D1557047

DATE DUE

MAY 3 1 2005		
APR 2 4 2009		
MAY 0 6 2009		

GAYLORD   PRINTED IN U.S.A.

OIL AND STATE IN THE MIDDLE EAST

OIL AND STATE
IN THE MIDDLE EAST

By George Lenczowski
University of California at Berkeley

CORNELL UNIVERSITY PRESS

Ithaca, New York

© 1960 by Cornell University

CORNELL UNIVERSITY PRESS

First published 1960

PRINTED IN THE UNITED STATES OF AMERICA
BY THE VAIL-BALLOU PRESS, INC.

60044

For John and Hubert

Foreword

THE idea of this book was born in 1951, when, under the impact of Dr. Mossadegh's crusade, Iran nationalized its foreign-owned oil industry. The author was fascinated by the dramatic clash of principles this nationalization engendered. On the one side was the oil company, basing its rights on the rule of law and on the progressive approach to industrial relations which it claimed to possess. On the other was an awakened nation, deeply persuaded of the justice of its cause and willing, in a moment of emotional outburst, to subordinate cold economic considerations to political and psychological priorities.

There was no doubt that the significance of the Iranian crisis transcended by far the boundaries of Iran proper. Regardless of the particular solution devised for the Iranian crisis, the struggle posed so many questions of a fundamental nature as to warrant a comprehensive inquiry into the totality of relations between the oil industry and the state in the Middle East. This conviction has been strengthened by the fact that, apart from a few monographs dealing with particular problems (such as the Iranian dispute), the works that have dealt with Middle Eastern oil as a whole have largely concentrated on the history of the oil companies. As a result a number of issues relevant to the understanding of the oil-state relationship in the Middle East have been touched upon but lightly, if at all. It was felt that a more systematic treatment was needed for such questions as the significance of Middle Eastern

FOREWORD

oil in the Free World's economy, the characteristics and legal status of concession agreements, territorial claims in submarine and desert areas, public opinion regarding Middle Eastern oil, company methods of handling government and public relations, and the whole sector of human and industrial relations, to mention some of the most important. It is on such insufficiently explored areas that this study aims to throw more light, within a broad analysis of the relationship between a privately owned Western enterprise and a sovereign state in the Middle East.

This being the objective, no attempt has been made to duplicate the work of others by writing a general history of oil concessions in the Middle East. Consequently the present volume has a topical rather than a chronological organization. While aiming at issues, the author felt compelled to supply a substantial body of descriptive material, without which a discussion of ideas would lack adequate foundation.

As for the anticipated audience, this book is addressed to the educated layman to whom certain basic geographical, political, and legal notions are not wholly alien. It is hoped, however, that the area specialist, the political scientist, and the business executive will also find enough in it to attract their attention and stimulate their thinking.

Research has been done in both the Middle East and the United States. Under a Ford Foundation fellowship the author spent the academic year 1952–1953 in the field to gather the preliminary materials for this study. This was followed by annual visits to the area from 1955 to 1958. In 1957 a grant was received from a special fund made available by the Rockefeller Foundation to the Department of Political Science of the University of California at Berkeley. This released the author from part of his teaching duties for the purpose of completing research and writing. Grateful appreciation is herewith expressed to both foundations and to the Department of Political Science for their assistance. Needless to say, these institutions are not responsible for the opinions contained in this volume, which are the author's own.

For permission to reproduce statistics pertaining to the production, reserves, and refining of oil in certain countries and for two

FOREWORD

quotations from Philip W. Thayer, ed., *Tensions in the Middle East* (Baltimore, 1958), thanks are given to the *Oil and Gas Journal* and to the Johns Hopkins Press, respectively.

The manuscript of this book was read by Dr. Halford L. Hoskins, Senior Specialist in International Relations, Legislative Reference Service, Library of Congress, whose critical remarks and constructive suggestions are hereby acknowledged with much gratitude.

G. L.

Berkeley, California
April 1959

Contents

FOREWORD vii

INTRODUCTION 1

PART ONE. *The General Setting*

I GROWTH OF OIL CONCESSIONS 9
   THE PRODUCING GROUP 9
   Iran 10
   Iraq 13
   Saudi Arabia 17
   Kuwait 19
   Other Oil-producing Countries 20
   THE TRANSIT GROUP 25

II MIDDLE EAST OIL IN THE EUROPEAN
   ECONOMY 27
   EUROPEAN CONSUMPTION TRENDS 28
   EUROPE'S REFINING CAPACITY 30
   TRANSIT ROUTES AND TANKER FLEETS . . . 31
   COST OF IMPORTS 34
   CONCLUDING REMARKS 35

CONTENTS

III OIL IN THE ECONOMY OF THE MIDDLE
     EAST 37
     REFINING, CONSUMPTION, AND NATIONAL TANKER FLEETS 41
     BENEFITS FOR TRANSIT COUNTRIES 43
     MIDDLE EAST DEVELOPMENT PROGRAMS 45
     Economic Developments in Kuwait 46
     Iraq's Development Plan 49
     Development Plan in Iran 53

PART TWO. *Companies and the Host Governments*

IV PATTERN OF CONCESSION AGREEMENTS . . 63
     STANDARD PROVISIONS 64
     DEPARTURES FROM STANDARD PATTERN 69
     Taxation 70
     Ownership 72
     Financial Safeguards 74
     The New Iranian Offshore Agreements 82
     The Arabian-Japanese Agreements 84
     SUMMARY 86

V LEGAL AND POLITICAL SAFEGUARDS . . . 88
     LEGISLATIVE SAFEGUARDS FOR HOST COUNTRIES . . . 90
     LEGAL STATUS OF CONCESSION AGREEMENTS . . . 94
     Challenges to the Validity of Concessions . . . 97
     A Search for Better Safeguards 101
     The Problem of Enforcement 104
     SELF-PROTECTION FOR THE COMPANIES IN TIMES OF
       CRISIS 107
     SELF-PROTECTION IN NORMAL TIMES 110

VI HANDLING OF GOVERNMENT RELATIONS
     BY THE COMPANIES 113
     ARAMCO'S GOVERNMENT RELATIONS ORGANIZATION . . 113
     IPC'S GOVERNMENT RELATIONS 117
     SPECIAL PROBLEMS: TECHNICAL ASSISTANCE . . . 120
     SPECIAL PROBLEMS: TRIBAL PROTECTION 121

CONTENTS

VII	TERRITORIAL CLAIMS: SUBMARINE AREAS	. 126
	CONTINENTAL SHELF PROCLAMATIONS	. 126
	UNCERTAINTIES AND ARBITRATIONS	. 129
	THE BAHREIN CONTROVERSY	. 134
VIII	TERRITORIAL CLAIMS: DESERT BORDERS	. 137
	THE BOUNDARY AGREEMENTS	. 137
	CONTROVERSY OVER SOUTHEASTERN BOUNDARIES	. 141
	THE BURAIMI DISPUTE	. 145
	THE QUESTION OF INNER OMAN	. 148
	CONCLUDING REMARKS	. 151
IX	PIPELINES: IN SEARCH OF A FORMULA	. 153
	EXIT MANDATES—ENTER ARAMCO	. 154
	Tapline and Syria: Drama of Ratification	. 155
	The First Revisions	. 160
	A Coup in Lebanon: Profit Sharing or Taxes?	. 160
	SPOTLIGHT ON IPC: NEW PIPELINES AND AGREEMENTS, 1953–1958	. 162
	TAPLINE'S PROFIT-SHARING FORMULA	. 164
X	INTERNATIONAL CONTROL: FROM BIG POWER PACTS TO PAN-ARAB PLANS	. 167
	ATTEMPTED REGULATION THROUGH ANGLO-AMERICAN AGREEMENTS	. 169
	A CRUSADE FOR UNITED NATIONS CONTROL	. 173
	OIL-SUPPORTED REGIONAL DEVELOPMENT	. 177
	PAN-ARAB OIL ACTIVITIES	. 187
	The Arab Oil Congress	. 196
	SUMMARY AND CONCLUSION	. 198

PART THREE. *Companies and the Public in the Host Countries*

XI	TYPICAL ATTITUDES: NATIONALIST THEMES	203
	ECONOMIC GRIEVANCES	. 209
	INDUSTRIAL GRIEVANCES	. 212

CONTENTS

POLITICAL GRIEVANCES	215
ARAB AND IRANIAN LITERATURE ON OIL	219
CONCLUDING REMARKS	224

XII OIL INDUSTRY'S REACTION 228
REFUTATION BY DEEDS 228
THE OIL INDUSTRY'S PUBLIC RELATIONS 235
Diversification of Media 240
Press Relations 242
Employee Magazines 245
SUMMARY 248

PART FOUR. *Companies and Employees in the Host Countries*

XIII OIL WORKERS: A NEW FORCE 253
LABOR LEGISLATION 254
IRANIAN INDUSTRIAL RELATIONS 261
IRAQI INDUSTRIAL RELATIONS 266
SAUDI ARABIAN INDUSTRIAL RELATIONS . . . 268
INDUSTRIAL RELATIONS IN THE TRANSIT COUNTRIES . 275

XIV PAN-ARAB LABOR MOVEMENT . . . 281
FALU'S CONSTITUTION AND OBJECTIVES 282
MEMBERSHIP AND UNIONIZATION DRIVES 284
FALU'S POLITICAL ACTIVITIES 287

XV THE HUMAN SIDE OF OPERATIONS . . 294
WORK AND WAGES 294
IN QUEST OF A DECENT LIVING 297
Food, Clothing, and Workmen's Compensation . . 297
Health Services 298
Housing 300
Saving for the Dark Hour 305
CAREER OPPORTUNITIES 306
Company Training Schemes 306
Advancement and Promotions for National Employees . 311

CONTENTS

PART FIVE. *A Case Study*

XVI REPERCUSSIONS OF THE SUEZ CRISIS . . . 319
    REACTIONS TO THE INVASION IN IRAQ AND LEBANON . 321
    SABOTAGE OF PIPELINES IN SYRIA 325
    Global and Regional Effects of Syrian Sabotage . . 326
    Consequences of Sabotage in Syria: Labor Dispute . . 328
    REACTIONS TO THE SUEZ CRISIS IN SAUDI ARABIA AND THE
        PERSIAN GULF 334
    THE TURKISH PIPELINE PLAN 338
    IRANIAN-TURKISH AGREEMENT 343
    PIPELINE SCHEMES IN ISRAEL AND EGYPT . . . 345

CONCLUSION 351

APPENDIX TABLES 361

BIBLIOGRAPHICAL NOTE 365

INDEX 371

Tables

1	Payments of the Consortium to Iran	40
2	Foreign exchange brought to Iran by the Consortium	40
3	Iraq development plan: Expenditures	51
4	Independent contractors servicing Aramco	232
5	Employment in oil industries in relation to general industrial employment	254
6	Level of annual earnings of national employees in IPC	295
7	Minimum daily wages in Aramco	295
8	Minimum daily wages in Iran under AIOC and the Consortium management	296
9	Company-operated major medical facilities in Iran and Saudi Arabia	299
10	Progress of housing under AIOC in Abadan	301
11	New housing constructed under the Consortium	302
12	Housing provided by the IPC group in Iraq	302
13	Houses constructed under IPC's home ownership scheme	303
14	Company-owned quarters for general employees of Aramco	304
15	Progress of training in the Iranian operating companies	308
16	Progress of training under Aramco	310
17	Iranian and foreign personnel of the operating companies in Iran	313
18	IPC's Iraqi and foreign personnel on the staff level in Iraq	313
19	Saudi and alien personnel in Aramco	314

TABLES

Appendix Tables

I   Crude oil production in the Middle East 361
II   Crude oil production of major world areas, 1958 . . . 362
III   Estimated oil revenues of certain countries in the Middle East 362
IV   Proved reserves of crude oil and refining capacity in the Middle East and North Africa in 1958 363
V   Tanker tonnage of major tanker-owning countries . . 363
VI   Tanker tonnage controlled by five major American oil companies at the end of 1957 364

Abbreviations

AIOC	Anglo-Iranian Oil Company
Aminoil	American Independent Oil Company
Aramco	Arabian American Oil Company
Bapco	Bahrein Petroleum Company
BPC	Basrah Petroleum Company
Consortium	Iranian Oil Participants
D'Arcy	Concession granted in Iran to William Knox D'Arcy
FALU	Federation of Arab Labor Unions
ICA	International Co-operative Alliance
ICFTU	International Confederation of Free Trade Unions
ILO	International Labor Organization
IPAC	Iran Pan American Oil Company
IPC	Iraq Petroleum Company
IRCAN	Iran Canada Oil Company
KOC	Kuwait Oil Company
MPC	Mosul Petroleum Company
NIOC	National Iranian Oil Company
OEEC	Organization for European Economic Co-operation
QPC	Qatar Petroleum Company
SIRIP	Iran Italian Petroleum Company
Tapline	Trans-Arabian Pipe Line Company
WFTU	World Federation of Trade Unions

Note on Weights and Measures

References to tons in this volume mean long tons (1 long ton = 1 metric ton multiplied by 1.0160).

With API (American Petroleum Institute) gravity of 30, 1 long ton of crude oil equals 7.310 barrels.

To convert "tons per year" to approximate "barrels per day" divide by 50 (example: 50,000,000 tons per year = 1,000,000 barrels per day).

To convert "barrels per day" to approximate "tons per year" multiply by 50 (example: 500,000 barrels per day = 25,000,000 tons per year).

Introduction

THE Middle East, like many another formerly colonial area in Asia and Africa, is now undergoing a dual revolutionary process: it is emancipating itself from Western dominance and it is changing its socioeconomic structure, with attendant political transformations. But although generally sharing the problems of other formerly colonial countries, the Middle East differs from them markedly in three specific respects. These are: first, its geographical location, which makes it one of the most strategic transit areas of the world for air, land, and maritime traffic; second, its possession of three-fourths of the proven world reserves of oil; and, third, its character as the arena of one of the most exacerbated and seemingly interminable political conflicts of this century—the Arab-Israeli imbroglio.

Seldom has the world witnessed a greater paradox than the existence of fabulous oil resources in countries whose people, at least in the beginning, were unable to exploit it for their own or anyone else's benefit. This inability was due to a number of factors such as unawareness of the presence of oil, lack of technical know-how, and absence of organizational skills and financial capacity to launch a major industrial enterprise. The missing skills had, therefore, to be supplied by the nationals of Western countries. The latter had to provide, at the outset, risk capital for exploration and development, to be followed later by steadily increasing investments to keep up with the growing demand for oil and with the rapid prog-

INTRODUCTION

ress of technology. But while their principal objective—that of producing oil—was being accomplished, their presence and activity in the midst of an alien society posed actual and potential political problems likely to complicate the achievements of purely technical and economic tasks.

No human activity occurs in a vacuum. Modern technological progress profoundly affects the society in which it takes place, and, conversely, social environment generates needs, cultures, habits, and attitudes that condition human inventiveness and search for technological improvements. In this continuing interaction there is a causal link between economic and technological phenomena and the political institutions and behavior of society. The influence is reciprocal, and the general sociopolitical rule seems to be that of balance. If economic and technical development requires modification in political institutions, the latter will have to be adjusted or a disturbing imbalance may occur. And, vice versa, a change in political institutions or policies may condition both the economy and the technology. In this process of change, interaction, and eventual balancing, situations arise in which maladjustment produces many social and political tensions. Depending on their nature and depth, these tensions may be eliminated by revolution or reform, or perhaps they may not be eliminated for long periods, thus causing a seemingly perpetual malaise in the life of a given society.

Although the phenomena so briefly described above are characteristic of every human society, they acquire particular intensity when two different societies come into close contact. And if it happens that a marked cultural, social, and economic lag exists between these two societies, the resulting tensions may acquire truly revolutionary proportions.

The development of the oil industry in the Middle East constitutes a vivid example of such interaction between two different types of societies, with attendant strains and stresses. On the one hand, it produces technical and economic change in the midst of societies subsisting on and institutionally geared to agrarian, trading, or pastoral economies. This change is rendered more abrupt by the fact that its motive force comes from the outside instead of from

INTRODUCTION

the native environment, as was the case in the industrial development of Western nations. On the other hand, the oil industry is being developed by foreign capital and management in areas experiencing political emancipation. Hence, however benevolent, enlightened, and socially progressive an industrial concern may be, it constitutes a substantial foreign element in the midst of a society whose general tendency is to free itself from dependence on external powers and influences.

Oil development in the Middle East is more than a "vivid example" of social interaction. In many ways the position of the oil industry is both unique and decisive in the social and political destinies of the Middle East. In the first place, it is the largest, and in most of these countries the only large, industry in operation. Consequently it is both the source of the first major problems of industrialization and the maker of precedents in the field of industrial relations. In the second place, profits derived from its operations provide such a high proportion of state revenues in the host countries as to overshadow all other sources of revenue and make these countries heavily dependent on the industry. In the third place, Middle Eastern oil has come to play a vital role in the peacetime economies and military requirements of major Western powers, supplying, as of 1956, over 80 per cent of the needs of Western Europe.

These three aspects determine the sociopolitical role played by the oil industry in the Middle East: socially, it is both a progressive and a disturbing factor; economically, it holds the key to fiscal balance and development programs; politically, it creates an ever-increasing interdependence between the West and the Middle East.

The Middle Eastern oil industry has a few other special characteristics. In the Middle East the oil industry operates on the basis of long-term concession agreements. This creates a special relationship between a foreign company and a host government. It poses the problems of negotiation and of validity of agreements, of their observance, modification, and revision. This special relationship is underscored by the elements of power entering it: one partner, the Middle Eastern state, represents the power of sovereignty; the other, the oil company, represents economic power of such magnitude as

INTRODUCTION

occasionally to dwarf the economic power of the host country. The company's home government may also have special interest in its operations—of a friendly or an unfriendly nature, depending on the circumstances. It may, for example, be concerned about the company's size and the existence or absence of the competitive factor. Or it may be interested in seeing that the company's operations are undisturbed and successful. In some cases the home government may even become a major shareholder and thus participate in the company's policy-making processes. Consequently an oil corporation cannot avoid being constantly aware of the political factor which its relationships with the host government and the home government engender.

The company's relations with the host country are not limited to dealings with the host government alone. In the nineteenth century or in some specific cases it may have been possible for a foreign concessionaire to deal with the ruler only, without regard to the feelings and attitudes of his subjects. But this could not have been considered a very safe procedure from the concessionaire's point of view. The tobacco concession in Persia in the 1890's is a case in point. There was no doubt as to the validity of the concession granting a British concern a monopoly of the production of and trade in tobacco. Nevertheless, public agitation, partly due to Russian hostility to the project, nullified the legal instrument, compelling its eventual abandonment. Today the role of public opinion in the political processes of practically every state is taken for granted. This means that the oil companies have to deal with the attitudes and reactions of the public in the host countries as well as transact their formal business with the governments. In fact, relations with the public may become so important as to overshadow official relations and, in some cases, may transform the latter into a mere function of the former. Public opinion is, of course, based on a variety of factors, some independent of and others dependent on the companies' policies and behavior. Companies' methods of handling labor and industrial relations are likely to influence public attitudes toward the oil industry materially, both because of the principles involved and because of the day-to-day contact between management and a sizable sector of the population.

INTRODUCTION

It will be the purpose of this study to describe and analyze the problems that arise out of the relationship between the foreign-owned oil industry and the state in the Middle East. Part One will give a historical introduction, followed by an account of the role that oil is playing in the economies of European and Middle Eastern countries. Parts Two, Three, and Four will be devoted to analysis of the oil industry's relations with three elements: the host governments, the public in the host countries, and the employees. Economic, legal, social, and political aspects of these relations will be examined with a view to evaluating their relative importance and, if possible, determining policy priorities. Part Five will present a case study which, it is hoped, will demonstrate the multifaceted character of the state-company relationship and the close interrelation of politics, mass psychology, and economics as factors affecting the position of the oil companies in the Middle East. Main attention will be paid to the major oil-producing countries that are independent and in which, consequently, state-company relations can follow their natural course without being artificially affected by the political will of an outside suzerain power. For this reason the British-protected Persian Gulf principalities will largely be left outside the scope of this study, except for occasional mention of their happenings as illustrative of more general trends in the area. Our inquiry will extend, however, to those countries through which oil is being regularly transported by pipeline inasmuch as they constitute a vital part of the total picture.

PART ONE

THE GENERAL SETTING

◈◈ CHAPTER I ◈◈

Growth of Oil Concessions

For the purposes of this study the Middle Eastern countries may be divided into two major groups: those which produce oil and those which serve as transit areas for the shipping of oil to foreign destinations. In the producing group four countries stand out, namely, Iran, Iraq, Saudi Arabia, and Kuwait, to name them in the chronological sequence of their development. Egypt, Bahrein, Qatar, Kuwait-Saudi Arabian Neutral Zone, Turkey, and Israel are minor producers, and only three of the latter, Bahrein, Qatar, and the Neutral Zone, are able to export their petroleum. In addition to the Middle East proper, Algeria and Libya, both linked to the area under our inquiry by profound ethnic bonds, have recently discovered oil. Consequently, they will be briefly reviewed in our survey of oil concessions. As for the transit countries, Egypt, Jordan, Syria, and Lebanon, special attention will be paid to the last two on account of their strategic position as territories providing both transit and terminal facilities.

THE PRODUCING GROUP

Companies conducting oil operations in the producing countries are, with a few exceptions, owned and managed by foreign interests. The various patterns of control are described under the countries concerned.

OIL AND STATE IN THE MIDDLE EAST

Iran

Chronologically the first producing country in the Middle East, Iran conducts its oil operations in three ways: (*a*) by an international Consortium on behalf of the state, (*b*) by the state directly, and (*c*) by foreign companies other than the Consortium, acting either on their own or in partnership with the state.

In the first category the state acts through the National Iranian Oil Company (NIOC), which in turn has delegated the so-called basic operations to an international Consortium formally known as Iranian Oil Participants, Ltd., incorporated in London. The Consortium is owned by the following interests:

British-Dutch 54%	British Petroleum Company (formerly Anglo-Iranian Oil Company)	40%
	Royal Dutch–Shell group	14%
French 6%	Compagnie Française des Pétroles	6%
American 40%	Standard Oil Company (New Jersey)	7%
	Standard Oil Company of California	7%
	Socony Mobil Oil Company	7%
	The Texas Company	7%
	Gulf Oil Corporation	7%
	Iricon Agency, Ltd.[1]	5%

Iranian Oil Participants, Ltd., operates in Iran through two affiliates, the Iranian Oil Exploration and Producing Company and the Iranian Oil Refining Company. The existing arrangement came into force on October 25, 1954, following a 3½-year deadlock produced by the nationalization of properties of the Anglo-Iranian Oil Company, former concessionaire, on March 20, 1951. The Anglo-Iranian (formerly Anglo-Persian) Oil Company had based its rights on the original concession granted to William Knox D'Arcy in 1901. This concession at first extended to the whole territory of Iran except for the five northern provinces of Azerbaijan, Gilan, Mazanderan, Gorgan, and Khorasan. As the result of a major revision of the concession in 1933 (which followed its unilateral can-

---

[1] Iricon Agency, Ltd., is owned by nine American oil companies in the following proportions: Richfield Oil Corp., 1.25%; American Independent Oil Co., 0.833%; and Standard Oil Co. (Ohio), Getty Oil Co., Signal Oil and Gas Co., The Atlantic Refining Co., Hancock Oil Co., Tide Water Associated Oil Co., and San Jacinto Petroleum Corp., 0.417% each.

GROWTH OF OIL CONCESSIONS

cellation by Reza Shah), the territory of the concession was reduced to 100,000 square miles located in southwestern Iran. Oil was first struck in 1908 in Masjid-i-Suleiman after seven years of stubborn efforts. Further wells were drilled in the adjoining area of Khuzistan, to be linked by a system of pipelines with the island of Abadan, where a major refinery and oil port were established. When, in 1913, the British Admiralty, then headed by Winston Churchill, decided to shift from coal to oil as fuel for the Royal Navy, Iranian oilfields acquired special importance as a principal supplier of petroleum for the British Empire. Iranian production steadily increased, assuming first place among the Middle Eastern countries and reaching, by 1950, the figure of 31.75 million tons a year. The next year, however, nationalization and the subsequent dispute brought about the virtual cessation of production, thereby permitting such newer producers as Saudi Arabia and Kuwait to replace Iran in its position of pre-eminence. Since the conclusion of the Consortium agreement in 1954, Iranian production has been making great forward strides, reaching in 1956 the rate of 32 million tons a year, which was equal to its highest figure prior to nationalization.

The Consortium, the foreign element in Iranian oil operations, is not a concessionaire but an operative agent on behalf of the Iranian state, the latter being the sole owner of its oil resources. With the establishment of the Consortium have come certain practical changes, two of which deserve special mention. One is that whereas the former concessionaire represented a single nationality, i.e., British, the new Consortium constitutes a multinational group. The other is that the so-called nonbasic operations—such as welfare and housing—have been assumed by the National Iranian Oil Company. It should be noted, however, that the basic industrial operations are conducted by the foreign-owned and foreign-managed companies and that in this sense the pattern of control by foreign elements has not radically changed.

In the second category operations are carried out directly by the NIOC in central Iran. Drilling (done by the Drilling and Exploration Company of Los Angeles) resulted, on August 26, 1956, in the dramatic discovery of oil in Qum. From well no. 5 oil burst forth in a 120-foot gusher, which ran unchecked for eighty-two days and inundated the surrounding desert with an estimated 5 million barrels, worth about $10,000,000 at current prices. Drilling was continued in 1956 and 1957, and new discoveries were made. The major problem for NIOC has been the organization of an adequate transportation system for what appears to be a major source of petroleum. The construction of a pipeline linking Qum with Alexandretta on the Turkish Mediterranean seaboard has been contemplated.

Following the adoption of a new oil law on July 29, 1957 (for details, see p. 91 below), Iran has again opened its doors to foreign oil companies, from which bids are being received for the areas outside the territory of the Consortium, both on land and offshore in the Persian Gulf. The precedent-setting development in this third category has been the agreement reached with the government-owned Italian concern Agip Mineraria in the summer of 1957 and ratified by the Iranian Parliament on August 12 within fourteen days after the passage of the new law. The agreement provided for the formation of an Iranian company, Iran-Italian Petroleum Company (SIRIP), owned equally by NIOC and Agip Mineraria, to conduct exploration and development in an area of about 23,000 square kilometers in three different zones. The first of these was located at the northern end of the Persian Gulf, partly on the coast but mostly offshore east of Abadan; the second lay on the eastern slopes of the central Zagros Mountains; and the third was on the coast of the Gulf of Oman, including an offshore area, in the proximity of the Pakistan border.

In the following year NIOC concluded two more agreements closely conforming to the pattern established by its agreement with Agip. The first, signed with the Pan American Petroleum Corporation (a wholly owned subsidiary of Standard Oil Company of Indiana) and promulgated by the shah on May 31, 1958, covered an offshore area of 16,000 square kilometers, divided into two parts: (1) approximately 1,000 square kilometers adjoining the north side of the Agip zone and (2) approximately 15,000 square kilometers adjoining the Agip zone on the south and extending to the median

GROWTH OF OIL CONCESSIONS

line of the Persian Gulf but excluding the islands of Kharg and Kargo, which were included in the Consortium area. The Pan American area was to be located beyond the 3-mile strip of territorial waters, which was also within the Consortium territory. According to the agreement, Pan American and NIOC were to set up a nonprofit Iranian corporation, Iran Pan American Oil Company (IPAC), whose task would be to conduct operations.

The next agreement was that concluded on June 22, 1958, between NIOC and Sapphire Petroleums, Ltd., a Canadian corporation. Almost identical in its main provisions (except for financial clauses) with the NIOC–Pan American agreement, it provided for two small areas of exploration, totaling 1,000 square kilometers and located also in the offshore zone. As in the previous agreements, a jointly owned company—in this case to be known as Iran Canada Oil Company (IRCAN)—was to be formed and its nationality was to be Iranian. Other similar agreements were expected to be concluded with respect to the remaining exploitable districts of Iranian territory as defined in the oil law of 1957.

Iraq

Chronologically the second major producing country in the Middle East, Iraq allows its oil resources to be exploited by the Iraq Petroleum Company (IPC) and the latter's affiliates, Basrah Petroleum Company and Mosul Petroleum Company. IPC is owned by British, Dutch, French, and American interests, in the following proportions:

British Petroleum	23.75%
Royal Dutch–Shell group	23.75%
Compagnie Française des Pétroles	23.75%
Standard Oil Company (New Jersey)	11.875%
Socony Mobil Oil Company	11.875%
Participations and Explorations Corporation (C. S. Gulbenkian Estate)	5%

IPC and its two affiliate companies operate on the basis of a revised concession agreement, concluded on February 3, 1952, and modified on March 24, 1955. The area of concession for all three companies comprises the whole of Iraq with the exception of 684 square miles near the Iranian border, which from 1925 to 1958 were subject to a concession held by Khanaqin Oil Company, an affiliate of British Petroleum.

The original concession was granted to IPC's predecessor, the Turkish Petroleum Company, on March 24, 1925, following protracted negotiations which were linked with the general peace settlement in the Middle East. The history of IPC's origins may be briefly summed up as follows: In 1914, shortly before the outbreak of the war, British and German interests joined to obtain an oil concession from the Ottoman government and for this purpose founded the Turkish Petroleum Company, incorporated in London. The over-all British share (with certain Dutch elements in it) was 75 per cent, distributed between the Anglo-Persian Oil Company (50 per cent) and the Royal Dutch–Shell group acting through its affiliate, the Anglo-Saxon Oil Company (25 per cent). The remaining 25 per cent was assigned to German interests represented by the Deutsche Bank. Although the concession, covering the provinces (vilayets) of Mosul and Baghdad, was granted in a note of the Ottoman grand vizier on June 28, 1914, it was never ratified owing to the outbreak of the war and thus never acquired sufficient legal validity to bind the Ottoman successor states after World War I.

The war and the partition of the Ottoman Empire, first through the secret wartime agreements and then during the peace settlement, posed the problem of the disposition both of the territories where oil was presumed to exist and of the rights to exploit these resources. With regard to the territorial dispositions, the main question revolved around the province of Mosul in northern Mesopotamia. According to the Sykes-Picot Agreement of May 16, 1916, Mosul was to be included in the projected French sphere of influence, which was to embrace Syria, Lebanon, and parts of Turkey proper. The remainder of Mesopotamia (Iraq) was to fall within the British sphere. Inasmuch, however, as the district of Mosul was occupied by the British in the course of military operations in Mesopotamia, Britain reopened the question of territorial division at the end of the war. A supplementary agreement concluded in December 1918, just prior to the beginning of the Peace Conference, by Prime Ministers Lloyd George and Clemenceau removed Mosul

GROWTH OF OIL CONCESSIONS

from the French sphere and placed it in the British instead. In return Britain pledged France a 25 per cent share in the exploitation of the oil resources of the region. This French share was to be obtained by the transfer of Germany's share in the Turkish Petroleum Company. These arrangements were formally confirmed at the San Remo Conference of April 24, 1920. This meeting of principal Allied powers, acting on behalf of the peace conferees in Paris, formally assigned the mandates in the Middle East to Britain and France, thus settling the territorial question. The oil was the subject of a simultaneous deal between Britain and France, known as the Long-Berenger Agreement, which confirmed the previous assignment of the 25 per cent share to France. The Treaty of Sèvres of August 10, 1920, which formalized the peace settlement with the defeated Ottoman Empire, incorporated the territorial provisions.

Despite the seeming finality of these dispositions, the latter were subjected to challenge, both in their territorial and their oil aspects. The territorial challenge came from renascent Kemalist Turkey, which, having liberated itself from foreign occupation and rejected the humiliating Sèvres Treaty, claimed Mosul province as an integral part of Turkish territory. This Turkish claim was given consideration in the Treaty of Lausanne of July 24, 1923, which declared that if Britain and Turkey did not reach agreement on Mosul within a specified period the matter should be presented to the League of Nations for decision. This was, indeed, what happened. Following prolonged maneuvers, the Council of the League on December 16, 1925, assigned the province of Mosul to Iraq, then under British administration, with the proviso that Iraq should continue under British mandate for the next twenty-five years and that 10 per cent of the oil output of the contested region should be made available to Turkey.

The challenge to the Franco-British dispositions concerning oil resources came from the United States. Expressing interest in business expansion in the Middle East, American oil interests raised objections to the arrangements whereby they were automatically excluded from participation in the Iraqi oil developments. Their cause was seconded by the Department of State, which, in the name of the Open Door principle, asked Britain to open her mandated territories, in this case Iraq, to penetration by American enterprise on an equal basis. After a lengthy exchange of notes understanding was finally reached both on government and on company levels to permit American participation in the exploitation of Iraqi oil. An agreement was signed in London on July 31, 1928, permitting American companies to participate in the Turkish Petroleum Company. The Anglo-Persian Oil Company (now British Petroleum), which initially was to hold nearly 50 per cent of the shares, now relinquished half of its holdings to the American group, thus giving the latter 23.75 per cent participation. The shares of Royal Dutch–Shell and the French company remained unchanged, each equaling that of the American group. The price that the American group had to pay for this share was an agreement not to seek separate concessions in the area roughly corresponding to the Asiatic territories of the defunct Ottoman Empire with the exception of Kuwait and the Khanaqin district of Iraq. This "Red Line Agreement" was designed to protect British interests against the possibility of expansion by the more dynamic American interests. This was in conformity with the general policy of exclusion practiced by Britain vis-à-vis foreign interests in the British-controlled areas, specifically in and around the Persian Gulf. The American group was originally composed of seven corporations, but following withdrawals and transfers of stock it was reduced to two: Standard Oil Company (New Jersey) and Socony-Vacuum Oil Company, each holding one-half of the total American share. These two companies were represented by their common affiliate, the Near East Development Corporation.

The newly constituted international group now changed its name to Iraq Petroleum Company, and the latter, on March 24, 1931, obtained from the government of Iraq a new concession, which remained valid until its revision in 1952. Its area included the territory of Iraq east of the Tigris River. The territory west of the Tigris became subject to two different concessions, one of which was initially held by the British Oil Development Company. The latter, having been for a brief time subjected to Italian majority control, was transferred in 1942 to the Mosul Petroleum Company, IPC's affiliate. The other concession, located south of the

GROWTH OF OIL CONCESSIONS

33rd parallel west of the Tigris, was granted in 1938 to another IPC subsidiary, the Basrah Petroleum Company.

Saudi Arabia

The oil industry in Saudi Arabia is operated by the Arabian American Oil Company (Aramco) on the basis of a concession first granted to Standard Oil Company of California on May 25, 1933, and later extended on July 21, 1939. Aramco is owned by four American corporations in the following proportions:

Standard Oil Company of California	30%
The Texas Company	30%
Standard Oil Company (New Jersey)	30%
Socony Mobil Oil Company	10%

The history of the Saudi oil industry is briefer and simpler than that of the oil industries of Iran and Iraq, yet it also contains certain episodes involving international complications. Initially there was a competition for concessionary rights between Iraq Petroleum Company and Standard of California. The latter won the contest by acting with greater speed and imagination and, in particular, by offering King Ibn Saud immediate payment in gold, as he demanded. Standard's success was significant internationally inasmuch as it introduced a newcomer into the area lying within the Red Line as laid out in the IPC agreement. Thus an American corporation not belonging to the IPC group penerated a region which the British element in that group wanted to keep out of bounds to outsiders, especially Americans.

Operations were actually carried on by the California Arabian Standard Oil Company, a subsidiary of Standard of California. In 1936 the Texas Company acquired a half-interest in the venture, thus increasing the number of American corporations acting independently within the Red Line area. The operating company's name was now changed to Arabian American Oil Company. The territory of the concession was substantially extended by a new agreement in 1939, and oil, first discovered in 1933, began to be produced in increasing quantities. The outbreak of the war, however, brought about a virtual suspension of production, which was not resumed on a major scale until 1945–1946. By that time it was clear that Saudi Arabia possessed vast oil reserves and that the growing demand for oil on world markets would call for substantial expansion of production and marketing. For this reason the two original participants, Standard of California and Texas, were willing to take in other American concerns which could provide additional capital and marketing facilities. Standard (New Jersey) and Socony expressed interest, but their French and other partners in IPC objected, invoking the Red Line Agreement as prohibitive of independent ventures by the members of their group in the area in question. Early in 1946 the two American corporations refused to reaffirm the validity of the Red Line Agreement, and in October of that year they denounced it as altogether null and void. Their argument was based on the changes brought about by the war, on the need for substantial increases in supplies of oil to meet postwar demands, and on the incompatibility of restrictive clauses with freedom of trade and the public interest. Having thus stated their position, in December 1946 Standard (New Jersey) and Socony Vacuum concluded an agreement with Standard of California and the Texas Company providing for their participation in Aramco.

In protest against this action the Compagnie Française des Pétroles, one of the chief partners in the Iraq Petroleum Company, filed a suit in February 1947 against the American group and other partners of IPC for violating the Red Line Agreement. Because of this legal complication formal agreement on the partnership in Aramco was delayed. Instead the four American companies agreed that the two original partners in Aramco would receive a bank loan to be guaranteed by Standard (New Jersey) and Socony Vacuum. This initial understanding was, however, replaced by a final agreement in March 1947 whereby Standard (New Jersey) acquired 30 per cent and Socony 10 per cent of Aramco's stock. The remaining 60 per cent was held in equal parts by the two original owners, Standard of California and Texas.

While this final agreement was being negotiated, the dispute between the French and the American group within IPC was transferred from legal litigation to private negotiation. On November 3, 1948, it ended in a settlement out of court. This settlement provided for the dissolution of the Red Line Agreement, thus removing

GROWTH OF OIL CONCESSIONS

the obstacles to participation by the two new partners in Saudi Arabian oil expansion. From that time on, this expansion was both rapid and spectacular. By 1950 Aramco's annual production of 25.9 million tons was second only to that of Iran, and, following the breakdown in Iranian production as the result of nationalization, it assumed first place, soon to be shared with that of Kuwait.

On December 30, 1950, Aramco and Saudi Arabia made a major change in the concession by concluding a fifty-fifty profit-sharing agreement. This agreement was of great importance inasmuch as it set a precedent for similar formulas in other oil-producing countries of the Middle East. A further step was the agreement of both parties on October 2, 1951, to apply the new formula before the payment of United States taxes rather than after, as had been done until then.

The Aramco concession in Saudi Arabia has played a special and rather unique role in the history of Middle Eastern oil on account of two factors. One was the purely American character of the company in an area hitherto closed to influences other than British; the other was the political disinterestedness of the American government in this particular region at the time when the concession was granted.

Kuwait

Oil exploitation in this British-protected principality is in the hands of the Kuwait Oil Company on the basis of a concession granted on December 23, 1934, and revised on December 1, 1951. The company is owned in equal shares by the Gulf Exploration Company and British Petroleum (formerly Anglo-Iranian), the latter acting through its affiliate, D'Arcy Kuwait Company, Ltd.

The granting of the concession was not a simple matter and in some respects was reminiscent of the earlier complications in Iraq between British and American interests. The concession was first sought in the 1920's by a British subject, Major Frank Holmes, acting in the name of the Eastern and General Syndicate, Ltd. This concern, which had been instrumental in obtaining other concessions in the Middle East, generally with a view to selling them, transferred its rights in 1923 to the Gulf Oil Corporation, an American company. The latter, however, soon encountered two major obstacles. On the one hand, the British government invoked the so-called nationality clause, i.e., a pledge made in 1913 by the ruler of Kuwait to exclude from oil concessions anybody who was not "appointed" by the British government, hence presumably British. On the other, as if to reinforce this legal argument, it developed that the Anglo-Iranian Oil Company also had a claim to the Kuwait concession on the basis of an earlier application and a subsequent agreement with the ruler.

It is obvious that if given free rein the ruler of Kuwait would have granted the American company the concessionary rights. Being bound to Britain by the exclusivity clauses of earlier treaties and pledges, he was pushed into a policy of contradictory promises, thus complicating the situation. As a result of Gulf's complaints, the State Department took up the defense of American interests and in a series of diplomatic *démarches* in London insisted on the recognition of Gulf's acquired rights. Eventually a compromise was reached in December 1934 between Gulf and Anglo-Iranian. Both agreed to share the concessionary rights in equal parts, and soon afterward a formal concession was awarded to their common affiliate, the Kuwait Oil Company. The company proved to be eminently successful in its explorations and struck oil in 1934 in the area of Burgan. The latter soon emerged as the world's largest single oilfield, with estimated reserves of 27.5 billion barrels, representing 17.80 per cent of the world total as of 1955.

Of the four major producers of oil, Kuwait is unique in that it is not a sovereign country but one virtually controlled by Britain. Moreover, its small size—20,000 square miles with a population of 250,000—provides a stark contrast with its wealth and its current oil revenues, which in 1958 were estimated at $415,000,000 a year.

Other Oil-producing Countries

The foregoing sections have dealt with the major producing areas. Oil is being extracted in lesser quantities in Egypt, Bahrein, Qatar, Kuwait–Saudi Arabian Neutral Zone, Turkey, and Israel. It has recently been discovered in Libya and Algeria, which, as Arab states, are likely to share some of the problems characteristic of the Middle East proper.

Of all these countries, *Egypt* has pioneered in developing an oil

GROWTH OF OIL CONCESSIONS

"regime" different from those in the area as a whole inasmuch as, instead of granting a major concession to a single company or group, it has divided its territory into relatively minor plots which it leases for exploration or exploitation to a number of enterprises, mostly foreign but some mixed or predominantly Egyptian. In 1955 no less than seven oil corporations were engaged in exploration for or production of oil in Egypt.

In *Bahrein* the basic political framework—the existence of a virtual British protectorate—resembles that prevailing in Kuwait, but the actual arrangements with oil industries vary. Bahrein's oil is being exploited by the Bahrein Petroleum Company (Bapco), which is owned in equal parts by Standard Oil of California and the Texas Company on the basis of a concession granted in 1930 and renewed on June 15, 1940. In view of the nationality clauses in the treaties linking Bahrein to Britain, this is a rather unusual situation: it represents exclusive American penetration in the most vital economic sector of the island principality while Britain continues her paramount political status to the exclusion of other powers (and even of their consuls).

The story of the Bahrein concession closely parallels the developments in Kuwait (in terms of time it preceded the latter), but with a different end result. The concession was initially obtained by Major Holmes, whose firm, the Eastern and General Syndicate, offered it for sale first to potential customers in Britain. Rebuffed on the British market, Holmes turned to Americans and found in Standard of California a willing buyer. At this juncture the British government stepped in, trying to keep an American company from entering a British-protected territory. Faced with this obstacle, Standard turned to the State Department for help, and the matter was transferred to the level of diplomatic negotiations. Invoking the Open Door principle, the State Department insisted on the right of American concerns to expand in Bahrein. The British government gave in reluctantly, and only on condition that the company be endowed with a British legal personality, that it have a British director, and that its operating personnel on Bahrein be predominantly British. These conditions were accepted, and with the approval of the ruler of Bahrein the Syndicate on June 12, 1930, transferred its concession to Bahrein Petroleum Company, Standard's subsidiary incorporated in Canada. Two years later oil began to be produced in commercial quantities. In 1935 half of Bapco's stock was purchased by the Texas Company. The concession originally covered 100,000 acres, but in 1940 it was extended to the whole of Bahrein.

In *Qatar,* in the sheikhdoms of the Trucial Coast, and in Muscat-Oman and the Aden Protectorate concessions or exploration permits were granted, in the 1930's, to various subsidiaries of Iraq Petroleum Company. Thus far oil has been found in Qatar only. The present concessionaire in Qatar is the Qatar Petroleum Company, an IPC subsidiary, whose annual production in 1958 exceeded 8 million tons.

In the *Kuwait–Saudi Arabian Neutral Zone* oil is being produced jointly by two small American firms, the American Independent Oil Company (Aminoil) and the Getty Oil Company on the basis of two separate concessions, each for an undivided half of the Zone. Aminoil obtained its concession on June 28, 1948, from Kuwait, and Getty (formerly Pacific Western) secured a concession from Saudi Arabia on February 20, 1949. Located in the vicinity of the famous Burgan field, the Neutral Zone was presumed to possess impressive reserves of oil, yet it took the exploration party nearly six years before it discovered oil in the area. Commercial production began in 1953.

The *Neutral Zone's offshore area* also became an object of solicitation by various oil companies. On December 10, 1957, Saudi Arabia granted a 40-year concession for its own undivided half to the Japan Petroleum Trading Company. On July 5, 1958, Kuwait followed suit by granting a concession to the same company for its half of the offshore area. The terms of both concessions were very similar, introducing the concept of integration in the company's operations, of which more will be said later.[2] The granting of these two concessions corresponded to the general increase in interest in the submarine areas of the Persian Gulf, as attested by the simultaneous agreements concluded between Iran and certain foreign companies. There was a strong belief that oil could be found in

[2] See p. 84 below.

GROWTH OF OIL CONCESSIONS

these areas in abundant quantities, a belief strengthened by Aramco's discoveries in the Safaniya and Manifa wells in 1957.

In *Turkey* oil was discovered in 1948 in the Ramandag district. As of mid-1955 the estimated reserves were among the smallest in the Middle East, not exceeding 80 million barrels, and production has been correspondingly modest (about 325,000 tons in 1958). Oil operations have been conducted by the state, in conformity with the general policy of *étatisme* of Kemalist Turkey. More recently, however, the Turkish Republic reverted to a more liberal policy in the economic field. This new trend found expression in the law of March 7, 1954, which set new regulations for the oil industry, encouraging free enterprise and, in particular, foreign concession seekers.

Israel is the latest comer in the Middle Eastern oil industry, having struck its first well in Helez (Huleikat) in 1955. Its production has been minor, amounting to 82,000 tons in 1958.

In *Algeria* oil operations are carried on by the Société Nationale de Recherche et d'Exploitation des Pétroles en Algérie (REPAL), whose stock is owned 50 per cent by the Algerian government and 48.45 per cent by the French government. The company struck oil in Hassi Messaoud in 1956. Following the construction of a 6-inch pipeline spanning the hundred miles between the oilfield and the nearest railway station in Touggourt, the first shipment of oil was made in late December 1957 to the port of Philippeville. Although the estimated reserves of more than 400 million tons augur well for the producing capacity of the fields, transportation is a problem. The latter has been complicated both because of the 400-mile length of the pipe-rail route and because of the costly security measures required to protect the shipments from attacks of the nationalist rebels.

In *Libya* no fewer than thirteen companies had been granted concessions by 1958, on the basis of a law of 1955, which, like the earlier Egyptian legislation, has encouraged multiple operations rather than exploitation of a large area by a single company. Oil was first struck in December 1957 in the Atshan area of the province of Fezzan, near the Algerian border, by Esso Standard (Libya), Inc., an affiliate of Jersey Standard. Commercial exploitation would require the building of a 560-mile-long pipeline to connect the oilfield with the port of Tripoli. Further encouraging finds in the north-central part of the country were made in 1958 by three American companies, Ohio Oil Company and its partners, Amerada Petroleum Corporation, and Continental Oil Company. The existence of oil was also ascertained by Socony Mobil Oil Company in central Libya some 100 miles south of the Mediterranean coast. The commercial value of these discoveries has not yet been conclusively established.

The degree of political emancipation of a producing country has been and is an important factor in the position of a concessionaire company. In a fully independent country the sovereign will of the state enters all phases of the relationship between the company and the government, from the conclusion of the agreement, through its implementation, to its termination. There is a definite link between a host country's political attitudes toward an oil company and the degree of its political emancipation at the time when the agreement was concluded. Certain countries, at present sovereign, were not fully independent at the time when a concession was first negotiated. This was clearly the case in Iraq, which was under British mandate when the concession was granted. By the same token, certain other countries, though legally sovereign, were subjected to such a degree of *de facto* foreign influence as to cause them to deny that a concession was granted of their own free will. In Iran this question has aroused much controversy between the government and the company.

In the dependent areas the conclusion of concession agreements did not necessarily represent the free consent of the native rulers inasmuch as Kuwait, Bahrein, and Qatar are virtual British dependencies. Britain's special position in those areas is secured by the so-called nationality and nonalienation clauses in the existing treaties. These clauses generally provide that no concessions for the exploitation of natural resources must be granted by these countries without Britain's consent. This arrangement differs, of course, from that in an outright colony, such as Aden, where it is not the native ruler, but the British authorities who negotiate and give their consent to the exploitation of resources. Granting con-

GROWTH OF OIL CONCESSIONS

tinued British political predominance in the Persian Gulf sheikhdoms, the practical and legal validity of these concession agreements seems to be assured. A different situation might arise, however, if British authority were to be weakened and the native rulers were to assert their independence.

Apart from the more or less sovereign quality of decisions made by producing countries, two other factors are apt to affect the position of the concessionaire companies. These are the form of government of the host country and the way oil revenues are used. A stable government, deriving its strength from well-ordered democratic political processes or from a deeply embedded authoritarian tradition is likely to be less erratic in its dealing with foreign interests than a weak government. And purposeful and constructive utilization of oil revenues by a host government may greatly help in creating a climate of public confidence in which the company's position will be less vulnerable than otherwise to criticism by discontented elements.

THE TRANSIT GROUP

Many of the points mentioned with regard to the producing states apply to the transit group as well. The sovereign quality of their decisions, the degree of their political emancipation, and the kind of governments they have—all are relevant to their relationship with foreign interests. But in contrast to the producing group, their national revenue is not so heavily dependent upon the income from oil and the basis for calculation of their revenue is much more open to controversy. Oil may be transported by open seas, by special waterways, or by pipelines. Countries through whose territories pipelines or waterways pass belong to the transit group. Syria, Lebanon, and Jordan serve as pipeline transit areas, and Egypt belongs to the transit category by virtue of the Suez Canal. So long as Israel cannot be used for the transit of Arabian oil because of the Arab countries' opposition, Syria and Egypt occupy the most strategic position in the transit group. Syria serves as a transit area for oil produced in Iraq and Saudi Arabia, two of the four major oil-producing countries of the Middle East. About

73 per cent of Iraq's 1956 oil exports depended on the availability of Syrian pipelines. As for Saudi Arabia, 35 per cent of its oil exports go through Syria. Oil from Iraq is transported to the Mediterranean through three pipelines. Two of them, 12 and 16 inches in diameter, were constructed prior to 1956 and lead to the Lebanese port of Tripoli. One, 30–32 inches in diameter, was opened in 1953 and terminates at Banias on the Syrian coast just north of Lebanon. During the British mandate in Palestine a fourth line conveyed Iraqi oil to Haifa, but since the establishment of the state of Israel Iraq has stopped the flow of oil through that line. In 1955 construction of a new 24-inch pipeline was begun, to be interrupted in 1956 as a result of hostilities in Egypt. These pipelines are owned and operated by Iraq Petroleum Company.

The pipeline carrying oil from Saudi Arabia to the Mediterranean crosses Jordan, Syria, and Lebanon, and Syria again figures as the indispensable transit area. The pipeline, 30–31 inches in diameter, was opened in 1950 and is operated by Trans-Arabian Pipe Line Company (Tapline), which is owned by Aramco's parent companies. The remainder of Saudi Arabian oil exports is shipped either by the Suez Canal (20 per cent) or by other maritime routes (44 per cent), the latter principally to destinations south and east of Suez. Egypt's role as a transit area will be better appreciated if we add that in addition to Saudi Arabian oil exports the Suez Canal handles 41 per cent of Iranian and 74 per cent of Kuwaiti oil exports.

The construction of pipelines necessitated concluding concession agreements that would guarantee the operating companies transit rights for a satisfactory length of time and protect them against discriminatory legislation. Eventually two sets of agreements were negotiated: one between Iraq Petroleum Company and Syria and Lebanon; the other between Trans-Arabian Pipe Line Company on the one hand and Syria, Lebanon, Jordan, and Saudi Arabia on the other. The provisions of these agreements and the problems to which they gave rise will be discussed later.

◈◈ CHAPTER II ◈◈

Middle East Oil in the European Economy

THE Suez crisis of 1956 brought about a sudden realization of Europe's heavy dependence on Middle Eastern oil. Although the initiated few had known the facts for a long time, the public at large was unexpectedly confronted with unfamiliar and perplexing data and events, which included the length of maritime routes through the Mediterranean and around the Cape, the throughput capacity of Syrian pipelines, the volume of oil traffic through the Canal, the tonnage of Western-controlled tanker fleets, oil rationing in Europe, the exchange difficulties of European governments, and the limitations on American ability to make up the sudden oil shortages of its partners in the Atlantic alliance.

The crisis and its multifarious repercussions—economic and political—will be examined in greater detail at the end of this volume. The purpose of this chapter is to supply a brief set of data, mostly of a statistical nature, likely to provide answers to the basic questions concerning Europe's dependence on Middle Eastern oil. How much oil does Europe consume annually? Does the consumption show an ascending trend and, if so, how steep? Who is the main supplier of oil for Europe in the postwar period? How does oil compare with other sources of energy—coal, gas, hydroelectric power, and possibly atomic energy—in filling Europe's needs for fuels? Is there an interrelationship between the growth of European industry and the consumption of oil? What are the differences in the dependences of individual countries on oil supplies? What is Europe's refining capacity and how much does Europe spend on her oil imports? Those are the essential questions, which, without going into a profound analysis of European economy, we will attempt to answer, as succinctly as possible, in the lines that follow.

EUROPEAN CONSUMPTION TRENDS

Europe outside the Soviet bloc has shown two simultaneous trends: a steadily increasing consumption of oil, both in absolute figures and in relation to other sources of energy, and an increasing dependence on the Middle East as a principal source for its oil supply.

European oil consumption before the war (1938) amounted to only 27 million tons a year. In the immediate postwar period (1947) it rose to 37 million tons a year. Since then it has grown at a rate of 13 per cent a year, reaching 100 million tons in 1955. In 1956 Europe's consumption was 115 million tons a year. The 13 per cent rate of increase was unusually high (as compared with 6 per cent in the United States and 11 per cent in the rest of the world) and could be attributed to the extraordinary pace of European recovery under the stimulation of the Marshall Plan. Yet according to a special study of the Organization for European Economic Cooperation [1] the rate of increase between 1955 and 1960 may be expected to amount to 9 per cent and between 1960 and 1975 to 6 per cent (equivalent to the current United States rate). At this pace European oil consumption would be 153 million tons a year in 1960 and 340 million tons in 1975. These figures are impressive. It

[1] L'Organisation Européenne de Cooperation Economique, *Le Pétrole: Perspectives européennes* (Paris, 1956). Many statistical data in this section are based on this study. The Organization for European Economic Co-operation (OEEC) comprises the following member countries: Austria, Belgium, Denmark, France, Germany, Greece, Iceland, Ireland, Italy, Luxembourg, the Netherlands, Norway, Portugal, Sweden, Switzerland, Turkey, and the United Kingdom. References to "Europe" in this section should be understood as relating to these countries. Inasmuch as Spain and Finland do not belong to the OEEC, the total figures for Europe outside the Soviet bloc should be revised upward to take account of these two countries' consumption.

should be pointed out, moreover, that they represent not only an increase in the consumption of oil in absolute figures but also a gradual displacement by oil of other sources of energy. According to studies made by the OEEC,[2] oil accounted for 8 per cent of the total European energy consumption in 1938. In 1949 this figure rose to 11 per cent and in 1955 to 18 per cent. In the latter year coal still accounted for 74 per cent of the total, and natural gas and hydroelectric energy claimed 1 per cent and 7 per cent, respectively.[3] Yet with definite natural limitations on the use of coal and hydroelectric energy and the difficulties in obtaining adequate manpower for expansion of the coal industry, it was inevitable that with the passage of time oil should claim an ever larger share in the total European energy consumption.

Actually consumption figures do not tell the whole truth about Europe's oil imports; the latter, because of a certain amount of re-exportation, are somewhat larger than the figures given above. In 1955 Europe's total gross imports amounted to 113 million tons of oil, of which 21 million came from the Western Hemisphere and 90 million from the Middle East. (Of the latter figure 88 million tons were crude oil and 2 million were refined products.) Thus the Middle East was responsible for about 80 per cent of European oil imports.

In analyzing Europe's growing oil needs, one should keep in mind three factors: the relationship between economic development and the consumption of oil, Europe's refining capacity, and transportation facilities. With regard to the first, an obvious causal link exists between European industrial production and the consumption of oil. Industrial production has been growing in the postwar period at a rate of 8 per cent a year. This has been reflected in the increased consumption of all forms of energy at a rate of 5 per cent a year. Within this total picture, oil, as mentioned earlier, accounted for a 13 per cent increase a year.

[2] OEEC, *Europe's Growing Needs of Energy: How Can They Be Met?* (Paris, 1956).
[3] Comparable figures for the United States in 1955 were as follows: oil, 44%; coal, 26%; natural gas, 25%; and hydroelectric energy, 5%.

EUROPE'S REFINING CAPACITY

Europe's increasing oil consumption was accompanied by a spectacular growth in refining facilities. In 1948 crude oil refined in Europe amounted to less than 20 million tons; in 1955 it reached 103 million tons, about 90 per cent of Europe's needs. The actual capacity of European refineries rose to 120 million tons. This expansion represented a total investment of between $1,750,000,000 and $2,250,000,000. Refineries necessitate auxiliary installations, which increased the investment some $500,000,000. If we add that it was also necessary to develop distribution facilities and means of transportation, it is probable that investments in these two fields have reached nearly $1,000,000,000. The general trend in the postwar period has been to construct the refineries close to the centers of consumption and away from the sources of production. This trend was accentuated after 1951, i.e., after the nationalization of oil in Iran and the resulting shutdown of the Abadan refinery. In the 1950's many new refineries were constructed, and the capacity of the existing ones was enlarged, both on the continent and in the British Isles. Especially pronounced was the rise of refining capacity in France, which with 33 million tons a year in the mid-1950's assumed first place in Europe. Britain was a close second, with a capacity of 31 million tons a year. The latter was strengthened by the construction of a big refinery in the British crown colony of Aden. Italy, with a capacity of 26 million tons a year, occupied third place, followed by Western Germany and the Netherlands, with annual refining capacities of 16.5 million and 14 million tons, respectively.

This marked increase in European refining capacity has had a doubly beneficial effect on Europe's economic position. First, it has permitted Europe to effect considerable savings in foreign exchange, a large portion of which had hitherto had to be spent on higher-priced imported refined products. Second, it has rendered Europe more independent of political uncertainties in the producing and transit areas. Europe's gain in this respect is obviously bound to affect the pattern of oil development in the Middle East. The latter, generally anxious to industrialize itself and to emerge from a raw-material-producing economy of a colonial type, has increasingly insisted on the construction of refineries near the sources of production. This conflict of interest has thus far been latent rather than actual because of the dynamic rise in production (the latter providing ever-mounting revenues to local governments) and also because of gradual construction of local refineries, which have had to be geared to the steadily rising Middle Eastern consumption.

TRANSIT ROUTES AND TANKER FLEETS

Transportation is another vital factor in Europe's demand for oil. Two elements have to be taken into account: one is the existence and serviceability of transit routes for Middle Eastern oil and the other is the availability of adequate tanker tonnage. In 1956 oil for Europe was being shipped by two major routes: the Suez Canal (65 million tons a year) and the Syrian pipelines, IPC and Tapline (36 million tons a year). The capacity of both routes was greater than the figures just given, inasmuch as the Suez Canal was handling a total of 77 million tons of oil a year (1.5 million barrels a day) and the combined throughput of IPC and Tapline pipelines amounted to 41 million tons a year (0.83 million barrels a day). The additional 12 million tons transported via Suez and 5 million tons pumped through the pipelines had the Western Hemisphere or Africa as their destinations. In case of emergency in Europe—and such an emergency occurred late in 1956 [4]—these "surplus" quantities could be redirected toward the European continent.

The steadily rising volume of oil shipments to the West necessitated an increase in the throughput capacity of the pipeline and the Suez Canal. Both the IPC and Tapline have taken steps to expand their lines, the IPC by planning for the construction of a new 24-inch pipeline with a throughput capacity of 9 million tons, linking Homs with Banias, and Tapline by launching a program aiming at the increase of its own capacity from 16 million tons a year (320,000 barrels a day) to 22 million tons a year (440,000

[4] See below, Chapter XVI, "Repercussions of the Suez Crisis."

barrels a day) by 1958.[5] Because of the Suez crisis late in 1956 IPC's expansion plans were suspended. Tapline, however, went ahead with its project, and in 1958 its work was completed.

Increase in the clearing capacity of the Suez Canal presented a complicated problem inasmuch as it involved not only the arduous technical task of deepening and widening the navigable channel but also of settling the issue of political control. By mid-1956 the Suez Canal was capable of handling vessels with a maximum draught of 35 feet.[6] A program of improvements prepared by the Suez Canal Company was aiming at an increase of the permissible draught to 36 feet in the first stage, to be followed by an increase to 40 feet in the succeeding phase. With a draught of about 36 feet the Canal could accommodate fully loaded tankers of up to slightly under 40,000 dead-weight tons. Larger tankers would have to pass only partly loaded, and passage of those above 65,000 tons would have to be subjected to strict limitations on account of the wash and suction effects on the banks of the Canal. Inasmuch as most tankers were designed to permit them to pass through the Canal, the majority of tankers under construction or on order in 1956 were less than 40,000 dead-weight tons. Because of both economic considerations and political uncertainties, the trend is now toward the construction of supertankers of 80,000 tons or more. Tankers like *Al-Malik Saud al-Awal,* of 46,550 dead-weight tons (with 344,000-barrel capacity), are beginning to make their appearance, carrying oil by routes other than the Suez Canal. In 1957 the newly commissioned *Universe Leader* beat all records with 85,500 dead-weight tons. Its capacity load is 590,000 barrels of oil. In the same year five tankers of 104,500 tons were ordered in Japanese shipyards by the National Bulk Carriers, Inc., of New York. Ships of this size could call on only a limited number of ports possessing adequate depth and berthing facilities. If kept busy, they not only would be less expensive than small tankers, especially of the standard 16,560-ton T2 type, but would also compete with the existing pipelines.

[5] United Nations, *Economic Developments in the Middle East, 1955–1956* (New York, 1957).

[6] Data in this section are largely drawn from OEEC, *Europe's Need for Oil: Implications and Lessons of the Suez Crisis* (Paris, 1958).

The adequacy of present tanker tonnage to supply the needs of Europe constitutes the second major aspect of the transportation picture. The world tanker fleet has experienced a spectacular expansion comparable to the production of oil itself. In 1939 the total number of tankers in the world amounted to 1,571 with an aggregate tonnage of 16.6 million tons. In 1945, despite heavy wartime losses (over 4 million gross tons, of which 2 million were lost in 1942 alone), the world tanker fleet reached the figure of 1,768, with a total tonnage of 21,668,000 tons, largely because of the intensive building program in the United States between 1942 and 1945. Most of the T2 tankers were constructed at that time; by 1955–1956 they accounted for about 25 per cent of the existing world tonnage. The world tanker fleet has expanded rapidly since the war. In 1956 it numbered 2,850 tankers, with an aggregate capacity of more than 44 million dead-weight tons, the average tanker capacity being a little less than 16,000 dead-weight tons. By the middle of 1956 all of this fleet was fully employed. About 25 per cent of it was used for shipments to the East Coast of the United States, close to 50 per cent was engaged in transporting oil to European ports, and the remainder was employed in other movements. There was no unemployed surplus of tankers in 1956, except for the United States Reserve Fleet, representing about fifty vessels. Consequently any distruption in tanker movements, necessitating longer voyages, was liable to produce serious complications and shortages in oil-consuming countries.

Steadily rising world production and consumption of oil will necessitate further increases in tanker tonnage in the years to come, regardless of new pipeline construction. According to a study by the OEEC, the aggregate tonnage of tankers under construction or on order in 1957 amounted to 30 million dead-weight tons (or 12.7 million gross tons), which was about 67 per cent of the then-existing world tanker tonnage.[7] These huge orders were due partly to the earlier-mentioned estimates of steadily rising consumption, partly to the desire of certain oil companies to lessen their de-

[7] OEEC, *Maritime Transport: A Study by the Maritime Transport Committee in May 1957* (Paris, 1957), p. 49. On Jan. 1, 1958, over 300 tankers were under construction with a total of nearly 5 million gross tons. Of these some 100 were of the super type (*New York Times*, Feb. 9, 1958).

pendence on the politically vulnerable pipelines and partly to the need of replacing many of the T2 tankers designed for wartime use, whose serviceability will end soon after 1960.

A vital consideration in the tanker situation is, obviously, the flag and ownership of the fleets. Most striking is the fact that the tonnage controlled by Russia and the Soviet bloc is negligible. On January 1, 1959, the combined Soviet, Chinese, and Polish Communist tanker fleet amounted to about 70 units of an aggregate weight of 821,000 dead-weight tons, equivalent to 1.53 per cent of the world tonnage of 53,500,000 dead-weight tons. The rest, i.e., over 98 per cent, was owned by the nations of the Free World. Of these, about 6 million tons were under United States registry and about 15 million tons under Liberian and Panamanian registry, most of the latter two representing American interests. More than half of the world fleet—about 27 million tons—flew the flag of European nations grouped in OEEC. Norway controlled 8,500,000 tons, closely followed by the United Kingdom with 8,000,000 tons. Other countries possessing large tanker fleets were France (2,600,000 tons), Italy (2,300,000 tons), Sweden (1,850,000 tons), Holland (1,700,000 tons), and Denmark (1,100,000 tons). Tanker fleets were partly owned by independent shipping interests. Ninety per cent of the tonnage, however, was either owned or controlled by the big oil companies, especially the seven major American, British, Dutch, and French companies.[8] Inasmuch as these companies control most of the Middle Eastern oil production—usually through affiliate companies sometimes owned jointly—it follows that movements of oil from the Middle East to European and other destinations are largely dependent on executive decisions emanating from these major companies.

COST OF IMPORTS

It is clear that the increasing dependence of Europe on oil imports necessitates considerable expenditure of foreign exchange,

[8] Distribution percentages of the Free World tanker tonnage in 1959 were approximately as follows: company-owned tankers, 40%; long-term charters for company use, 36%; consecutive-voyage charters (1–2 years' duration) for company use, 14%; for a total owned and controlled by oil companies of 90%. On-the-spot tonnage was 10%.

thus affecting the totality of European payment balances. The OEEC Oil Committee has estimated that the total cost of delivered oil for member nations was $2,050,000,000 in 1955 and will be $3,150,000,000 in 1960, and $8,000,000,000 in 1975.[9] To this approximately 10 per cent must be added for the cost of bunkering the European-registered tankers. The figures thus obtained represent the total expenditures. From the point of view of the European balance of payments, any European receipts would represent a credit, thus offsetting the basic imbalance. It has been estimated that the receipts for freight carried under European flags amounted to $590,000,000 in 1955 and will amount to $760,000,000 in 1960 and $1,590,000,000 in 1975, which would equal 75 per cent of the total freight cost for oil supplied to Europe. To this we may add various expenditures made by the companies in Europe, including substantial local purchases. A good example is the Arabian American Oil Company, which in order to reduce its expenditures in dollar areas has set up a special affiliate, The Aramco Overseas Company, with headquarters in The Hague for the purpose of making purchases on the European and other nondollar markets. Arrangements of this kind obviously tend to lessen the acuteness of the payment balance problem for European importers of Middle Eastern oil.

CONCLUDING REMARKS

This general review of the role of Middle Eastern oil in the European economy would be incomplete without at least mentioning the role that these oil imports play in the life of certain individual countries. The dependence of various European countries on oil as a source of energy is far from uniform. Measured in hard-coal equivalents, the consumption of petroleum products varied in 1954–1955 from 82 per cent of the total energy consumption in Greece to 9 per cent in Germany. Corresponding percentages were 13 for the United Kingdom, 22 for France, 32 for Italy, 40 for Sweden, and 24 for Norway and the Netherlands.[10] These figures are obviously not stationary, and whereas the average increase of oil

[9] These figures are from OEEC, *Le Pétrole*, p. 79.
[10] OEEC, *Europe's Growing Needs of Energy*, pp. 75 ff.

consumption in Europe has been, as mentioned earlier, 13 per cent a year in the postwar period, percentages for individual countries have varied considerably, showing in some cases, especially in highly industrialized countries, spectacular increases. Thus, despite the relatively modest consumption of oil in the over-all energy picture in Germany, the latter's rate of consumption increase was 20 per cent a year in the postwar period. In Italy the increase was even more rapid, amounting to 120 per cent between 1950 and 1955.

It should be evident from these figures that not only have imports of Middle Eastern oil played a vital role in the postwar development of European economy, but that they will steadily expand their role in the over-all European energy consumption, thus forging ever-stronger links between the economic and strategic destinies of Europe and the Middle East.

◈◈ CHAPTER III ◈◈

Oil in the Economy of the Middle East

IN the preceding chapter we sketched the role of Middle Eastern oil in European economy. It is legitimate to ask at this juncture: How important is oil to the Middle East itself? What does it represent in terms of investment, government revenues, and employment? Is it conducive to the change from a primitive to more advanced levels of economy? Are the producing countries satisfied with the extracting operation or do they develop refining facilities as well? Do they acquire and increase their own tanker fleets? How about the transit states: Do they share in the oil-generated benefits and, if so, to what extent? And are the oil revenues being spent wisely on development programs or are they being hoarded or wasted on unessential luxuries?

To answer these questions, let us begin with the volume of production. In 1958 the Middle East was producing oil at the rate of 215 million tons a year (4.3 million barrels a day), which represented about 24 per cent of the world total of 885 million tons a year (17.7 million barrels a day).[1] The comparable figures for 1956, the year ending with the Suez crisis, were 190 million tons a year (3.8 million barrels a day) for the Middle East, which constituted 22 per cent of the world total of 840 million tons.[2] The rapidly ex-

[1] *Oil and Gas Journal,* Dec. 29, 1958.
[2] Actual Middle Eastern production in 1956 amounted to 170 million tons, the reduction being due to the Suez crisis which disrupted production in the last two

panding oil industry necessitated large investments of capital, estimated to have reached a total of $2,440,000,000 in the 10-year period between 1946 and 1955. The annual rate of investment in 1955 and 1956 was about $210,000,000, or about one-third less than it had been in the immediate postwar years but still not negligible. This investment was distributed as follows:

Production	$ 980,000,000
Marketing	75,000,000
Pipelines	587,000,000
Refineries	465,000,000
Other	333,000,000
Total	$2,440,000,000

The steadily increasing production, combined with the larger share in company profits obtained by the producing countries after 1951–1952, had its effect on the rise of government revenues. Direct revenues collected by the governments of the producing countries from the oil operations rose from about $500,000,000 in 1953 and $680,000,000 in 1954 to about $880,000,000 in 1955 and $940,-000,000 in 1956. The estimated figure for 1958 amounted to slightly over $1,274,000,000. The region did not benefit from oil operations in a uniform fashion. Foremost among the beneficiaries were the four major producing countries, Kuwait, Saudi Arabia, Iraq, and Iran. Of these only Iraq and Iran regularly publish state budgets, and in both countries operating oil companies have released figures of their payments to the governments. In 1955 Iraq's oil revenues amounted to $207,000,000. As a result of the drop in production due to sabotage of Syrian pipelines in November 1956, these revenues decreased to $193,000,000 in 1956 and to $137,000,000 in 1957. Following repair of the pipeline, the figure for 1958 was $235,000,000, thus indicating a return to the normal pattern. By contrast, Iran's revenues from oil rose from $90,000,000 in 1955, to $214,000,000 in 1957, and to $246,000,000 in 1958, showing an up-

---

months of the year. Of the total produced in the world, the output of the Free World (i.e., of countries other than Russia, eastern Europe, and Communist China) amounted to 780 million tons. Data in this section are mainly based on UN, *Economic Developments in M.E.*

ward trend stronger than the Middle Eastern average as Iran resumed its prominent place among the producing states. Kuwait's revenue in 1958 was estimated at about $415,000,000 and Saudi Arabia's at about $300,000,000, and the two lesser producers, Qatar and Bahrein, received $57,000,000 and $11,000,000, respectively.

Proceeds from oil accounted for an estimated 90 per cent of Kuwait's total government revenue and an estimated 85 per cent of the Saudi Arabian national budget. Since the conclusion of the Consortium agreement oil money has become an ever-greater proportion of the Iranian state budget. In 1955 it accounted for 41 per cent of the total state receipts, in 1956 for 45 per cent, and in 1957 for 51 per cent. These figures represent a substantial advance over the pre-Mossadegh period when Iran's revenues from oil oscillated between one-eighth and one-fifth of all government revenues.[3] In Iraq oil revenues were approximately 70 per cent of the total receipts of the government in 1955 and 1956.

In addition to these direct payments, the producing countries obtained a number of other financial benefits stemming from local purchases by the operating companies, payroll expenditures for national employees, awards of contracts to local entrepreneurs, local spending by the companies' foreign personnel, payments of customs duties for the companies' dutiable imports, and certain other payments such as, for instance, contributions to social insurance

[3] The exact percentage is somewhat difficult to ascertain in view of the discrepancy between the data supplied by the Iranian government and those emanating from the Anglo-Iranian Oil Company. Official Iranian sources asserted that at no time during the 40-year period of oil production did royalty payments exceed 15 per cent of total government revenues. (See "Iran Presents Its Case for Nationalization," *Oil Forum*, March, 1952; hereafter referred to simply as "Iran Presents Its Case.") More specifically the percentage of oil revenue in the total receipts was stated to be 12 per cent in 1950. (See United Nations, *Review of Economic Conditions in the Middle East, 1951–1952* [New York, 1953], p. 77.) The AIOC claimed that in 1950 its payments of about £16,000,000 to Iran should be compared to the total national budget of about £90,000,000. This would represent 17.8 per cent of the total. (See *The Anglo-Iranian Oil Company and Iran: A Description of the Company's Contributions to Iran's Revenue and National Economy; and of Its Welfare Activities for Employees in Iran* [n.p., 1951], p. 3.) No attempt to reconcile the differences between these figures was made by either party.

schemes. As an example, Tables 1 and 2 give the figures on these additional benefits published by the international Consortium operating in Iran:

Table 1. Payments of the Consortium to Iran (in round figures)

	1955	1956	1957
Paid to government and NIOC (for stated payments) *	£36,500,000	£55,000,000	£77,400,000
Paid to companies' Iranian staff and labor and to social insurance fund	20,500,000	19,500,000	20,900,000
Purchases on local market and payments to local contractors	3,000,000	2,500,000	3,500,000
Total	£60,000,000	£77,000,000	£101,800,000

Source: Iranian Oil Exploration and Producing Company and Iranian Oil Refining Company, *The Iranian Oil Operating Companies: The First Year of Activity* (London, 1955). See also *Iranian Oil Operating Companies* (n.p., 1956, 1957).
* Stated payments include income taxes on trading and operating companies, taxes on salaries, custom duties, and the like.

Table 2. Foreign exchange brought to Iran by the Consortium (in round figures)

	1955	1956	1957
Payments to government and NIOC (stated payments and companies' income tax)	£35,000,000	£54,000,000	£76,400,000
Other payments	18,000,000	16,000,000	15,000,000
Total foreign exchange	£53,000,000	£70,000,000	£91,400,000
Foreign exchange earned by Iran from nonpetroleum exports	£42,000,000		£38,000,000

Source: See Table 1.

The Middle East benefited in many ways other than financial from the companies' operations. Notable among these benefits was the substantial employment of local manpower. In 1956 the Iraqi oil industry was employing close to 11,500 nationals, the Saudi Arabian 13,000, the Kuwaiti 6,000, and the Iranian 45,000. Of the lesser producing countries, Bahrein's corresponding employment figure was 6,000 and Qatar's 2,000. Among the transit countries, the manpower used on the pipelines and in the refineries totaled 3,600 persons in Syria and 2,800 in Lebanon. These figures refer to the nationals of the countries concerned who were employed directly by the oil companies in exploration, production, refining, or pipeline operations. To this should be added less stable but not negligible figures of laborers employed by local contractors for jobs connected with the oil industry. Local distributing operations, frequently conducted by different marketing organizations, gave employment to additional thousands of workers. Substantial numbers of foreigners were also employed by the producing and pipeline companies. In 1957 this personnel was estimated, for the whole of the Middle East, at some 250,000 individuals. Their purchasing power, substantially higher than that of most people in the area, acted as a considerable stimulus to local economy.

REFINING, CONSUMPTION, AND NATIONAL TANKER FLEETS

Among the advantages the growing oil industry was bestowing upon the Middle East was development of the area's refining capacity. This is noteworthy in view of the trend mentioned earlier and observable in Europe to increase the refining facilities located close to the major markets. By mid-1958 the Middle East possessed twenty refineries and topping plants of an aggregate capacity of over 75 million tons. A considerable part of this capacity is not utilized, especially in the Abadan and Haifa refineries; in the fall of 1956 actual refining operations were being carried on at a rate of 50 million tons per year and in the fall of 1957 at about 55 million tons. The 1956 figure was an increase of about 35 per cent over the preceding year, which, in turn, exceeded by 23 per cent the refining operations of 1954. In 1956 the Middle East not only returned to, but even slightly surpassed, the aggregate refining done in 1950 when the Abadan refinery was in full operation. The increases were due to reactivation and overhauling of the Abadan refinery, increased production at the Aden refinery, the launching of three new refineries—Daura in Iraq, Sidon in Lebanon, and Batman in Turkey—and the expansion of the Bahrein and Kuwait refineries. With the completion of its new plant early in 1958 the Kuwait refinery increased its capacity from 1.5 million tons to 11 million tons. Further increases in refining capacity were planned for the late 1950's and the early 1960's. These included construction of new refineries in Alexandria (capacity 200,000 tons) and Cairo (2 million tons), formation of a refinery company in Jordan with a capital of JD (Jordan dinars) 3,000,000 (approximately $8,400,000), construction of a government refinery in Homs (Syria) under a contract awarded to a Czechoslovak firm (750,000 tons), expansion of the capacity of the Batman refinery from 300,000 tons to 380,000 tons, and construction of another refinery in Izmir (750,000 tons).

The growth of the region's refining capacity was due partly to the companies' policies and partly to the desire of the host governments to achieve higher economic standards by emerging from the raw-material-producing stage into the manufacturing stage. Three incentives were operating: a desire for the greater revenue rendered possible by the higher prices commanded by refined products, a general desire for industrialization, and a need to supply local populations with adequate amounts of refined petroleum products. Local demand for these products has been growing at the rate of 13.6 per cent a year since 1950. In 1955 the Middle East consumed a total of 10.82 million tons of major refined petroleum products. Egypt alone accounted for about one-third of this amount, with Iran, Israel, Turkey, and Iraq each claiming between 10 and 13 per cent of the total. As could be expected in a region rich in oil and relatively poor or underdeveloped as to other energy resources, oil accounted in 1954 for 69.4 per cent of the total energy consumption. This figure would be considerably higher if we excluded Turkey, the only country that consumed more solid fuels than oil (about four times more). Other countries of the region relied overwhelmingly, and in some cases exclusively, on oil as a source of energy.

Another by-product of the oil industry in the Middle East was the stimulation it gave certain countries to develop their own tanker fleets. Egypt, which in 1955 possessed two tankers with a combined capacity of 26,000 dead-weight tons, spent £E (Egyptian

OIL IN MIDDLE EAST ECONOMY

pounds) 700,000 (approximately $1,960,000) in 1955–1956 on the purchase of tankers while earmarking £E1,500,000 for further acquisitions in 1956–1957. In Iran orders were placed by a private group for the construction of two 32,000-ton tankers in Holland, to be chartered by the National Iranian Oil Company. In Iraq studies were undertaken looking toward the creation of a tanker fleet of 100,000 tons. By 1956 Turkey had acquired eight tankers with an aggregate capacity of 95,000 dead-weight tons, and in Kuwait a government-sponsored company was formed, with the co-operation of the oil companies, for the purpose of establishing a tanker fleet. Its capital was reported to be £3,700,000. Although not directly connected with the oil operations in the Middle East because of the Arab boycott, Israel began developing its own tanker fleet, primarily of course, to import oil from abroad. The first tanker was acquired in 1954. By mid-1956 Israel's total tanker fleet consisted of four tankers with a combined tonnage of 54,000. Three 19,500-ton tankers were expected to be delivered by German shipyards in 1958, with further acquisitions scheduled thereafter.[4]

BENEFITS FOR TRANSIT COUNTRIES

Although the producing countries benefited far more than the transit countries in terms of cash revenue and additional advantages, the latter were by no means deprived of the benefits generated by the oil industry. Pipeline conventions usually provided not only for transit and terminal fees, but also for payments for protection and certain other services.

In 1956 and thereafter Syria's revenue from IPC pipeline operation was to amount to $18,332,000, according to the upward revision of fees in December 1955. This estimate was based on the assumption that the throughput via Syria would be 26 million tons of crude, of which 18.5 million would go through Banias. Syria's actual earnings fell to about $13,440,000 in 1956 as a result of the disruption of IPC's pipeline traffic in the wake of the Suez crisis, but even the latter figure was considerably in excess of $4,420,000, which Syria had previously been receiving. At the same

[4] Data in this section are drawn from UN, *Economic Developments in M.E.*

time Syria's revenue from Tapline amounted to about $1,160,000 a year. The total of $15,000,000 obtained from the pipeline systems represented about 16 per cent of the total Syrian government revenue in 1956. This was expected to increase with the anticipated conclusion of a revised agreement with Tapline.[5]

Lebanon's cash benefits from the pipelines crossing its territory were not impressive prior to 1958, largely because of delays in reaching definite agreements with IPC and Tapline. According to tentative proposals advanced in the course of negotiations with IPC in March 1958, Lebanon's revenue was expected to increase from $980,000 to $2,980,000 a year, on the assumption of 7.5 million tons throughput, of which 7 million tons would go through the port of Tripoli. Simultaneously the government was receiving about $1,250,000 a year from Tapline, a figure expected to be at least doubled when a new agreement with Tapline was concluded. In 1956 Lebanon's total revenue from pipeline operations (about $2,200,000) amounted to 4.6 per cent of the total state revenue. This was very modest in comparison with Syria. Lebanon, however, enjoyed a number of supplementary benefits, such as location of Tapline's headquarters in its territory, considerable purchases of food and other locally produced necessities by Aramco, Tapline, and IPC, chandlering for tankers calling at the Sidon and Tripoli terminals, tourist spending by oil company employees, and business opportunities for various contractors in servicing the pipelines, the terminals, and the refineries of Tripoli and Zahrani.

To conclude this section, mention may be made of the benefits accruing to Egypt from the transit of oil through the Suez Canal. The Suez traffic showed a spectacular rise from 7.8 million tons of cargo in 1900 to 30.6 million in 1947 and 107.5 million in 1955. Although oil accounted for only 14 per cent of the total tonnage transited in 1930, its proportion rose to 45 per cent in 1947 and to 62 per cent in 1955. In 1956, prior to the crisis of October and November, the Canal traffic amounted to 122 million tons a year,

---

[5] Calculations made on the basis of data contained in *Recueil des lois syriennes et de la législation financière: Exposé du Ministre des Finances sur la situation économique et financière de la Syrie en 1956* (Damascus, 1957) and UN, *Economic Developments in M.E.*

of which oil cargoes accounted for 63 per cent and, more particularly, Middle East oil exports for 61 per cent. It is thus clear that the steadily increasing tanker traffic has been primarily responsible for the impressive upward trend in the aggregate figures of traffic. Before the nationalization of the Suez Canal Company on July 26, 1956, Egypt benefited from the Canal operations in two ways: from its share of the transit dues paid by the company and from payments for various services and goods supplied to the transiting ships during their brief stops in the Canal Zone. Some of these services were rendered by the authorities, thus adding to the direct payments received by the Egyptian government. Others, including catering to tourists, were provided by private interests. This made it exceedingly difficult to calculate the additional benefits accruing to the Egyptian economy as a whole. In 1955 direct payments of the company to the Egyptian government amounted to about £E6,000,000, while the payments to the government-provided services and goods reached £E5,000,000, adding up to a total of about £E11,000,000 ($30,800,000). This represented approximately 5.7 per cent of the budget revenue.

With the nationalization of the Canal, Egypt's revenue from transit fees was expected to rise to £E50,000,000 ($140,000,000) a year. This figure was not, however, immediately realized, largely because many ships in transit continued to pay their fees to the account of the Suez Canal Company in Paris rather than directly to Egypt—a situation accepted by the Egyptian government pending final settlement with the nationalized company. In any event it was evident that oil cargoes would continue to be a major source of government revenue from the Suez Canal and, with anticipated increases in Middle Eastern production and European consumption, would probably claim an even larger share of the total Suez traffic.

MIDDLE EAST DEVELOPMENT PROGRAMS

By far the greatest advantage derived by the producing and transit countries from oil operations was financial support for impressive development schemes. Of the four principal producing countries, Iraq, Iran, and Kuwait made major efforts to direct their oil revenues toward development plans. Ironically, financial capacity based on oil income in these three countries was in inverse proportion to size and needs. Kuwait, with the highest revenue, had to cater to the needs of a small area inhabited by only 250,000 people. Iraq, whose oil revenue was about one-third lower than that of Kuwait, had to take care of about 5 million inhabitants. It was, however, blessed with potentially fertile soil and an abundance of water in its two major rivers. Iran's position was the least favorable of the three, especially in the immediate post-Mossadegh era, inasmuch as its oil revenues, amounting to about one-half of Kuwait's, had to serve a country of the combined area of France, Britain, Spain, and Italy, inhabited by some 17 million people.

Economic Developments in Kuwait

Of these three countries, Kuwait is an absolute patriarchal monarchy in which no clear distinction has thus far been made between the ruler's private purse and the public treasury. With the sudden influx of wealth from oil—i.e., after the fifty-fifty profit-sharing agreement of 1951—an attempt is being made to transform the town of Kuwait into a model city and to establish a welfare state within a short space of time.[6] Oil revenue has accordingly been roughly divided into three parts, of which one is to pay current expenditures, one is to serve the needs of development, and one is to be invested abroad. This division was made without prior establishment of sound administrative and budgeting practices, and as a result considerable confusion ensued. Typical of the initial chaotic state of affairs was the arrangement between the government and five major British engineering firms—each in partnership with an arbitrarily selected Kuwaiti merchant—whereby those firms were to execute a variety of projects according to a plan prepared by a British expert.[7] The monetary arrangements were obviously incompatible with sound public financial practices: the firms in question were to be paid for the work done on the basis of cost plus 15 per cent, with no ceiling on expenditures. This led, as

[6] For a general description, see Sir Rupert Hay, "The Impact of the Oil Industry on the Persian Gulf Shaykhdoms," *Middle East Journal*, Autumn, 1955.

[7] For details, see Elizabeth Monroe, "The Shaikhdom of Kuwait," *International Affairs*, July, 1954.

could be expected, to much waste, expressed in part in unnecessarily lavish and elaborate structures contrasting with drab native buildings. Moreover, the privileged position of the five local partners in these ventures caused considerable discontent among other merchants. This combination of financial and social factors led the government to cancel its agreements rather abruptly with the five contracting firms, which brought about a slowdown, if not a complete standstill, in the development program in 1954–1955.

Even these lessons of mismanagement have not impressed upon the government the need for instituting an adequate budgetary system which would be both strictly adhered to and published. Consequently in figuring out Kuwait's public expeditures one is obliged to rely on estimates. According to the latter,[8] out of its total income of about $280,000,000 in 1956 (deriving mostly from oil royalties and income taxes paid by the oil companies), Kuwait's government expenditures accounted for about $160,000,000, the remainder or $120,000,000 being invested by the ruler in the London stock and securities market. Twenty per cent of the expenditures went for salaries, and the remaining 80 per cent or about $130,000,000 were channeled into the economic development program.

This program, administered by the Development Board, began moving forward again after the 1954–1955 interruption. Among the projects scheduled before 1954 were fifty-five public schools, one of which, the secondary school at Shuwaikh occupying spacious grounds over 100 acres in extent and capable of accommodating 1,200 pupils, was designed to become a university later. The student population of Kuwait rose to 20,000 in 1956, served by over 1,000 teachers, brought in mostly from Egypt and Palestine. Notable among the educational institutions was the Technical College at Shuwaikh, which was to provide training for 600 students in engineering and other technical skills. In the field of public health the government's aim was to provide medical treatment for the whole population free of charge. A hospital was opened in Sulaibakhat in 1954,

[8] U.S. Department of Commerce, Bureau of Foreign Commerce, World Trade Information Service, *Economic Reports*, pt. 1, no. 57–32, "Economic Developments in Kuwait, 1956."

and another general hospital, with 750 beds, was under construction in 1956–1957. Among the early public utility projects was a sea-water distillation plant—the largest in the world—capable of providing 2 million gallons of drinking water a day and designed for an eventual output of 5 million gallons. This has been in operation since 1952. In 1955 the first main power station with a total capacity of 30,000 kilowatts was completed.[9]

The newer projects under the development plan, i.e., those completed or begun in 1956–1957, encompassed such a variety of works as a sand-lime brick factory, a concrete factory, twelve new schools, 1,500 low-income-group housing units, a second municipal power station, street construction and demolition under the town-planning scheme, expansion of the distillation plant, port dredging and pier construction, asphalting of airport runways, construction of a road to Basra, new municipal and justice buildings, and a new state-owned hotel.

All in all Kuwait's development was progressing at a spectacular rate. The sheikhdom's per capita income, approximately $1,500 a year, was among the highest in the world, and the ratio of 1 motor vehicle per 12 inhabitants (as compared with 1 for every 34 persons in Italy) testified to the higher living standards achieved by the population. There is no doubt that, in contrast to other oil-producing states, Kuwait is relatively free of many perplexing problems and is in a position to minister to the needs of its small population. Consequently even the mistakes committed in administering the huge revenues cannot prevent considerable progress from being achieved. Moreover, it should be pointed out that although Kuwait's ruler is following paternalistic practices, he is genuinely desirous of lifting the living standards of his people and providing them with adequate social services, and his personal habits are free from lavish ostentation. Kuwait's real challenge lies not so much in its ability to develop itself as in its potential role in developing the resources of the Middle East on a regional basis. This problem will be discussed later.

[9] For details on this phase of development, see Kuwait Oil Company, *The Story of Kuwait* (London, 1957), pp. 66 ff.

Iraq's Development Plan

Co-ordinated planning for the development of Iraq's economy began as early as 1950, i.e., two years before the revision of IPC's concession agreement that appreciably raised Iraq's oil revenues. In that year legislation was passed establishing a Development Board, providing for a 5-year program, and allocating all oil revenues to the board. A year later the first 5-year program was formally enacted. It called for the expenditure between 1951 and 1955 of ID (Iraqi dinars) 66,000,000 ($185,000,000).[10] The requirements of the government budget, however, and the general domestic situation made it politically inadvisable to practice too much austerity. Consequently in 1952 the basic development law of 1950 was amended retroactively to 1951, to allocate 70 per cent of oil revenues to the board, the remaining 30 per cent being made available to fill anticipated gaps in the ordinary budget. In the meantime two important events further influenced Iraq's development planning. One was the submission in February 1952 of a report by a World Bank mission which had earlier been invited to survey Iraq's economy and present recommendations. The other was the conclusion in the same month of a revised agreement with Iraq Petroleum Company, retroactive to January 1951, thanks to which Iraq's oil revenues rose from about $15,000,000 in 1950 to about $110,000,000 in 1952, with further anticipated increases. Iraq, as one observer has put it, has "come into money." Consequently the original 5-year plan was converted in March 1952 into a 6-year plan, 1951–1956, providing for a total expenditure of ID155,000,000 ($435,000,000).

The new broadened plan required certain organizational adjustments. In 1953 the basic development law was replaced by a new law, which created a Ministry of Development and provided for a new organization of the Development Board.[11] Henceforth the board was to be composed of ten members, of whom three were to

---

[10] One ID equals $2.80 in U.S. currency.

[11] This was Law no. 27 of 1953 for the Development Board and the Ministry of Development, which established the system now in existence. Its text can be found in *Al-Waqai' al-'Iraqiyah*, no. 2280 of May 13, 1953.

be cabinet ministers and seven—to be called executive members—were to be appointed for 5-year periods by the Council of Ministers. Among the seven executive members one was to be a vice-chairman, one an irrigation specialist, and one a financial expert. Cabinet representation on the board was to include the Prime Minister as chairman and the Ministers of Development and Finance. Before long the two expert positions were filled by an American (Wesley Nelson) and a Briton (Sir Edington Miller) for irrigation and finance, respectively. The guiding principle was to make the board independent of political pressures and fluctuations, and this was achieved remarkably well. The Ministry of Development was divided into eleven component units, of which the so-called Technical Committees were charged with the bulk of the work. These were the First Technical Committee for irrigation, water drainage and storage, and flood control; the Second Technical Committee for roads, bridges, communications, public works, and housing; the Third Technical Committee for industrial, electrical, and mining projects; the Fourth Technical Committee for agriculture, forestry, and artesian wells; and, established in 1956, the Fifth Technical Committee for housing.

Conscious of the magnitude of its task the Development Board sought the advice of foreign experts. At the request of the government an American consulting firm, Arthur D. Little, Inc., prepared an industrial survey report, recommending among other things that Iraq should concentrate first on export industry, especially petrochemicals, to assist in generating increased purchasing power for a domestic consumer goods industry. At the invitation of the Development Board, Lord Salter prepared in April 1955 a report on the timing and balance of the different projects of the board and on their co-ordination with the actions of other authorities, with an eye to the impact of these projects upon the general economy of the country.[12] In addition, the United States Operations Mission (Point Four) was at hand to help Iraq with technical know-how.

Soon the steadily rising oil revenues allowed much bolder planning than in the initial stages. In 1955 a new 5-year program was

[12] Iraq Development Board, *The Development of Iraq: A Plan of Action*, by Lord Salter (n.p., 1955).

promulgated, providing for a total expenditure of ID304,000,000 ($850,000,000) between 1955 and 1959. This program was barely under way, however, when it was amended by Law no. 54 of 1956, which called for an expenditure of ID500,000,000 ($1,400,000,000) in the six years of 1955–1960. Table 3 shows the division of funds and their planned increases from the 1955 level to that of 1956.

Table 3. Iraq development plan, 1955–1956: Expenditures (in millions of Iraqi dinars)

		Law 43 of 1955 (1955–59)	Law 54 of 1956 (1955–60)	% of total
	Main projects			
I.	Administration studies and organization expenditures	5.5	7.5	1.5
II.	Flood control, irrigation, and drainage	108	153.5	30.8
III.	Roads, bridges, and communications	74	124.5	24.9
IV.	Main public buildings	20.5	37	7.4
V.	Summer resorts and rest houses	2	2.5	0.5
VI.	Housing	6	24	4.8
VII.	Industry, mining and electrification	43.5	67	13.4
VIII.	Development of animal, plant, and underground water resources	6.5	14.5	2.8
	Total main projects	266	430.5	86.1
	Minor projects			
IX.	Public buildings and institutes	32	59.5	11.9
X.	Miscellaneous projects	6	10	2.0
	Total minor projects	38	69.5	13.9
	Grand total	304	500	100.0

Source: Government of Iraq, Development Board and the Ministry of Development, *Law No. 54 for the Year 1956 Amending the Law of the General Programme of the Projects of the Development Board and the Ministry of Development No. 43 for the Year 1955* (Baghdad, 1956).

As will be seen from Table 3, the up-dated development plan called for the total expenditure of ID500,000,000 during the six years 1955–1960. To finance the plan revenues estimated at ID390,-000,000 were expected, of which ID60,000,000 were received in 1955. The remainder was expected to come in during the subsequent five years at the rate of about ID66,000,000 a year. This would leave a total deficit of ID110,000,000 or 22 per cent. However, the Development Board possessed a cash balance exceeding ID50,000,000 in unspent funds of previous years and, in addition, expected to effect savings of no less than 15 per cent of the estimated expenditures. Added together, these sums appeared sufficient to close the apparent gap.

Started in 1951–1952, the development program began to show the first major results in 1955–1956. Under the supervision of the First Technical Committee for flood control, irrigation and drainage, two major flood control structures were inagurated: the Wadi Tharthar and the Habbaniya projects. The first consisted of a dam constructed at Samarra, about sixty miles north of Baghdad, which harnessed the waters of the Tigris River, diverting them in case of need toward the extensive depression known as Wadi Tharthar. The project took four years to complete and cost ID11,000,000. Its principal objective was to prevent the flooding of major areas in south-central Iraq including Baghdad, which in 1954 had experienced the worst flood in forty-seven years, resulting in damage estimated at ID25,000,000. The second project provided a dam at Ramadi on the Euphrates River, whose waters in case of flood were to be diverted to a desert depression through Habbaniya Lake. Completed late in 1955, this project took five years and cost over ID4,000,000. Although primarily designed as flood control measures, both projects were eventually expected to serve irrigation programs also. Water stored in the Wadi Tharthar project would be able to irrigate 14 million acres of land. The greater Mussayib irrigation project was completed in 1956, adding some 250,000 dunams [13] to Iraq's cultivable areas.

The results were also encouraging as to road and bridge building. Two major bridges, at Kufa and Hindiya on the Euphrates, were completed in 1956. Work on roads to link the major urban centers of Iraq was also in progress in 1956.

The mid-fifties saw considerable addition to the country's indus-

[13] One dunam equals 0.62 acres.

trial capacity. New constructions included a cotton textile mill and two cement plants, and work was advanced on a sugar refinery and an asphalt refinery. Moreover, work was completed in 1956 on the petroleum refinery at Daura near Baghdad, with a capacity of 24,000 barrels a day.

Work conducted under the Fourth Technical Committee for agriculture embraced a successful smut control program, which according to estimates increased Iraq's 1956 grain crop 10 to 15 per cent over the previous year, as well as the drilling of numerous water wells.

A significant development also took place in the field of housing for low- and middle-income groups. With the creation of a special Housing Committee in the Ministry of Development, work was begun on the construction of nearly 3,000 dwellings, which, added to the 2,500 houses being built under the supervision of the Ministry of Social Affairs and to the housing program undertaken by the army for its personnel, were likely to relieve the housing shortage in the rapidly growing urban centers of Iraq.

All in all, thanks to the high level of oil revenues, Iraq at the beginning of 1958 could look forward to an era of spectacular economic expansion, provided it avoided political upheavals likely to affect the judicious use of its oil money. The revolution that occurred on July 14, 1958, shook the country to its foundations, ushering in an altogether new era socially and politically. Ostensibly dedicated to progress and reform, the new leaders bitterly attacked the prerevolutionary development program as failing to respond to the needs of the people and promised sweeping changes in its emphasis and execution. With the resulting cancellation of many contracts (especially those with foreign firms), the plan suffered a serious setback in 1958–1959. How much of this was due to the dislocation bound to occur in any immediate postrevolutionary period and how much to more profound causes (such as the rise of Communist influence and basic unwillingness to do business with the West) is difficult as yet to determine.

Development Plan in Iran

As early as February 15, 1949, the Iranian Majlis enacted a bill providing for a 7-year development plan and establishing a Plan Organization for its execution. The plan called for a total expenditure of Rls. 21,000,000,000 ($651,000,000 at the official rate of 1 rial equals 3.1 cents), divided into seven major installments ranging from approximately Rls. 2,000,000,000 ($62,000,000) in the first year to Rls. 3,600,000,000 ($111,600,000) in the last. With the exception of the first six months, the entire revenue derived from the oil royalties was to be assigned to the implementation of the plan.[14] The allocation of funds in the plan was as follows:

	Millions of rials
Agriculture	5,250
Roads, railway, harbors, airfields	5,000
Industries and mines	3,000
Petroleum (creation of an Iranian Company)	1,000
Posts, telegraph, and telephone	750
Social and municipal reforms	6,000
Total	21,000

The plan was based on two consecutive reports by foreign experts. The first of these was submitted to the Iranian government on August 2, 1947, by the Morrison-Knudson Company of Boise, Idaho. It was followed on August 22, 1949, by a 5-volume report presented by Overseas Consultants, Inc., a concern grouping eleven American engineering firms.[15] Unfortunately for Iran the execution of the plan encountered many difficulties, partly economic and partly political. At the time when the plan was being launched, Iran's oil royalties from the Anglo-Iranian Oil Company did not exceed $45,000,000 a year and could not alone provide sufficient financing for the plan. Hopes for a substantial loan from the United States or the International Bank of Reconstruction and Development did not materialize, and negotiations with AIOC for an upward revision of royalties ended in deadlock. The mortal blow to the plan was, however, delivered by the nationalization of the oil industry in 1951, which put a halt to virtually all production and exports of oil.

[14] S. Rezazadeh Shafaq, "The Iranian Seven Year Development Plan, Background and Organization," *Middle East Journal*, Jan., 1950, pp. 100 ff. For the text of the plan, see *Rooznameh Rasmi Keshvar Shahanshahi Iran* (Official Gazette), no. 1170, dated Bahman 30, 1327 (Feb. 19, 1949).

[15] *Report on Seven Year Development Plan for the Plan Organization of the Imperial Government of Iran*, 5 vols. (New York, 1949).

Following the conclusion of the Consortium agreement and the resumption of oil production, the plan was resuscitated. On March 13, 1956, the shah promulgated a new bill, enacting a second seven-year development plan which provided for a total expenditure of Rls. 70,000,000,000 ($921,000,000 at the new rate of 76 rials to a dollar) between 1956 and 1962. The latter sum was composed of two amounts: Rls. 17,200,000,000 which had been approved prior to the new development plan and allocated for programs already in progress and Rls. 52,800,000,000 for new projects.[16] The rising oil revenues, however, soon allowed somewhat bolder planning. Consequently in October 1956 the planned expenditures were revised to a total of Rls. 81,188,000,000 ($1,068,000,000) and the plan was rephased to expend less in its first four years of operation and more in the last three. According to this revised version, the planned allocations were to be as follows:

	Millions of rials	*Percentage of total*
Agriculture and irrigation	19,854	26.0
Communications	30,136	32.6
Industries and mines	11,388	15.1
Social development and education	19,810	26.3
Total	81,188	100.0

In principle the financing of the new plan was to be based entirely on oil revenues, of which approximately 60 per cent were to be transferred to the Plan Organization during the first three years (up to March 20, 1958), and 80 per cent thereafter.[17] This principle,

[16] Official Gazette, no. 3253, dated Farvar 22, 1335 (April 11, 1956), with amendment, *ibid.*, no. 3521, dated Esfand 13, 1335 (March 4, 1957).

[17] Article 8 of the Plan Law said in this connection: "Credit for development operations will be derived from the oil revenues. The total of funds collected by the Ministry of Finance as income tax in accordance with the Oil Agreement of 7 Aban 1333 [Oct. 29, 1954] and the income received by the N.I.O.C. from trading companies as stated payment (12.5 per cent of the posted price) will be divided among the Ministry of Finance, the N.I.O.C., and the Plan Organization as follows:

"For the years 1334, 1335, and 1336: (*a*) The N.I.O.C. will place at the disposal of the Ministry of Finance any surplus remaining from the stated payment after having covered the authorized expenditures foreseen in that Company's charter. (*b*) The Ministry of Finance will utilize the surplus in question plus 10 per cent of the total oil income (including income tax and stated payment) for covering the expenditures for the maintenance and management of available Government non-

however, was subject to two reservations: (*a*) "in case that the total revenue from oil should exceed $144 million in the year 1335 [March 1956–March 1957] and $188 million in each one of the years 1336 and 1337 [March 1957–March 1959], the surplus [was to be] placed at the disposal of the Ministry of Finance . . . in order to cover the essential needs of the Government"; [18] (*b*) should the revenue from oil be less than the plan expenditures for the years 1334, 1335, and 1336 (March 1955–March 1958), the government was authorized by law to make up the shortage in the Organization's requirements by raising loans from domestic and foreign institutions up to the amount of $240,000,000.

To implement the plan the law provided for the Plan Organization a corporate personality and fiscal independence. Similar to development legislation in Iraq, the law aimed at giving the Organization a nonpolitical character and making it separate from existing government departments. Accordingly the Plan Organization was to consist of a managing director, a high council, and a control board. The managing director was to be appointed by imperial decree for a period of three years on nomination by the Council of Ministers; he was to be eligible for reappointment. His term of office could be terminated by reason of death, resignation, or incompetence. The high council was to consist of seven members to be appointed for the full term of the plan. They could be removed only for the

---

profit establishments such as roads, health and educational establishments, and the like, as well as for establishments of that kind which will be created and turned over to the Government for management as a result of the execution of the programs. (*c*) The balance of the total oil income (including income tax and stated payment) shall be paid to the Plan Organization for Plan operations.

"In the ensuing years until the end of the term of the Plan: (*a*) 80 per cent of the total oil income (including income tax and stated payment) shall be paid to the Plan Organization. (*b*) From the remaining 20 per cent, after having covered the authorized expenditures foreseen in the N.I.O.C. charter, the surplus will be placed at the disposal of the Ministry of Finance for the expenditures for the maintenance and management of non-profit establishments."

The stated payment from the Consortium in 1956 amounted to about 28% of the total oil revenue. This, added to the 10% mentioned in the above-quoted law, amounted to 38% of the total oil revenue which was to be transferred to the government and the NIOC for their current operations. The remaining 62% were earmarked for the Plan Organization.

[18] Quoted from Art. 8, Note 2, of the Plan Law of 1956.

same reasons as the director. The high council was entrusted with the approval of plans, projects, cost estimates, executive regulations, contracts, and supervision of all functions performed by the Plan Organization. The control board was to be composed of six members, of whom three were to be appointed by the Senate and three by the Majlis from a list of persons submitted by the government. Its function was to consist primarily in auditing the Organization's expenditures.

The law attached special importance to the procedures to be followed by the Plan Organization. The latter was to "pay particular attention to the question of facilitating the investment of private capital in productive enterprises and also make adequate efforts for the safeguarding and strengthening of small urban and rural handicraft industries." Moreover, the operations of the plan were to be divided into four classes according to the method of execution:

1. Operations which fall within the current routine of the ministries and government agencies. Such operations were to be executed by those ministries and agencies under the supervision of the Plan Organization.

2. Operations capable of being performed by existing government agencies but requiring the services of consulting engineers in their planning stage. These were to be executed by the existing agencies, with the proviso, however, that manufacturers and contractors were to be recommended by the consulting engineers and approved by the Plan Organization. The execution of the operations was to be controlled by the consulting engineers as representatives of the Plan Organization.

3. Operations for the execution and planning of which no machinery exists within the government. Such operations were to be performed by specially created agencies, and the consulting engineers were to be responsible for blueprinting and supervision on behalf of the Plan Organization.

4. Operations which can be executed by individuals and private institutions. In such cases the Plan Organization was enjoined to encourage the investment of private capital and to facilitate its task by formulating projects, giving directives and technical aid, as well

as granting or procuring credit from internal and foreign sources. For the exercise of such functions the Plan Organization was authorized to establish a bank or an agency fully financed by the Plan Organization itself.

No less important were the provisions of the law providing for co-operation between the Plan Organization and municipal authorities. Municipalities were encouraged to undertake electric power plants and public utility projects by a clause in the law which said that one-half of the cost of such projects would be paid by the Plan Organization. The latter was, moreover, empowered to extend loans to poor municipalities as well as to such private organizations or individuals as should desire to undertake these projects in lieu of a municipality.

The activity and financial estimates of the Plan Organization were to be subject to review and approval by the Joint Plan Committee of the Majlis and the Senate. The committee's authority was repeatedly stressed by the law.

As will be seen from the preceding paragraphs, the Iranian Plan Law was more detailed and complex than its Iraqi counterpart. Despite its intention to keep the Plan Organization free of political pressures and immune from day-to-day government interference, it did a good deal to complicate the Organization's task. For one thing, the Organization was not free to choose the method of execution of a given project and was compelled in some cases to rely on the services of existing government agencies whose competence and efficiency could justifiably be questioned. For another, the Joint Plan Committee of both Houses was, perhaps, given too much power of intervention to be quite healthy for the progress of the Organization's work. And, thirdly, the independent tenure of the managing director and members of the high council was placed in jeopardy at the outset by the law's reference to "incompetence" as a legitimate cause for dismissal. Furthermore, while availing itself of the services of consulting engineers, Iran, in contrast to Iraq, did not appoint any foreign experts to the high council. This no doubt flattered the Iranians, yet it also deprived Iran of the advice and executive capacities of impartial foreign experts at the highest level of decision making.

The task of the Plan Organization was, by and large, more formidable than that of the Development Board in Iraq. Apart from the basic factors of a larger area and population, Iran was suffering from two evils, neither of which was present in Iraq to the same degree. One was the continuous budgetary deficit, which in the Persian years 1333, 1334, and 1335 (corresponding roughly to 1954, 1955, and 1956) amounted to $50,000,000, $36,000,000, and $53,000,000, respectively. Consequently Iran was faced with a constant threat lest its current budgetary requirements throw its elaborate development planning into confusion, thus inflicting economic and moral harm upon the country. The other evil was the psychological "hangover" of the Mossadegh era, which despite the new start under the aegis of the constructively-minded shah could still be detected in various manifestations of suspicion, pessimism, and negativism. The limiting, by law, of the Plan Organization's authority to planning and supervisory functions, instead of giving it a free hand in the execution of projects, was symptomatic of this trend. Lacking a strong institutional foundation, the Organization had perforce to rely more on personalities. Fortunately for the country the managing director, Abol Hassan Ebtehaj, was a man of unusual abilities and strength of character who, with the support of the shah, was capable of carrying on his arduous duties despite continuous sniping from many disaffected critics. Strong pressure was put on him to embark on many projects which would have an immediate impact on Iran's economy and on living standards. Various interests insisted that priority be given to projects in their districts. The director rather successfully resisted them, insisting on careful studies by consulting engineers before any major project was undertaken. Consequently the first years (1955–1957) were largely devoted to studies and estimates, with relatively little actual constructive activity. Nevertheless, despite its initial slowness the plan was gaining momentum. In 1956 a major road-building program was begun, and in May 1957 the shah inaugurated a new branch of the Transiranian Railway linking Teheran with Meshed. Another branch, connecting the capital with Tabriz, was opened in 1958.

These achievements did not suffice, however, to protect the Plan director from the criticisms and jealousies which his vast powers stimulated. On February 15, 1959, the Majlis passed a bill consisting of a single article, which transferred all the director's powers to the prime minister, who was given the choice of managing the Plan Organization directly or by an appointee of his own. The objective of the bill was to transfer to regular government departments most of the tasks performed by the Organization and to restrict the latter to technical supervision. Faced with such a drastic change, Ebtehaj tendered his resignation. His departure and the passing of the new law meant a definite repudiation of the original concept of the plan to make it an independent and nonpolitical venture. What effects this change will have on Iran's development program it is as yet too early to say.

It is evident that, in Iran as in Iraq, oil permitted the government to launch a major development plan likely to alter its economy substantially. Both plans, moreover, were apt to generate a number of additional benefits. Free enterprise would probably be stimulated; new patterns of co-operation between central and local government were evolving; training in engineering and administrative skills was being advanced; the growth of urban centers was gradually draining off the semiemployed surplus population of the villages, presenting a challenge to landowners to offer better wages and living conditions to agricultural workers; and social consciousness and increased civic responsibility were gradually replacing apathy and cynicism.

PART TWO

COMPANIES AND THE HOST

GOVERNMENTS

CHAPTER IV

Pattern of Concession Agreements

FOREIGN oil corporations in the Middle East operate, without exception, on the basis of special concession agreements by which they obtain exclusive right to exploit the petroleum resources in a defined area for a long period of time. In return for this privilege they undertake to explore for and produce oil and to pay the host country royalties and certain other sums in connection with specific phases of their operations.

Although each concession contains special stipulations that distinguish it from others, a number of provisions are common to practically all concessions granted by the producing countries. A brief review of these recurring features may be helpful in determining the general character of the concession agreements.[1]

---

[1] Texts of the concession agreements discussed in this section may be found in the following publications: D'Arcy concession, 1901—J. C. Hurewitz, *Diplomacy in the Near and Middle East, A Documentary Record: 1535–1914* (Princeton, 1956), I, 249 ff. (hereafter referred to as Hurewitz); AIOC concession, 1933—Hurewitz, II, 188 ff.; IPC concession, 1931—Iraq Petroleum Co., Ltd., *Agreement Concluded on 24th March, 1931, Revising the Convention Made on the 14th March, 1925 with the Government of Iraq* (n.p., n.d.); Aramco concession, 1933;—*Umm al-Qura* (Mecca, July 14 and 21, 1933), and also in a booklet, *Ittifaqiyat Sharikat Istithmar al-Petrol wa Mustakhrajatihi wa al-Ma'adin* (Agreements of Companies for the Exploitation of Petroleum and Its Products and Minerals), (Mecca, 1359 A.H.); Consortium agreement, 1954—Hurewitz, I, 348 ff.

STANDARD PROVISIONS

1. The privileges granted to companies include, as a rule, the exclusive right to explore, prospect, extract, refine, and export crude oil and related matters (such as natural gas) within the area of the concession. These privileges do not necessarily include the exclusive right to transport oil and oil products, which operation may be subject to separate agreements (IPC, 1; AIOC, 1; Aramco, 1 and 30; Consortium, 4).[2]

2. The initial documents of concessions (in contrast to revised agreements) generally included a time limit within which exploration must begin, failing which the concession might be revoked by the host government (IPC, 4 and 5; D'Arcy, 16; Aramco, 6 and 7).

3. Concessions are usually granted for extended periods. Thus the original D'Arcy concession in Iran was to last sixty years. The revised agreement with Anglo-Iranian Oil Company provided also for a sixty-year period from the date of the revision. The agreement with the Consortium is technically an exception to the general rule, providing for twenty-five years with extensions up to fifteen years. But this agreement is in many ways a continuation of the original D'Arcy concession, and it will provide for ninety-three years of foreign exploitation of Iranian oil under special agreements. Similar longevity is characteristic of the concessions of the IPC group in Iraq, all of which are to last for seventy-five years. Aramco's concession stipulates sixty years, and the Kuwait Oil Company's concession seventy-five (D'Arcy, 1; IPC, 2; AIOC, 26; Aramco, 1; Consortium, 49).

4. The area of the concession is generally very extensive. In some cases it embraces the whole country, for example, in Kuwait, Bahrein, and Qatar. In others it covers a major region which, either because of its geological formation or because of its political character, constitutes a well-defined territorial entity. Thus the original D'Arcy concession embraced the territory of Iran with the exception of the five northern provinces (Azerbaijan, Gilan, Mazanderan, Gorgan, and Khorasan) which were traditionally regarded as subject to Rus-

[2] These references in the text to the concession agreements will contain the name of the company and the relevant article of the concession.

CONCESSION AGREEMENTS

sian influence. The original IPC concession considered the province of Mosul as most likely to contain oil, but eventually the whole territory of Iraq, with the minor exception of the Khanaqin District (the so-called Transferred Territory) was covered by the concessions of the IPC group.

Although the concession agreements generally do not provide an exact definition of "territory," in practice "territory" has come to mean land territory and territorial waters but not the continental shelf. For the latter separate concessions have been granted, in some cases to different companies than those operating the mainland concessions (D'Arcy, 6; IPC, 3; AIOC, 2; Aramco, 2; Consortium, 2).

5. Companies are generally granted the right to establish their own systems of transportation and communication for the efficient conduct of their operations. Among these facilities radio, telegraph, and telephone as well as railroads, vessels, and airplanes are usually specified in the concession agreements (IPC, 19 and 20; AIOC, 5; Aramco, 19; Consortium, 4B).

These agreements also provide for the right of the host country to requisition these facilities in case of war or national emergency, usually on payment of fair compensation (IPC, 19 and 20; AIOC, 5).

6. For the purpose of conducting their operations the companies are entitled to acquire land in the host countries, either from the state or from private individuals. If it is state land, concessions provide either for gratis assignment of land to the company or for sale at a specially reduced price. The price of land purchased from private persons should conform to the prices prevalent in a given district at the time of purchase. Holy places, monuments, and historical sites are exempted from these provisions (IPC, 21; AIOC, 4; Aramco, 25; Consortium, 7).

7. Companies are generally required to supply the host governments certain specified amounts of oil and oil products for local consumption. This clause is usually accompanied by a proviso that quantities thus supplied should not be subject to royalty calculation (IPC, 14 and 15; AIOC, 19; Aramco, 19; Consortium, 14 and 15).

8. Companies are permitted to establish subsidiary companies for the conduct of specific phases of their operations. The profits of such subsidiaries are included in the royalty calculations (D'Arcy,

9; IPC, 33; AIOC, introductory definitions; Aramco, 32; Consortium, 1P, 1Q, and 18B).

9. Host governments usually reserve the right to be represented in their dealings with companies by specially appointed delegates whose salaries and expenses are to be borne by the companies. Such representatives are given the right of inspection of the companies' operations and accounts (IPC, 16; AIOC, 15; Aramco, 1950, 7).

10. As a matter of general principle unskilled laborers employed by companies have to be nationals of the host country. As for managers, engineers, technicians, other skilled workmen, and clerks, they may be aliens if qualified personnel cannot be found in the host country. Agreements frequently obligate the companies to train local nationals in the skills required for these positions (IPC, 29; AIOC, 15; Aramco, 23; Consortium, 5F).

11. Companies undertake to present every year to the host governments reports of their operations, including data on the discovery of new oil deposits and geological plans and records. Such information is to be treated as confidential by the host governments (IPC, 8; AIOC, 13 and 14; Consortium, 4G; Aramco, 26).

12. Should new shares be issued to the public, companies are obliged to open subscription lists in the host countries (AIOC, 18). In some cases concessions specify the percentages of new issues which are to be made available to the public in the host country. In the case of Aramco it is 20 per cent (Aramco, 32).

13. Companies are generally either exempted from all direct and indirect taxation (AIOC, 6D and 11; Aramco, 21) or are assured that such taxes as may be applied shall not be different or higher than those imposed upon other industrial undertakings (IPC, 27).

14. Concessions exempt companies from customs duties and licenses for the importation of all machinery, equipment, and materials necessary for the conduct of their operations (IPC, 28; AIOC, 6C; Consortium, 34; Aramco, 21).

15. In case of disputes between the host countries and the companies concerning their rights and obligations under concession agreements, the latter provide for arbitration. Arbitration proceedings usually are expected to include the nomination of an arbitrator

CONCESSION AGREEMENTS

by each party and the choice of an umpire by the arbitrators. If the latter cannot agree on an umpire, either party may request the president of the World Court to appoint one. Should either party fail to appoint an arbitrator, the president of the World Court may be requested to appoint a sole arbitrator to deal with the case (AIOC, 22; Aramco, 30; Consortium, 44; IPC, with slight deviation from this pattern, 40).

16. Financial arrangements in major concessions usually include the following provisions: (*a*) payment of a lump sum to the host government at the time when the concession is granted (the so-called bonus payment); (*b*) payment of a dead rent during the period of exploration; and (*c*) payment of royalties calculated either on the basis of the volume of oil produced, sold, or exported, or on the basis of profits, or both. At the risk of oversimplification but with the view of obtaining a clearer idea of general trends in this respect, we may say that the financial history of the concessions is divided into three major periods: the first, from the inception of the first concessions (AIOC, IPC) to the early 1930's; the second, from the early 1930's to the early 1950's; and the third, from the early 1950's onward.

The first of these periods was characterized by agreements which provided for rather modest royalties, the latter to be based either exclusively on a fixed rate per ton of crude produced (thus IPC; the basic rate was 4 gold shillings) or on the company's profits (the original D'Arcy concession provided for 16 per cent of the annual net profits).

The opening of the second period was marked by two major revisions of the existing concessions and the grant of a new concession. In 1931 IPC's concession was revised to provide for a minimum payment of £400,000 in gold a year even if actual production did not warrant such a payment on the basis of the previously agreed fixed rate per ton. In 1933 the revised AIOC concession provided for a combined system based on a fixed rate per volume sold (4 shillings a ton) and on a percentage of profit (20 per cent of dividends in excess of £671,250), with the stipulation of a minimum payment of £750,000 a year regardless of the volume of production and the

67

amount of profits.[3] Almost simultaneously the new concession operated by the California Arabian Oil Company in Saudi Arabia (precursor of Aramco) made a provision for royalties of 4 gold shillings a ton of oil produced. Thus the prevalent pattern during the twenty years between the early 1930's and the early 1950's was payment at a fixed rate (4 gold shillings) per ton of oil produced or sold. Payments above this standard rate were provided for only in the revised AIOC concession, partly because the latter followed a strong Iranian campaign to increase the income from oil and partly because the Iranian operation was the oldest and the one that had resulted for many years past in the actual production of oil. Consequently the Iranian government had an empirical basis on which to base its demands. As for the IPC, its concession, though based on earlier agreements with the Turkish Petroleum Company, did not yet reflect a producing operation in full strength, major exports from Iraq having started only in 1935. Similar considerations applied to the new concession in Saudi Arabia.

Great expansion in the production of oil after World War II led the producing countries to clamor for major increases in payments so as to obtain a fair share in the spectacular profits the concessionaire companies were deriving from their operations. Iran was the first to begin negotiations, with AIOC, shortly after the end of the war, but the first major change, which affected the rest of the Middle East, was made in Saudi Arabia. On December 31, 1950, Aramco and the Saudi Arabian government adopted a fifty-fifty profit-sharing formula, which had earlier been introduced in Venezuela. This formula, based on net profits after the payment of United States taxes, was later readjusted to provide for a division of profits before American tax payments. In February 1952 Iraq followed suit, and in 1954 the Consortium agreement with Iran also provided for the same equal sharing of profits. The revised concession in Kuwait

---

[3] In addition, in lieu of taxes the company was to pay during the first 15 years 9 pence for each of the first 6 million tons of petroleum and 6 pence for each ton in excess of 6 million, the total amount not to go below £225,000. During the following 15 years, the company was to pay 1 shilling for each of the first 6 million tons and 9 pence in excess of the figure of 6 million tons, the total amount never to be less than £300,000.

of 1951 likewise followed this pattern. In each case details vary, but the basic pattern is the same.

17. The existing major concessions do not provide for a partnership between the host government and the company but for a foreign-owned and foreign-managed enterprise. IPC's concession in Iraq is the only one to stipulate the appointment, by the government, of "a Director to the Board of the Company, who [is to] enjoy the same rights and privileges and receive the same emoluments from the Company as the other Directors" (IPC, 35). The original concessions of Aramco and the Anglo-Iranian Oil Company made no mention of native directors. A change, however, was introduced with the conclusion of new or revised agreements in the 1950's. According to the agreement of February 1952, the government of Saudi Arabia is authorized to nominate two members to Aramco's Board of Directors. The company, on its part, undertakes to have the said nominees elected to the board. The government delayed nominations for seven years. In 1959, however, its first two nominees were made members of the board.

The Iranian Consortium agreement of 1954 likewise introduced the innovation of government-appointed directors, to serve on the boards of the two operating companies. The Consortium members agreed "that two of the members of the Board of each Company shall be persons nominated by NIOC" (Consortium, 2D).

Although it may be open to question whether the presence on a board of one or two national directors is likely to influence the companies' basic decisions, the right to appoint them undeniably constitutes an important gain for the host governments, which can thus keep reasonably well informed of the companies' operations, plans, and policies.

DEPARTURES FROM STANDARD PATTERN

The standard pattern of concessions has just been given. From this pattern a number of departures have been made in the third period previously referred to. These departures may be classified into two basic categories: (*a*) revisions and changes affecting the still-existing older concessions and (*b*) departures contained in the new agreements, especially those recently concluded by Iran with certain foreign companies for the development of its offshore areas as well as those signed by Japanese interests with respect to the offshore territory of the Neutral Zone. In the first category three major issues likely to modify the standard relationship between the host countries and the companies had come up. These issues were taxation, ownership, and financial safeguards.

Taxation

It has been a deeply embedded principle for the companies to insist on complete exemption from taxation. The reasoning behind this principle was as follows: there is a basic difference between the operations of an oil company in the West and in the Middle East. The latter is an underdeveloped area under a variety of political systems, most of which are of recent vintage. Inasmuch as the native industrial undertakings are either of minor proportions or nonexistent, it is clear that a foreign-owned oil company is likely to become the largest single industrial enterprise in a given country and that its profits, granting the success of its operations, will exceed the profits of any other enterprise in such a country. Consequently submission to taxation would be tantamount to acceptance of possible discrimination, inasmuch as a legally enacted progressive tax system might in practice apply to one target only—a big foreign corporation.

This sounds like a valid argument. But its strength was somewhat diminished when, in some cases, notably in Iran, it could be proved that if the company had been subjected to existing tax rates (even without any discriminatory features) the benefits to the host government would have been many times greater than those realized on the basis of the stipulated royalties. Consequently a strong feeling developed in certain Middle Eastern countries that either the companies should be subjected to taxation or, at least, that the royalties to the host governments should not be less than the income to be derived from taxation, should the latter be applied.

Despite the fact that Iran was the country displaying the most pronounced "revisionist" mood in this respect, Saudi Arabia made the first move. On November 2, 1950, a royal decree imposed an income tax on individuals and corporations, the latter having to pay 20 per cent of their net profits. This was followed by another decree, published on December 27 and 29, 1950, instituting an additional income tax of 50 per cent "on companies engaged in the production of petroleum or other hydrocarbons." These decrees obviously went counter to the stipulations of the concession and called for some clarification of the future relationship between the company and the government. This clarification was promptly achieved by the agreement of December 30, 1950, whereby Aramco submitted to the principle of taxation, but the government undertook that the total of income taxes "and all other taxes, royalties, rentals and exactions of the Government for any year would not exceed fifty per cent of the net operating profits of the company, these profits to be calculated after the payment of foreign [i.e., mostly American] income taxes." [4] Thus the Saudi-Aramco agreement represented both a departure from the old practice and a compromise—a departure in that it introduced the principle of taxation and a compromise in that it limited it to a certain maximum. Although in purely financial terms the new agreement meant only a fifty-fifty profit-sharing formula, politically it had a deeper meaning inasmuch as it provided a new basis for exacting payments from the company, while introducing the element of national sovereignty in the form of tax legislation.

Once brought out into the open, the issue of taxation not only reappeared in other producing countries but affected the transit conventions as well. In December 1951 the Kuwait concession was revised along the Saudi lines. In 1952–1953 certain economists in Lebanon suggested that the Trans-Arabian Pipe Line Company and IPC be subjected to income taxes. Ultimately, after a lengthy controversy between the government of Lebanon and IPC, the Lebanese Parliament on June 29, 1956, passed a law imposing income taxes on all concessionary companies operating in Lebanon. The law elicited protests from Tapline and IPC as contrary to the terms of their concessions and has not yet (April 1959) been carried out. (A fuller examination of this dispute will be found in Chapter IX, p. 163 below.) This new trend in government-company relations also

[4] Hurewitz, II, 315 ff.

found expression in Syria, where on November 29, 1955, a revised agreement was signed between the government and IPC, providing for increased payments to the former. The new agreement, theoretically at least, recognized the principle of taxation by providing that royalties and other dues would be paid by the company "in lieu of taxation."[5]

The taxation principle found its way likewise into the Iranian-Consortium agreement of October 29, 1954, which in Article 28 recognized the income tax liability of the trading companies and the operating companies, with the reservation, however, that this liability should not exceed the rates provided for in the law existing on August 1, 1954 (i.e., the Income Tax Act of 1949) and that the law itself would be amended to conform to those provisions of the agreement which deal with the expenses relating to the earning of income by the said companies.

Ownership

In addition to its tax feature, the Consortium agreement contained another departure from the standard concession pattern. The agreement followed, as we know, the nationalization of the oil industry by Iran and the resulting 2½-year-long deadlock. In many ways the new agreement conforms to the basic standard of concessions. The principal practical difference is that the agreement distinguishes between the basic and the nonbasic operations, the latter being entrusted to the National Iranian Oil Company as an agency empowered by law to operate the oil industry in the name of the state. In Article 17B the agreement mentions the following nonbasic operations: housing estates, maintenance of roads used by the public, medical and health services, operation of food-supply system, canteens, restaurants, and clothing stores, industrial and technical training and education, guarding of property, welfare facilities, public transport, communal water and electricity supplies, other public services, such further services as may be agreed upon between NIOC and the relevant operating company, and also "the provision, mainte-

[5] Syrian Republic, Ministry of Finance, *Conventions Made Between the Government of the Syrian Republic and the Iraq Petroleum Company, Ltd., 29 November, 1955. Approved by Law No. 128 dated 6th December, 1955* (Damascus, n.d.).

nance and administration of warehouse and other auxiliary services, insofar as such auxiliary services are required in connection with the services above mentioned."

The transfer, through the Consortium agreement, of the nonbasic operations from a foreign to a national organization constitutes a significant political development. On the one hand, it relieves the foreign company of a whole sector of responsibility in which it usually is most vulnerable to nationalist criticism. On the other, it presents the host government with a challenge to live up to what the public and the government itself have been demanding from the concessionaire in the way of various social services. In this sense the new arrangement is apt to remove an important element of potential friction between the host country and the foreign concessionaire. But while this is so, another clause in the agreement says that "NIOC shall perform and carry out non-basic operations with due regard to economy and efficiency and in such a manner as to meet the reasonable requirements of the Operating Companies." A further clause stipulates that the NIOC should prepare its budget of nonbasic operations in consultation with the operating companies and that the capital expenditures and operating expenses included in the budget should be subject to agreement by both parties. Moreover, other sections of the agreement describe in detail how both the NIOC and the operating companies should supply funds for the execution of these operations. These provisions are understandable and necessary. They considerably enlarge, however, the area of day-to-day co-operation between the national and the foreign organizations, with all the attendant risks of such a relationship.

Although the Consortium agreement has introduced a new formula with regard to the ownership pattern, the most striking departure from the established pattern has undoubtedly been made by NIOC in launching a nationally managed operation in the Qum area. To be sure, the ability of the Iranian state to operate an important oilfield has not yet been fully tested. Certain technical operations, such as drilling, have from the beginning been entrusted to foreign experts. Adequate organization of transportation, whether by pipeline or by other means, has yet to be set up both in its technical and financial aspects. And it remains to be seen how soon the Iranian state will be in a position to add to the basic producing functions the refining and marketing operations.

Financial Safeguards

As time passed, certain host governments found out that the concession agreements did not fully safeguard their financial interests. This was true because agreement on rates and minimum amounts of royalties was not sufficient to maximize the benefits that the host country might derive from a given concession. If the concession document passed over certain matters or dealt with them vaguely, opportunities arose to interpret it in a way deemed detrimental to its interests by the host country. The host governments discovered that there were a number of ways in which their benefits could be limited, and they accordingly insisted that new or revised agreements should provide safeguards. Some of the ways in which host countries' benefits were limited follow:

1. Nonexpansion of production. For various reasons, sometimes due to company policies and sometimes not, production in certain countries was held at relatively low levels. This was true at one time or another in Iraq, Saudi Arabia, and Iran. In each case the cause was different. Because the operating companies are frequently owned, at least in part, by the same holding corporations, and because the latter may often have the last word as to the operating companies' policies, it is possible for the holding corporations to decide that emphasis should be placed on production in one rather than another country.

This was the grievance of Iraq against IPC in the immediate post-World War II period. Iraq complained that, instead of increasing production and developing the pipeline system, the elements holding the dominant position in IPC favored expansion in Iran and the Arabian Peninsula. Consequently in the revised IPC concession of February 1952 Iraq insisted on and obtained safeguards in this respect.

In Saudi Arabia during World War II there was such a marked decrease in production as to cause the government serious financial difficulties. Here, however, it could not be attributed to the company, but rather to the war emergency, which prompted the American government to set up a system of priorities with respect to steel and other equipment. Under this system Iran, with its well-established Abadan refinery, was granted priorities, while deliveries to Saudi Arabia were virtually suspended, reducing the production to an almost symbolic minimum. Although there is no doubt that Aramco had no influence whatsoever on this decision, a host government might not always accept such practices with equanimity.

As for Iran, the virtual stoppage of its production in the period between nationalization and the Consortium agreement permitted other producing countries to take its place in the world markets. Consequently, when negotiating the agreement with the Consortium, the Iranians were particularly anxious to insert provisions guaranteeing not only a minimum production in the opening stages, but also a progressive expansion, and such provisions were included in Articles 20 and 21. Article 20A contained a guarantee by the Consortium members that if the effective date (i.e., the date on which the agreement comes into force) was on or before October 1, 1954, the aggregate of oil exported or delivered to a refinery or to NIOC, should not be less than 17,500,000 cubic meters in 1955, 27,500,000 cubic meters in 1956, and 35,000,000 cubic meters in 1957. Should the effective date be after October 1, 1954, the quantities so guaranteed were to apply to the first three annual periods commencing three months after the effective date. Paragraph B of the same article further made this significant statement:

Following the above annual period in which the guaranteed quantity of 35 million cubic meters shall have been attained, it would be the policy of the Consortium members, assuming favorable operating and economic conditions in Iran, to adjust the quantity so attained in such a manner as would reasonably reflect the trend of supply and demand for Middle East crude oil.

Article 21 of the agreement provided further safeguards in this respect, proclaiming that "the Consortium members will strive for, without guaranteeing" to assure exports of a certain minimum tonnage of refined products during the three first years following the effective date of the agreement.

2. *Taxation by home governments.* Whenever payments are calculated on the basis of profits, the question arises as to the extent of

permissible deductible expenses. Is taxation by the company's home government to be considered such a deductible expense? For a long time, i.e., until the early 1950's, it was so considered, but the only company to which this practice applied was the Anglo-Iranian Oil Company, one part of its payments being based on profits. It obviously did not affect either IPC or Aramco inasmuch as neither of the two had a profit formula as a basis of payments until late in 1950. When, however, Aramco concluded its profit-sharing agreement with Saudi Arabia in December 1950, taxes to the United States were recognized as a deductible expense. The formula was quite explicit:

In no case shall the total of such taxes i.e. Saudi Arabian taxes, royalties, rentals and exactions of the Government for any year exceed fifty per cent (50%) of the gross income of Aramco, after such gross income has been reduced by Aramco's cost of operation, including losses and depreciation, and by income taxes, if any, payable to any foreign country.

This new formula, however, proved short-lived. Anxious to increase its income from oil and rejecting the right of a foreign government to have a prior claim to taxation on profits derived from an operation conducted in Saudi territory, the Saudi Arabian government insisted on and obtained, on October 2, 1951, a new agreement, whereby it would tax the company before the payment of foreign income taxes. A similar formula was subsequently adopted in the revised IPC agreement of February 3, 1952, which introduced a profit-sharing system instead of the old fixed rate on units of production.

As for Iran, where profits (or rather dividends) formed the basis for at least a part of the payments, taxation by the British government constituted one of the main grievances. For a number of years the amounts paid by AIOC to the British Treasury exceeded many times over the total received by Iran.[6] This rather spectacular dis-

[6] During the 40 years of operations (1911–1951) Iran received a total of £113,000,000 in payments. During the same period the British government received an estimated £250,000,000 in taxes from AIOC and its subsidiary companies. In the last year before nationalization (1950) Iran's royalties had amounted to about £16,000,000, while taxes paid to the British government reached the sum of £50,500,000. In addition to the taxes, the British government was receiving its share of dividends as owner of the majority stock of the company.

proportion was one of the principal factors contributing to the violent emotions released in Iran in 1951 at the time of nationalization. Eventually, in the Consortium agreement the remedy was sought not in the "before taxes" profit-sharing formula, but in the actual subjection of the operating companies to an agreed maximum of Iranian taxation. (See p. 72 above.)

3. Arbitrary increases in expenses. The earlier concession documents did not go into elaborate detail as to permissible costs and expenses. For example Article 10 of the original D'Arcy concession in Iran stated that the concessionaire would pay the "Government annually a sum equal to 16 per cent of the annual net profits of any company or companies that may be formed" to exploit the oil resources, but it did not define the meaning of the term "net profits." Nor did the revised concession granted to AIOC in 1933 contain any definitions of this kind. Although the latter agreement was much more elaborate than the preceding one, it changed the basis of payments to the government from a percentage of the profits to a combination of fixed rate per unit of volume and a percentage of dividends paid to stockholders. Needless to say, net profits and dividends actually paid are two different things and do not necessarily equal each other. Although the amount of profit can be reasonably foreseen if due account is taken of gross income, allowable expenses, and the general trends on the market, the distribution of dividends is subject to the company's policy decisions and, perhaps, to the home government's regulations. The history of the AIOC concession is illuminating on this point, inasmuch as not only did the company channel into reserve funds and further investments large sums to which the Iranian government objected, but the British government during World War II adopted dividend-restricting regulations, thus decreasing Iran's income, despite an actual increase in the company's profits. For all these reasons Middle Eastern governments seeking revised agreements or concluding new ones in the postwar period displayed a tendency not only to adopt the profit-sharing formula in lieu of the old arrangements, but also to specify with greater precision than hitherto the meaning of "net profits."

4. Pricing. The profits of the operating companies ultimately depend on the prices they charge for oil. Leaving aside the question of whether or not the structure and character of the international oil industry permit unimpeded competition in this respect,[7] the fact remains that there are many ways to determine prices. In the first place, prices will differ according to the point at which oil is being delivered to customers. In the cases of Iraq and Saudi Arabia the choice may be between delivery in the ports of the Persian Gulf or at the end of the pipelines in the eastern Mediterranean. In the second place, oil may be sold at the same point at different prices to different customers. What is involved here is the much-debated problem of discounts to special purchasers. This problem arose at one time or another with respect to practically every concession. It was standard practice for AIOC to grant special discounts to the British government, whose navy was a major customer of the company. The rate and amount of these discounts were never made public in the company's reports, nor were they revealed to the Iranian government, but they were far from negligible and must have affected both the profits and the dividends distributed by the company. Aramco has had a policy of granting a 5 to 10 per cent discount to its parent companies, depending on the amount of oil sold. Although the concession agreement did not specify how prices were to be set, Saudi Arabian officials felt that their government was being deprived of a legitimate share of profits by the policy of discounts. Discussions between the government and the company eventually led to an additional agreement in 1953 that abolished the discounts.

5. *Nonintegrated operations.* Oil company operations comprise a number of phases and activities beginning with exploration and ending in marketing. As a rule a concessionary company does not engage in all of these activities. Its functions are generally limited to exploration, extraction, and refining, with transportation and marketing excluded from its operations. In some cases the latter functions are performed by the concessionaire's subsidiary or affiliate companies; in others it is the parent companies which undertake these tasks. Two major problems are likely to emerge from these

[7] For an official view of this question, see United States, Federal Trade Commission, *The International Petroleum Cartel, Staff Report Submitted to the Subcommittee on Monopoly of the Select Committee on Small Business* (82nd Cong., 2nd Sess., Committee Print no. 6; Washington, D.C., 1952).

practices. One of them pertains to the role of subsidiary companies. Should the profits of such companies be included in the over-all profits of the concessionaire company and, consequently, should they be subject to royalty calculations? Although concession agreements are generally unanimous in considering such profits as subject to royalty provisions, yet in one notable case a dispute has arisen between the concessionaire company and the government with regard to the interpretation of these provisions. This was between AIOC and the Iranian government, the latter reproaching the company for not including the profits of one of its subsidiaries, the British Tanker Company, with its own profits. So far no formulas have been devised with respect to subsidiaries in the new or revised concession agreements. The latter generally contain similar provisions on this subject to those included in the older agreements. But it may be taken for granted that, having been alerted to this problem, the host governments will henceforth jealously watch their rights in this respect.

The second problem, somewhat related to the first, is the degree to which the concessionaire company should limit its functions to the basic exploration and production activities, while leaving transportation, most of the refining, and marketing to others who are not its subsidiaries. Although it is clear that there may be obvious technical, organizational, or economic limitations upon a concessionaire's capacity to perform these other functions, host countries are and will continue to be interested in "integrating" as much as possible the companies' operations with a view to maximizing their benefits from them. Refining is a subject of particular solicitude—not only that the company should include it in its normal operations, but also that it should take place within the territory of the host country. Refined products command a higher price than the unrefined, and a refinery is likely to offer numerous employment opportunities in the host country. Moreover, although the extraction of oil is undoubtedly an industrial operation, it nevertheless represents a primary raw-materials-producing phase, characteristic of the colonial economies. In the eyes of the host governments only the refining of crude products truly adds to the industrialization of their underdeveloped countries and thus assures desired economic progress. A number of concession agreements provide for the erection of refineries by the concessionaire companies, and the general trend is toward a greater emphasis on this point.

Although understandable, a host country's desire for the development of local refining facilities cannot be viewed in isolation from other economic aspects, more especially from its influence on demand. Possession of its own refining facilities is also insisted upon by many a consumer country for economic, fiscal, and strategic reasons. Such countries are primarily interested in buying crude oil, and if they find that only refined products are offered in one place they may turn elsewhere for crude. Thus expansion of refining by a producing country does not automatically assure greater benefits.

As for transportation, a distinction may be made between tanker and the pipeline operations. A tanker organization may exist prior to, or independently of, the principal oil-producing operation; hence it is difficult for the host government to insist on integration of tanker transportation into the normal activities of the concessionaire. If the concessionaire company is owned by holding companies which engage in tanker operations of their own, it will be in the interest of the latter to continue and to expand their own transportation system, thus removing it from the profit-sharing operations of the concessionaire. A pipeline operation, however, usually follows rather than precedes the principal producing activity and may be said to be linked to it by much stronger ties than is a tanker operation. Moreover, to the extent to which the pipelines are located in the territory of the producing country, the latter will tend to regard them as part of the whole operation covered by the concession. Consequently, if the company or its parent or subsidiary organizations derive any profit from the pipeline, the host government will understandably endeavor to benefit from it as well.

A concrete example is supplied by the so-called Sidon price dispute between Aramco and Saudi Arabia. The dispute arose out of a claim by the Saudi Arabian government that the company's profits from crude transported by pipeline to the Mediterranean should be calculated not on the basis of the posted price in the Persian Gulf ($2.08 a barrel in 1957), as had hitherto been the practice, but on the basis of the price obtained at the pipeline terminal in Sidon,

CONCESSION AGREEMENTS

Lebanon ($2.59 a barrel). In other words the Saudi government laid claim to one-half the profits realized by the operation of Tapline, which had been created for the purpose of effecting savings in transportation costs for Aramco's parent companies. This claim was first formulated by the Saudi government on November 28, 1956. The company denied its validity on two grounds: (*a*) Aramco's concession agreement subjected its profits from production but not from transportation to the profit-sharing formula; (*b*) the pipeline operation was carried out by Tapline, which was not Aramco's subsidiary but a subsidiary of Aramco's parent companies; hence Aramco had no title to the profits thus realized. In reply the Saudis pointed out that crude transported by pipeline to Sidon was formally owned by Aramco until the moment of loading on tankers. Further negotiations proving inconclusive, the Saudi government on August 6, 1957, demanded the payment within one month of approximately $103,-000,000, which was said to represent 50 per cent of the savings realized by Aramco since the beginning of Tapline operation in 1951. Faced with this ultimatum, the company decided to submit the matter to arbitration and on November 2, 1957, named its arbitrator. This procedure was accepted by the government, which appointed its own arbitrator on December 1, 1957.

There is no doubt that behind the technicalities of this dispute lay the fundamental desire of the host country to see Aramco as an integrated oil operation and not merely as a producing agent for its four parent companies.[8] In purely business terms there was no reason why Aramco's parent companies should want to part with their hard-won advantage by transferring their marketing organization to Aramco itself. Moreover, as a long-range proposition such a transfer might contain germs of danger by depriving the parent companies of an element of strength which, failing other adequate safeguards, they might need in their relations with a sovereign host country.

Regardless of the outcome of arbitration in the Sidon price dispute

[8] See in this connection interviews with Sheikh Abdullah Tariki, director general of petroleum and mineral affairs in Saudi Arabia, by Sam Pope Brewer reported in "Saudi Arabia Seeks Better Aramco Deal," *New York Times*, March 6, 1957, and by Wanda Jablonski in "Terms for New Saudi Concessions Spelled Out," *Petroleum Week*, June 20, 1958.

it was certain that Saudi Arabia, or any other host country, would encounter serious opposition by the concessionaire companies to the idea of integrated operations. What was, however, not easily obtainable from old concessions could be tried when negotiating new ones. This became evident with the conclusion by Saudi Arabia of the earlier-mentioned agreement with a Japanese concern for the development of the offshore areas in the Neutral Zone. Salient features of this agreement will be reviewed in a subsequent section.

The New Iranian Offshore Agreements

The three agreements which NIOC concluded in 1957 and 1958 with Agip Mineraria, Pan American Petroleum Corporation, and Sapphire Petroleums, Ltd., differed considerably from standard concessions in the Middle East and from the Consortium agreement in particular.[9] The main features these agreements had in common were as follows:

1. They provided for a partnership (described as a noncorporate joint structure) between NIOC and a foreign company to set up an operating company of Iranian nationality. Profits from such a venture were to be taxed at the rate of 50 per cent by the Iranian state, with the remainder to be divided in equal parts between the two partners. The Iranian state was to receive, directly or indirectly, 75 per cent of the profits.

2. Exploration expenses until the discovery of oil in commercial quantities were to be borne by the foreign (i.e., non-Iranian) partner, who assumed the obligation to spend a specified amount within a prescribed time limit. In all three agreements this period was set at twelve years. The amounts to be spent by the three companies were as follows: Agip Mineraria, $22,000,000; Pan-American, $82,000,000; and Sapphire, $18,000,000.

3. The primary term, applied separately to each commercially exploitable field, was to be twenty-five years from the date of commencement of commercial production (defined as the sale and de-

---

[9] For the text of the agreement of Iran with Agip Mineraria, see *Petroleum Times* (suppl. to vol. LXI, no. 1572), Nov. 8, 1957; with Pan American Petroleum Corporation, see *Platt's Oilgram News Service* (Chicago ed., suppl. to vol. 36, no. 97), May 19, 1958; and with Sapphire Petroleums, see *Iranian Official Journal*, Shahrivar 13, 1337 (Sept. 4, 1958).

CONCESSION AGREEMENTS

livery of 100,000 cubic meters from the area concerned). This term was to be followed by three 5-year renewals if the non-Iranian partner so desired.

4. On the second and third renewals of the agreement the foreign partner was to accept the terms of other agreements considered on the whole more favorable to Iran than the existing agreement. This may, perhaps, be designated as "the most-favored-nation clause" within the specific context of oil agreements. Of the older concession agreements, only that of IPC in Iraq contained a provision reminiscent of this clause.

5. Strict limitations were imposed on the employment of foreigners by the operating companies, with provisions for the training and progressive Iranization of the personnel. Foreign employees were not to exceed 2 per cent of the total personnel at the end of ten years or 49 per cent in top executive positions.

6. The boards of directors of the operating companies were to consist of Iranians and non-Iranians in equal numbers. The chairman in each case was to be nominated from among the Iranian directors, with the vice-chairman and managing director from the non-Iranians.

7. All three agreements provided for the gradual relinquishment of areas in which no oil was exploited within specified time limits.

These were the main common features of the agreements. The principal difference among them lay in provisions regarding bonus payments. Whereas the first and the third (i.e., the Agip and the Sapphire) agreements made no mention of bonuses, the second (i.e., the Pan American) provided for a $25,000,000 bonus payable to Iran within thirty days of the effective date. This bonus was, however, to be amortized by Pan American in the course of operations. Furthermore, the Pan American and the Sapphire agreements contained a provision that preferential consideration was to be given to Iranian flag tankers. The NIOC-Agip agreement did not contain such a stipulation.

The agreement between NIOC and Agip Mineraria, a company wholly owned by the government of Italy, reopened the question of participation by a foreign government in the development of Middle Eastern oil resources. This is a delicate question both in the Middle East and in the West. In the Middle East the general trend has been to shun agreements with government-owned foreign concerns in order to avoid dangerous political entanglements. This was expressed not only in the older Iranian complaints about the British government's controlling voice in the affairs of AIOC, but also in the more recent resolutions of the Arab League which urged its member states not to grant oil concessions to any company which was owned even in part by a foreign government.[10]

In the West the prevailing practice was to leave oil ventures to free enterprise. By entering the field hitherto worked by private corporations, the Italian government opened the door to possible international complications, since its instrument, Agip Mineraria, does not possess any significant marketing outlets for crude oil production. It was feared in Western oil circles that, by virtue of its official status, the company would induce the Italian government to discriminate in its favor by compelling refineries operating in Italy to process its crude oil. Thus far crude for Italy's refineries had come mostly from members of the Iranian Consortium, who would thus be the first to suffer from such a discrimination. Any decrease or nonexpansion in the Consortium's production would in turn affect the interests of Iran as a recipient of half of the profits. Furthermore, criticism was voiced in Western oil circles that the Italian government, whose financial position had to be bolstered by outside aid, was allowing its funds to be used in a speculative venture. Inasmuch as the government was borrowing money for different purposes, it was suggested that the lending institutions might want to question the diversion of funds into other channels than those originally intended.

These criticisms were vigorously rejected by leaders of the government-controlled Italian oil enterprises, who accused the major Western oil corporations of undue pressures and exclusiveness, while reaffirming their determination to go ahead with their plans for the development of Iranian offshore areas.

The Arabian-Japanese Agreements

The two agreements concluded between the Japan Petroleum Co., Ltd., on the one hand and the governments of Saudi Arabia

[10] See p. 190 below.

and Kuwait on the other for the exploitation of the Neutral Zone's offshore areas likewise introduced significant innovations into the pattern of concessions.[11] The main features of these agreements were as follows:

1. Integration. The company was to exercise its concession rights through a subsidiary, Japan Arabian Oil Company, which would operate as an integrated entity, carrying on production, refining, transportation, and marketing. This was undoubtedly the most significant departure from the existing pattern.

2. Payments. The financial clauses called for payments (composed of rents, royalties, and taxes) totaling not less than 56 per cent of the company's net income to Saudi Arabia (presumably from the half of its operations) and 57 per cent to Kuwait. Taxes were to be paid to the two signatory governments on all operations, i.e., inside and outside Saudi Arabia and Kuwait. Foreign taxes on income derived wholly outside the host country were to be deductible except those levied on tanker operations.

3. Personnel. The agreements provided for strong representation of the host countries on the company board and management committees. Saudi Arabia and Kuwait together were to nominate one-third of the board membership. The chronologically earlier agreement with Saudi Arabia also stipulated that Saudis, if available, should constitute not less than 70 per cent of the work force inside Saudi Arabia and the concession area and not less than 30 per cent outside of it. If Saudi employees were not available, preference was to be given to (*a*) citizens of other Arab League states, (*b*) citizens of other Arab states, and (*c*) citizens of other friendly states. Equal compensation was to be given to Saudis and non-Saudis with substantially similar responsibilities.

4. Supervision of expenses. Expenditures and purchases were to be examined and supervised by committees composed equally of representatives of the host governments of the company. This again was an innovation.

5. Refining obligation. Each agreement provided for the con-

[11] For the text of the agreement between Japan Petroleum Co. and Saudi Arabia see *Platt's Oilgram News Service* (N.Y. ed., spec. suppl. to vol. 36, no. 26a), Feb. 6, 1958; and between Japan Petroleum Co. and Kuwait, see *Petroleum Times*, July 18, 1958.

struction of a refinery when production reached a certain level. This level was defined as 30,000 barrels a day in both the Kuwait and Saudi Arabia agreements. Moreover the Kuwait agreement provided for the establishment of a Research Institute in Kuwait once the above-mentioned level of production was attained.

6. *Diplomatic noninterference.* The company renounced any right to diplomatic recourse in case of disputes. Moreover no direct or indirect interest was to be acquired in its stock by a foreign government or government entity. This provision conformed to the Arab League recommendations on this subject and differed markedly from the Iranian approach as exemplified by the Agip agreement.

7. *Time limit and relinquishment.* Both concessions were granted for a term of forty years beginning with the discovery of oil in commercial quantities and both provided for a gradual relinquishment of unexploited areas at stated intervals.

SUMMARY

The foregoing review of the concession agreements indicates that the era of simple and brief documents belongs irrevocably to the past. The growth of the oil industry in the Middle East has been accompanied by growth in the complexity of the legal foundations upon which it is built. This complexity is partly due to the desire of the concessionaire companies to define with the maximum clarity their rights and obligations, but even more to the policies of the host governments. After having undergone a dual process of emancipation and education, the governments are insisting both on a higher price to be paid for the exploitation of their oil resources and on a number of additional advantages and safeguards. Such special provisions as a guarantee of minimum production, calculation of profits before rather than after the home government's taxation, participation of the host country's nationals on the boards of the companies, and insistence on career opportunities for national employees are illustrative of the new trends, which reflect greater economic awareness and increased social consciousness on the part of the host governments.

The growing demands of the host states for a better deal have resulted, as we have seen, in a series of revisions of the older concessions and also in the conclusion of recent agreements that have set new patterns of nearly revolutionary significance. Foremost among the latter are high initial bonus payments, abandonment of the fifty-fifty formula in favor of higher payments, and the concept of integrated operations—all accepted by some recent concessionaires. Expectation of a higher share in profits on the part of the host governments is partly due to the fact that the value of much of the territory around the Persian Gulf has been upgraded in view of the presence of oil as demonstrated by those companies which had taken the earlier risks of exploration. Partly it is due to changing concepts, under the general impact of nationalist trends, of what is fair and proper.

To a conservatively minded person the ascending expectations and demands of the host countries may seem somewhat disturbing. The most-favored-nation formula has informally (and in the IPC concession of 1952 formally) been the rule in the oil business in the Middle East. The question, therefore, is asked: Are these new agreements with small independent newcomers going to affect the older patterns? And if so, to what extent and by what methods? Is this going to be a process of gradual adjustment or of abrupt crises? And will respect for the sanctity of contracts prove to be a mitigating factor in the face of growing nationalist pressures? Having invested heavily in the development of their industry, the oil companies obviously have to think of protecting these investments by devising the best possible safeguards. The concession document is the most important legal safeguard that a company possesses. Naturally, therefore, it will tend to insist on its binding force as the main ingredient of the company-government relationship. The host government on the other hand is in many respects an unequal partner: economically it may be weaker than a concessionaire company, but it usually possesses the attribute of sovereignty, which places it in a position of superior strength vis-à-vis a private corporation. This inequality of partners carries within itself a germ of considerable tension. Once we speak of sovereignty, we introduce the element of political power. Some basic issues arising in this connection will be examined in the next chapter.

◇◇ CHAPTER V ◇◇

Legal and Political Safeguards

APART from regulating the economic relationship between the company and the host government, the concession agreement represents also a political fact of major importance. Obviously it is not a matter of indifference to whom, for how long, and on what terms a major national resource is going to be entrusted for exploitation. Consequently not only is the decision to grant a concession made by the highest policy-making elements in the host country, but in modern times special legal safeguards have been provided against arbitrary or lighthearted granting of such concessions.[1]

These safeguards can, in the first place, be found in the constitu-

[1] A vivid example of the recklessness with which Eastern rulers in the past disposed of the wealth of their countries is supplied by the concession granted in 1872 by Nasr ed-Din Shah of Iran to a naturalized British subject, Baron de Reuter. The concession, of 70 years' duration, gave De Reuter the right to construct a railroad joining the Caspian Sea with the Persian Gulf and any branch line judged convenient by the concessionaire; a streetcar monopoly in the whole of Iran, with the privilege of claiming rights of way and any land necessary for this purpose gratis; the right to extract from public domains, without payment, all the materials necessary for the construction of the streetcar system; the exclusive privilege of exploiting in the whole of Iran all the state forests, of extracting coal, iron, copper, lead, oil, and any other minerals with the exception of gold, silver, and precious stones, and of exploiting through the construction of dams, wells, reservoirs, and canals all the water resources of the country that had not yet become the subject of other concessions. Reuter's concession represented a virtual bartering away of most of the wealth of the country to a single foreigner. For details, see M. Nakhai, *L'Evolution politique de l'Iran* (Brussels, 1938), pp. 21–22.

tional laws of the host countries. The Iranian Fundamental Law of December 30, 1906, explicitly provides in Article 24 that "the granting of commercial, industrial, agricultural and other concessions, irrespective of whether they be to Persian or foreign subjects, shall be subject to the approval of the National Consultative Assembly." [2] The Turkish Constitution likewise deals with this matter in Article 26, stating that "the Grand National Assembly directly exercises such functions as . . . approving or annulling contracts and concessions involving financial obligations. . . ." [3] Similar provisions may be found in the Lebanese Constitution of May 23, 1926 (Article 89),[4] the Iraqi Organic Law of March 21, 1925 (Article 94),[5] and the Syrian Constitution of September 5, 1950 (Article 146).[6] The Egyptian Constitution of January 16, 1956, does not explicitly vest the right of approving the concessions in the legislature but states that "law will arrange the regulations and procedure regarding the grant of concessions connected with the exploitation of natural resources and public utilities" (Article 98).[7] The provisional Constitution of the United Arab Republic of March 5, 1958, omits explicit reference to concessions. Its Article 68, however, recognizes the validity of "all the dispositions established by the legislations in force in each of the two provinces of Egypt and Syria" at the time of the proclamation of the new Constitution, and Article 29 may be interpreted as indirectly referring to concessions by declaring that "the Government cannot contract a loan or engage itself in a project affecting the funds of the Treasury . . . without the approval of the National Assembly." [8] The Constitution of Jordan vests the right of granting concessions in the Council of Ministers acting with the sanction of the king (Article 77).[9]

[2] Helen Miller Davis, *Constitutions, Electoral Laws, Treaties of States in the Near and Middle East* (rev. ed.; Durham, N.C., 1954), p. 110.
[3] *Ibid.*, p. 454.   [4] *Ibid.*, p. 304.   [5] *Ibid.*, p. 169.   [6] *Ibid.*, p. 429.
[7] *Middle Eastern Affairs*, Feb., 1956, p. 74.
[8] Arab Information Center, *Basic Documents of the Arab Unifications* (New York, 1958), pp. 14 and 20.
[9] Davis, *op. cit.*, p. 251.

LEGISLATIVE SAFEGUARDS FOR HOST COUNTRIES

Apart from these basic constitutional provisions a number of Middle Eastern countries have adopted special legislation regulating the exploitation of mineral resources. Acting under the impact of a threat that Iran's remaining oil resources might fall prey to Soviet ambitions, the Iranian Majlis passed, on December 2, 1944, a law which forbade any cabinet minister to enter into negotiations for an oil concession or to grant it without prior authorization of the Majlis. The author of the law was Deputy Mohammed Mossadegh, who subsequently emerged as the main proponent of a radical policy toward the Anglo-Iranian Oil Company. It was on his initiative that the Majlis adopted and the Senate confirmed on March 15 and 20, 1951, respectively, the famous nationalization law which provided that "the oil industry throughout all parts of the country, without exception, be nationalized; that is to say, all operations of exploration, extraction and exploitation shall be carried out by the Government." [10] This law was followed by another, promulgated on May 2, 1951, and entitled "Law Regulating Nationalization of the Oil Industry," which, in nine articles, provided for the methods and organization which should govern the transfer of the oil industry from private to state ownership.[11] In conformance with these laws a "Statute of the National Iranian Oil Company" was drafted early in 1955. According to it the company was "to engage in the petroleum industry (throughout the territory and the continental shelf) in any of its phases, including search and exploration for petroleum, natural gas and other hydrocarbons; extraction, refining, manufacturing, processing, transporting and storage" and "any other operations required for the commercialization, distribution, sale and export of the said substances and their by-products and derivatives." For the purpose of carrying out the foregoing operations the company was authorized "to associate and cooperate with any companies in Iran or abroad in all matters relating to petroleum." [12] Although this

---

[10] Iranian Embassy, Washington D.C., *Some Documents on the Nationalization of the Oil Industry in Iran* [Washington, n.d.], p. 2.

[11] *Ibid.*

[12] From a mimeographed text by courtesy of the National Iranian Oil Company.

LEGAL AND POLITICAL SAFEGUARDS

statute has not yet been formally approved by Parliament, it has nevertheless become an effective document, being implemented in practice.

Iranian oil legislation underwent a radical change with the passage on July 29, 1957, of a new Petroleum Act by Parliament. Reversing the Mossadegh law of 1944, the new bill authorized the government (acting through the National Iranian Oil Company) to enter into agreements with foreign concerns. Three types of enterprises were foreseen: (*a*) a "mixed organization" possessing a single juridical personality, to be owned in part by the NIOC and in part by another company; (*b*) a "joint structure," whose ownership would likewise be shared by NIOC and another company but which would not constitute a single juridical person; and (*c*) an enterprise to be operated independently by a company other than NIOC. In mixed organizations not less than 30 per cent of the ownership interest was to be held by NIOC. On the assumption that taxes received by the state would equal 50 per cent of the profits of such an enterprise, the total part accruing to Iran would be 65 per cent. While this appeared like a rigid a priori requirement, it should be pointed out that it applied to only one type of the contemplated organizations, the other two types being free of such stiff terms. The 65 per cent participation of the state in mixed ventures should probably be regarded as the goal to which the state was tending. Furthermore, the law provided for the proclamation of territorial zones for exploration and exploitation, for which bids from prospective operators would be received. Agreements were not to exceed twenty-five years in duration, with the possibility of three 5-year renewals.

Other provisions dealt in detail with the size of districts open to exploration, bonus payments, rentals, the sharing of costs between NIOC and other companies, taxation, and related problems. Rather significantly Article 4 of the law proclaimed that "authority to act as an operator or contractor can be granted with respect to foreign persons only in cases where, under the current laws and the economic regime of the foreign country concerned, Iranian persons would be permitted to engage in economic activities and in particular to enter into operating arrangements similar to those envisaged by this Law, in the territory of such foreign country." This was interpreted as virtual exclusion of the Soviet bloc countries from the development of Iranian oil resources.

The general tenor of the law was to avoid the control of Iranian oil resources by a single large corporation by using the device of granting restricted areas for exploitation, in no case larger than 16,000 square kilometers, with a maximum of five districts to be operated by the same company. At least one-third of the total exploitable territory was to be conserved at all times as national reserves. Moreover, no doubt in deference to the national sensibilities stirred up during the Mossadegh era, the law carefully avoided the term "concession" in its references to agreements with foreign concerns.[13]

Turkey, though not a major producer, has also enacted special legislation dealing with oil affairs. For many years the Turkish state, dedicated as it was to the principle of *étatisme,* was unwilling to allow foreign capital to explore and develop the potential oil resources of the country. This attitude found its expression in a 1929 law that entrusted a special government agency with the task of exploration and exploitation. The advent to power in the postwar period of the Democratic Party with its emphasis on free enterprise brought about a new policy in oil affairs. On March 7, 1954, the Grand National Assembly passed a law that authorized the granting of exploration permits and concessions, stating the terms in advance.[14] The international oil industry welcomed the new law as a positive step, and within a few months a number of foreign corporations applied for and obtained exploration permits.

In some countries, notably Egypt, the legal basis for oil operations is established not only by special oil laws, but also by general legislation regulating the activities of foreign corporations. On July 20, 1947, the Egyptian Parliament adopted a rather stringent Companies' Law providing minimums for Egyptian representation or participation on boards of directors, as employees or laborers, and in salaries, wages, and stock.[15] An amendment to this law, passed

---
[13] *Ibid.*

[14] *New York Times,* March 8, 1954, and *Oil Forum,* May, 1954.

[15] The figures were as follows: 40% for directors; 75% for employees; 90% for laborers; 65% of salaries; 80% of wages; and 29% of stock. See *Egyptian Economic and Political Review,* Nov., 1954, p. 27.

in 1948, raised the percentage of stock to be held by Egyptians to 51 per cent. This legislation was promptly followed on August 12, 1948, by a special law on mines and quarries, restricting licences for oil mining to Egyptian companies only and determining the rate of royalties to be paid to Egypt.[16] All this discouraged the foreign oil corporations operating in Egypt, several of which either abandoned or suspended their exploration activities.[17] Following the coup of 1952, however, the revolutionary government of Egypt enacted new legislation designed to attract foreign capital. A law of July 30, 1952, reversed the ratio of stock to be held by the Egyptians, allowing foreign companies to hold up to 51 per cent. The basic law on mines and quarries, however, remained unchanged, making it virtually impossible for a foreign concern to obtain a large-scale concession covering a substantial part of Egyptian territory. Conseqently oil operations in Egypt are carried out by a number of companies in fairly restricted areas.

In Syria legislation affecting foreign oil companies may also be divided into two categories. One, general, is exemplified by Law no. 151 of March 3, 1952, which proclaims that all foreign corporations must have as their chief representative in Syria a Syrian citizen. Another, specific, is expressed by Law no. 7 of December 21, 1953, called the "Law on the Mines," which in Article 9 states that exploration permits and concessions can be granted only to persons of Syrian nationality or to companies, provided (*a*) that the latter are founded in Syria and have their legal domiciles in it; (*b*) that in granting a permit or a concession due attention is paid to considerations of national defense and the safeguarding of the country's freedom and its financial and economic interests. The law enumerates certain basic provisions which any concession must contain and establishes a variety of fees and taxes to be paid by the concessionaires. It also proclaims that, in addition to the above, every act of concession shall determine the rate of royalties and that this rate shall not be less than one-half of the net profits from exploitation. The Ministry of Public Works and Communications is designated as an

[16] *Ibid.*, p. 26.
[17] Standard Oil Co. (N.J.) withdrew from Egypt in 1949, leaving behind it its affiliate sales organization, Esso Standard (Near East). Socony-Vacuum and Anglo-Egyptian Oilfields, Ltd., decided to suspend their exploration operations.

authority competent to deal with and control matters relating to mining. The law is explicit in giving the Minister of Public Works the right to grant survey permits. Full exploration permits are to be granted by decree, which presumably is a presidential prerogative. The law does not state clearly which is the authority empowered to grant concessions for exploitation. The Minister of Public Works is to be aided, in an advisory capacity, by the Mining Council, to be composed of representatives of various departments and experts. The law has been supplemented by Presidential Decree no. 294 of January 30, 1954, explicitly permitting foreign companies to seek permits and concessions in Syria.[18]

LEGAL STATUS OF CONCESSION AGREEMENTS

It is thus within the framework of constitutional provisions and the general or specific legislation that foreign concessions operate. So long as such laws are antecedent to the concessions, no major legal conflicts are likely to arise. But if they are enacted after the granting of a concession, and if the two are mutually incompatible, as was the case in Iran, a need arises to determine which legal act —the contract or the law—should enjoy primacy. Actually the issue may be broadened to include all breaches of the contract by the host state, whether by executive or legislative acts. The question that may be posed is therefore: Does law recognize the validity of a contract concluded between a sovereign state and a foreign national and can a state be held responsible for breaches of such a contract?

In reply to this question, it is perhaps proper to point out that the principle of *pacta sunt servanda* has been firmly embedded in the laws and customs of civilized states. In the West it has been surrounded by an aura of special sanctity, and much of the progress of Western civilization can be ascribed to consistent respect for this principle. The Islamic system of law did not develop a general

[18] Both laws are given in *Al-Jumhuriyat as-Suriyah, Wizarat al-Ashghal al-'Ammah wa al-Muasalah, Mudiriyat al-Ma'adin wa al-Maqal'i, At-Tashri'u al-Khass bi al-Manajim wa al-Maqal'i* (Republic of Syria, Ministry of Public Works and Communications, Directorate of Mines and Quarries, *Legislation on Mines and Quarries,* Damascus, 1954).

LEGAL AND POLITICAL SAFEGUARDS

theory of contracts comparable to that of the West.[19] Nevertheless, it did proclaim the duty of honoring agreements, in conformity with injunctions of the Koran and the Traditions.[20] Paradoxically, however, whereas the principle of *pacta sunt servanda* has been generally applied to contracts between private persons on the one hand and to treaties between states on the other, its application vis-à-vis agreements between a state and a private person (i.e., in the sector where the difference in power is likely to result in greater abuses than in other sectors) has not gained similar unquestioned acceptance. True enough, national laws have begun to distinguish between the sovereign and nonsovereign acts of a state and in due course hold the state responsible for the latter category, thus giving private individuals and corporations the right of redress in case of a breach of contract by the government.

But, while the binding force of contracts between a state and a private person has been thus reaffirmed, it appears that this rule is firmly accepted only in national laws. The question therefore arises: Can foreign concessionary companies enjoy the added protection of international law? Is a contract concluded between a sovereign and a foreigner, individual or corporate, binding under international law? Thus far unanimous consent has not been obtained on this point. Broadly speaking, two major trends of thought are discernible in this respect. One is that, as a law-making sovereign, the state has the right to alter the laws it makes and, if the public welfare so requires, it can lawfully cancel its contracts with individuals without incurring any international responsibility. This

---

[19] For an illuminating discussion of contracts in Islamic law, see Subhi Mahmassani, "Transactions in the Sharia," *Law in the Middle East*, ed. M. Khadduri and H. J. Liebesny (Washington, D.C., 1955), vol. I.

[20] The following passages in the Koran have frequently been invoked to demonstrate the Moslem duty to honor pledges and agreements:

Sura V, Verse 1: "Oh ye who believe, fulfill your pledges."

Sura V, Verse 91: "Honor your covenant with God, when ye enter into covenant with Him; and violate not your oaths, after ratification thereof, since you have made God a witness over you. Verily God knoweth that which ye do."

Sura XVI, Verse 94: "And take not your oaths to practice deception between yourselves."

Sura IX, Verse 4: "So fulfill your engagements with them to the end of their term: for God loveth the righteous."

thesis has been partly based on the old concepts that only states and not individuals are subjects of international law and that states (being sovereign) enjoy immunity from suits in foreign countries.

The other trend is to affirm that under international law the state is definitely bound to honor its contractual obligations toward aliens. Supporters of this thesis point out, first, that, "international law today recognizes that individuals and other subjects are directly entitled to international rights . . . , [that] the alien has been internationally recognized as a legal person independent of his state; [and that] he is a true subject of international rights." [21] Similarly the old concept of sovereign jurisdictional immunity has been gradually abandoned. In 1948 the United States Supreme Court went so far as to proclaim that "the principle of sovereign immunity is an archaic hangover not consonant with modern morality." [22] The thesis of the binding force of such contracts under international law has been upheld by a number of court decisions, individual text writers, and group studies. In the latter category (which represents systematic efforts of teams of scholars with a view to the codification of international law) the Harvard Law School's "Research in International Law" declared in 1929: "A State is responsible if an injury to an alien results from its non-performance of a contractual obligation which it owes to the alien, if local remedies have been exhausted without adequate redress." [23] Similarly the Preparatory Committee to the Hague Conference for the Codification of International Law in 1930 drafted the following statement for discussion: "A State is responsible for damage suffered by a foreigner as the result of the enactment of legislation which directly infringes rights derived by the foreigner from a concession granted or a contract made by the State." [24]

Despite these not infrequent assertions of the binding force of

---

[21] Report of the Special Rapporteur on "International Responsibility" of the United Nations International Law Commission, UN Doc. no. A/CN.4/96, pp. 51, 58. Quoted by Lowell Wadmond in an unpublished paper, "The Sanctity of Contract between a Sovereign and a Foreign National," read at the meeting of the American Bar Association, London, July 26, 1957, p. 10.

[22] *Larson v. Domestic-Foreign Corp.*, 337 U.S. 682, 703, 704 (1948). Quoted by Wadmond, *op. cit.*, p. 12.

[23] Wadmond, *op. cit.*, p. 41. [24] *Ibid.*, p. 45.

contracts in international law and the general evolution toward limitation of sovereign state powers, the matter has not yet fully emerged from the realm of controversy. It should be pointed out that, the Harvard studies and the work of the Preparatory Committee notwithstanding, neither the Hague Conference of 1930 nor the Institute of International Law (which considered this matter at its session in 1952) have succeeded in reaching agreement on the responsibility of states and the treatment of foreign property. More recently the International Law Commission of the United Nations has been working toward the codification of international law on the same subject, but the preparatory report submitted by its *rapporteur,* Señor García-Amador, has suggested that the responsibility of states occurs only if a breach of contract with an alien "(a) is not justified on grounds of public interest or of the economic necessity of the State; (b) involves discrimination between nationals and aliens to the detriment of the latter; or (c) involves a 'denial of justice,'" the latter being defined as a lack of genuine judicial redress.[25] These reservations are so far-reaching as to narrow down to almost nothing (if not to abolish altogether) the area in which the responsibility of states for a breach of contract could be applied.

Challenges to the Validity of Concessions

In addition to these fundamental considerations concerning the binding force of contracts in international law, the validity of concession agreements has sometimes been challenged by the host governments on the grounds of national emancipation, immaturity, and domestic revolutionary change. The argument of national emancipation runs as follows: Certain Middle Eastern countries initially granted oil concessions at a time when they were either formally subjected to foreign domination (such as under a mandate, protectorate, or exclusive treaty relationship) or greatly dependent, *de facto* if not *de jure,* on foreign influence. Only when foreign control or influence was removed, could the governments of liberated countries act freely with a view to promoting their national interests. They should not, therefore, be penalized by inheriting the obligations contracted by their imperialist predecessors. Here again this actual or potential plea by the host governments has been aided by

[25] *Ibid.,* pp. 48 ff.

lack of a firm principle in international law. The problem of succession in the law of nations has been a matter of controversy. According to Oppenheim, "the recent practice of states . . . tends to establish as a rule of international law the duty of a successor state . . . to respect the acquired rights of private persons whether proprietary, contractual, or concessionary." "There is much to be said in favour of the view," he adds, "that, if before the extinction of the State which granted the concessions every act necessary for vesting them in the holder had been performed, they would survive the extinction and bind the absorbing State. But every case must be studied on its merits, and it is difficult to lay down a general principle." [26] Similarly the Permanent Court of International Justice, while upholding the principle of succession in some notable cases, did not formulate any absolute and comprehensive rule in this respect.[27]

The second argument—that of immaturity—may be summed up as follows: Even if the concession was granted freely, without any foreign pressure or foreign sovereign decision, still the general experience and development of the host country should be taken into account when evaluating the validity of the concession agreement. This argument has not, to the writer's knowledge, been given express formulation by any host government. Yet it has been asserted informally more than once that at the time when foreign interests offered to exploit the potential oil resources of the host country, the latter was so retarded, its leaders so inexperienced in terms of modern economics, and its fiscal needs so compelling that a relatively small advance payment with a promise of what then appeared as reasonable royalties was sufficient to induce those leaders to barter away their precious natural resources without detailed investigation. But circumstances have now changed. With the passage of time rulers of the host countries have learned a good deal both about the value of oil as well as about the terms that other countries have obtained when granting concessions. They have also learned about

[26] L. Oppenheim, *International Law* (4th ed.; London, New York, Toronto, 1928), I, 168 n. 1 and p. 169.

[27] The cases in question were those of the *German Settlers in Poland* and *Mavrommatis Palestine Concessions*. See J. L. Brierly, *The Law of Nations* (4th ed.; Oxford, 1949), p. 139.

the companies' profits, which have many times surpassed the original investment. Consequently, in fairness and justice, they expected that the terms of the concessions would be changed to conform to the changed circumstances, failing which they were prepared to enact, unilaterally, legislation which would vindicate their views and requirements.

In contrast to the argument concerning succession, this one is hardly admissible in the eyes of the law. Conceivably this reasoning could be based on either of two legal rules, namely, those of fraud in the making or *clausula rebus sic stantibus*. But if the first of these were to be invoked, evidence of actual fraud would have to be supplied, which would be an extremely difficult thing to do. In fact, in view of the character of mineral concession agreements, in which the concessionaire undertakes the whole risk of finding or not finding mineral deposits, it could be ruled out in practice. As for *clausula rebus sic stantibus,* this doctrine, known both in private and in international law, proclaims that a contract is valid only so long as the circumstances upon which it is based have not changed. To invoke this doctrine in the situation described above would be to suggest a farfetched and misleading construction. The basis of the doctrine *rebus sic stantibus* is that a treaty or a contract is terminated if the foundation upon which it rests disappears. Therefore, to quote one authority,

> not every important change of circumstances will put an end to the obligations of a treaty. The *clausula* is not a principle enabling the law to relieve from obligations merely because new and unforeseen circumstances have made them unexpectedly burdensome to the party bound, or because some consideration of equity suggests that it would be fair and reasonable to give such relief.[28]

Thus it is clear that it is futile to attempt a legal justification of such unilateral actions on the part of host countries. The matter clearly lies outside the scope of the law. The law being primarily an instrument for the maintenance of the *status quo* cannot provide a solution to a situation which is essentially political in nature. All

---

[28] Brierly, *op. cit.*, p. 245. For a clarifying discussion of the doctrine *rebus sic stantibus*, see also Oppenheim, *op. cit.*, pp. 746–753.

that law can do is to insist that unilateral action aiming at the abrogation of a concession or its modification is illegal. The change in circumstances to which the host countries are wont to refer is integrally linked with the general problem of change in the political structure of the world. Thus far no satisfactory answer has been found, either legally or politically, to the question of peaceful change in international relations.[29] In certain instances where consensual approval has been given to a change effected by one party, the appearance of respect for legality has been preserved, but in reality this was a violation of contract followed by formal acceptance of the change by the parties concerned, and usually accompanied by a solemn reassertion of the sanctity of contractual obligations.[30]

The third and last argument is put forward by governments issuing from revolution or from some sort of popular upheaval. This argument is to the effect that the decisions (including laws, treaties, and contracts) of the old government did not represent the will of the people and therefore should be considered null and void. This type of argument gained fairly wide currency, especially in the 1950's, when revolutionary upheavals of various depths and dimensions occurred in such countries as Iran, Egypt, Lebanon, Syria, and Iraq. The fact that the new regimes in those countries did not necessarily represent an advance in electoral or representative processes (which would be the only objective way of measuring the degree of popular representation) did not seem to dampen the zeal of the new rulers in asserting that they, in contradistinction to the old regime, truly reflected the will of the masses. The basis of this assertion was to be found in the sociopolitical shift that had, in some instances, taken place as a result of the change of government. While the old regime was often representative of the dominant wealthy and conservative classes, the new one frequently brought to power elements representing the have-not strata, whether the lower middle class or the intelligentsia. The thesis of the new rulers was that

---

[29] For a useful discussion of this question, see Frederick S. Dunn, *Peaceful Change* (New York, 1937), and Hans J. Morgenthau, *Politics among Nations* (New York, 1948), pp. 350 ff.

[30] Such was the case in the unilateral Russian denunciation of the Black Sea demilitarization clauses of the Treaty of Paris of 1856 and the subsequent recognition of this step by the powers at the London Conference of 1871.

the old group, being corrupt, selfish, and linked to foreigners by a variety of interlocking interests, was bound by virtue of its character to make agreements detrimental to the welfare of the country as a whole.

Although in some cases such an argument may have considerable moral appeal, international law leaves no doubt as to its illegality. A state is "bound internationally by the acts of the person or persons who in actual fact constitute its government." [31] Obligations undertaken by a government bind the state on whose behalf such a government is acting, and they remain attached to the state regardless of later changes in government. To introduce the element of sociopolitical distinction or to debate the representative character of a given government with a view to determining the binding force of engagements it has contracted would produce such chaotic conditions in the field of contracts as to make them meaningless from the legal point of view. To give an example, between 1949 and 1954 Syria had five *coups d'état*. Following each *coup* a newly emerged government invariably claimed to be the only true representative of the people's will. No matter what mental acrobatics were to be exercised in such situations, no satisfactory solution as to the validity of obligations could be found if it were to be based on the character of the governments that had contracted them.

To recapitulate: The treatment of contracts between a sovereign and a foreign national in international law shows paradoxical gaps and uncertainties. While the validity of contracts is firmly upheld by the law as against challenges based on immaturity and revolution and fairly firmly as against those based on nonsuccession, the law fails to establish a firm rule about the binding force of such contracts as against executive or legislative breaches committed by the host country. Thus the kernel of the problem is left without a fully satisfactory answer.

A Search for Better Safeguards

The difficulty is compounded by the fact that there is no international court with compulsory jurisdiction over such cases. Both the Permanent Court of International Justice and its successor, the International Court of Justice, have based their competence on vol-

[31] Brierly, *op. cit.*, p. 130.

untary acceptance of jurisdiction by the parties to disputes, and whenever a doubt arose—as when Iran nationalized its oil industry—these courts have tended to limit their own jurisdictional power.

On the other hand, logic, equity, and general principles of law among civilized nations firmly uphold the principle of *pacta sunt servanda*. To quote one authority: "If the government is free to have recourse to its governmental power to escape its obligations under a contract with a private individual, then doing business with governments on a contractual basis becomes a wholly uncertain thing, so far as the private contractor is concerned." [32]

For this reason the companies have invariably insisted on provisions stipulating that in cases of disputes with the host governments their concession agreements were to be submitted to arbitration. Such provisions usually call for the application of principles of law recognized by civilized nations in general or of those applied by the World Court at The Hague.[33] The inclusion of such provisions in the concession agreements reflects the companies' conviction that if a dispute goes before an impartial international body the latter would naturally take the concession agreement as a point of departure and pronounce against the abuse of sovereign power by a host government. Host governments have usually submitted to arbitration. In a few cases, however, especially if the stakes were high, they have refused to accept arbitration, despite clear provisions for such in the concession agreements. A notable case was Iran, which in its dispute with AIOC in 1951–1952 both rejected arbitration and pleaded against the competence of the World Court, asserting that the matter fell fully within its domestic jurisdiction.[34]

The companies have steadily opposed the concept of domestic jurisdiction and the exclusive application of the host country's national laws to such cases, for understandable reasons. The general juridico-political climate of the host countries has not inspired confidence in the impartiality of their judicial processes, especially where the interests of the foreigner clashed with those of the state.

[32] F. S. Dunn, *The Protection of Nationals* (Baltimore, 1932), p. 165.

[33] Thus Art. 22F of the AIOC concession, Art. 46 of the Consortium agreement, and Art. 28 of the Aramco concession.

[34] For a detailed discussion, see Alan W. Ford, *The Anglo-Iranian Oil Dispute of 1951–52* (Berkeley, 1954), pp. 164 ff.

LEGAL AND POLITICAL SAFEGUARDS

This has been due partly to the traditional *mores* of the society and partly to the virtual subordination of judicial to executive authority in most of the Middle Eastern countries. Because of the authoritarian patterns of government—whether feudal-patriarchal or military-dictatorial—this subordination is not likely to disappear in the foreseeable future. The numerous political trials staged in Egypt, Syria, and Iraq in 1955–1958 bear eloquent testimony to this point.

The insufficiency of international law in the domain of contracts has stimulated much thinking and searching for remedies. The work under scholarly auspices, such as the Harvard group or the Institute of International Law, has already been mentioned. More recently suggestions have been made that concession agreements should be reinforced by intergovernmental treaties specifically concluded for that purpose. These suggestions originated in the pipeline crisis in Syria in 1956 and the resultant plans to construct alternate pipeline systems under international guarantees.[35] Although commendable in many respects, such suggestions are not likely to provide either a general remedy (which would call for a full clarification of the legal principle involved) or a solution for those concessions already in operation, reinforcement of which by new treaties would surely be resisted by the host countries.

In 1957 a German study group attempted to break the deadlock by submitting new and original proposals. Acting on the assumption that "the economic development of the less advanced countries of the world, particularly in Asia and Africa, calls for a continually increasing measure of assistance on the part of the Western industrial nations," the Society to Advance the Protection of Foreign Investments has proposed that an international convention for the mutual protection of private property rights in foreign countries be concluded as a multilateral instrument. The society's proposals have been embodied in a memorandum, accompanied by a draft convention with commentaries.[36] The preamble of the convention invokes Article 17 of the "Universal Declaration of Human Rights" made at the United Nations in 1948, which stipulated that (*a*) every-

[35] For a more detailed discussion of the crisis, see Chapter XVI below.
[36] *International Convention for the Mutual Protection of Private Property Rights in Foreign Countries,* draft presented by Gesellschaft zur Forderung des Schutzes von Auslandsinvestitionen e.V., Cologne, Nov., 1957.

one has the right to own property alone as well as in association with others and (*b*) no one shall be arbitrarily deprived of his property. The high lights of the draft convention are a prohibition of expropriation of alien property for thirty years after investment, respect for alien property even at a time of armed conflict between contracting parties, and provisions for both an international court and an arbitration committee to deal with possible litigations.

Although these proposals have attracted favorable attention in Western industrial and banking circles, thus far no official action has been taken to adopt them formally.

The Problem of Enforcement

The foregoing review of the legal status of the concession agreements leads one to observe that in the middle of the twentieth century, in an era when the "one world" concept has gained wide currency, a paradoxical imbalance exists in international relations. On the one hand, we witness the gradual liquidation of Western colonial empires and the appearance of the United Nations with its emphasis on the equality of nations and the pacific settlement of disputes. The use of force is decried as incompatible with a decent world order, and self-help is outlawed except for self-defense in case of aggression. On the other, the principle of noninterference in the internal affairs of independent states seems to enhance the freedom of individual governments to act as they please within their own borders under the protective cloak of national sovereignty. In other words, while the machinery for the maintenance of peace among states has been improved, the standards of respect for universally recognized legal principles, especially in relations between the state and the alien individual, have deteriorated. Under these circumstances such documents as the Universal Declaration of Human Rights are bound to remain idealistic expressions of good intentions rather than mark a real advance in the protection of an individual vis-à-vis the power of the state.

The complicating element in this general picture of imbalance is the fact that major investments have been made by persons and corporate bodies of the strong and advanced Western states in underdeveloped countries undergoing a process of emancipation from the former's control. The latter often tend to disregard the distinc-

LEGAL AND POLITICAL SAFEGUARDS

tion between a legitimate business enterprise, established solely on the basis of commercial merit, and an old-fashioned, imperially supported enterprise geared to unilateral exploitation of a colony. For this reason it has become doubly difficult in this age of formal condemnation of resort to force for the advanced Western states to intervene effectively with a view to protecting their nationals' interests in foreign, underdeveloped areas. Should such intervention or a threat of it occur, the host countries would tend to equate the West's defense of its legitimate interests with its economic or political imperialism.

The inadequacy of modern enforcement procedures poses a problem equal in gravity to the insufficiency of the substantive rules of international law. Even if we assume that a host government, having accepted arbitration or international judicial proceedings, has been adjudged guilty of the breach of contract, how can it be compelled to redress the wrongs committed if it refuses to obey the verdict? If the dispute is brought before the United Nations Security Council, it is doubtful whether an adequate solution can be found, because of the power of veto vested in the permanent members of the Council. If it is brought before the General Assembly, a majority of whose members represent either ex-colonial or underdeveloped states, it is equally doubtful whether the majority would sanction collective action against one of their own kind. Under the circumstances the limitation of self-help procedure solely to action in self-defense, as provided by Article 51 of the Charter, seems to put a premium on unlawful actions by sovereign states, provided they abstain from the technical act of aggression. This is a sad state of affairs and one not conducive to the encouragement of investments in underdeveloped countries. In this connection it may be observed that the nationalization of the Suez Canal by Egypt in 1956 has acted as a warning reminder, stimulating new thinking about the sanctity of international obligations. It has in fact led some early supporters of the United Nations to revise rather drastically their views on the usefulness of this organization and on the employment of force to safeguard one's legitimate interests. To quote Professor Eagleton:

> If the United Nations does not, or can not protect a member and that member is not allowed to defend itself, the aggrieved state is left with no

means of redress. This would mean that law-abiding states would be left at the mercy of law-breaking states and that law breakers would have sanctuary in the United Nations and could not be touched by the use of force. . . . Consequently I cannot agree with [the proposition] that we must put peace above law and justice. This would mean that the law-breakers would rule the world.[37]

In a similar vein, Professor Goodhart points out:

It will . . . be necessary to determine whether the limitation of force to self-defense against an armed attack is a reasonable and practical provision in a world in which the United Nations has not itself been able to carry out its duty to prevent threats of aggression and other breaches of the peace. There is, unfortunately, much to be said for Mr. Acheson's [38] view that moral pressure alone may not prove adequate against those who are prepared to use physical force. It is not certain how the United Nations can continue to be effective if it insists that so impractical a doctrine is an essential part of its existence.[39]

To sum up: Although the present legal safeguards against arbitrary moves by the host countries are inadequate, the fact remains that current international rules and institutions (supported by the presently prevailing attitude in Washington) have considerably lessened the likelihood that states will use force in defense of their nationals' interests abroad. Consequently the military hazards resulting from unilateral action by the host countries have considerably decreased. By contrast, the political and economic hazards have perhaps increased in comparison with the situation a quarter of a century ago. This is because the national economies and state budgets of the host countries have become heavily dependent on oil revenues. Any major disturbance in the steady flow of these revenues is apt to produce serious economic dislocations, and this in turn may easily affect the social and political stability of the host countries. It is not an exaggeration to say that, by adopting a unilateral course of

[37] Clyde Eagleton, "The United Nations and the Suez Crisis," in *Tensions in the Middle East*, ed. Philip W. Thayer (Baltimore, 1958), pp. 281–282.

[38] The reference is to Dean Acheson's article, "Foreign Policy and Presidential Moralism," *The Reporter*, XVI, no. 9 (1957), 10–13.

[39] A. L. Goodhart, "Some Legal Aspects of the Suez Situation," in *Tensions in M.E.*, p. 259.

action which seriously injures the interests of a concessionaire, a host government is apt to court economic and political disaster.

SELF-PROTECTION FOR THE COMPANIES IN TIMES OF CRISIS

Because companies today cannot rely to any great extent on the military might of their home countries, and because diplomatic support of their interests may prove inadequate for lack of military backing, they have to protect themselves largely by their own means. In this connection we may distinguish between a time of crisis and a normal time. In time of crisis the companies might avoid irreparable damage if they have the ability (*a*) to suspend for a lengthy period the totality of their operations in the host country without financial collapse, (*b*) to avoid exclusive dependence on the host country as a provider of raw material or transit rights, and (*c*) to secure a "hands-off" attitude on the part of their major competitors.

Each of these conditions merits a brief comment, which may help in clarifying certain points at issue between the companies and the host countries. The first condition, that of financial ability to withstand the shock of the total suspension of operations, is the most obvious and the one without which any company engaging in foreign operations would be courting suicide. But such an ability exists only if the company in question has ample financial resources of its own or if it has pooled its resources with those of other companies. Both aspects have a connotation of bigness which frequently provides a major point of criticism of the companies both at home and abroad. Yet both the magnitude of the operation itself, in technical and economic terms, and the political risk involved make bigness an almost indispensable attribute of the concessionaire enterprise.

The second condition—ability to avoid undue dependence on the host country as a provider of raw material or transit rights—overlaps the first one in many respects but is not identical with it. Actually fulfillment of the second condition means that the first one may be somewhat less rigorous. If a company possesses alternatives in terms both of access to crude oil and of ability to use different ways and routes for exportation than those provided by a given host country, its freedom of maneuver is so much the greater. But to achieve this capacity a company must not only possess the previously mentioned feature of bigness, but also spread its operations over a number of countries. This is what has happened with a majority of the foreign corporations holding concessions in the Middle East. Either these corporations themselves or the holding companies which own them carry on substantial operations in more than one country. To cite a few examples, Iraq Petroleum Company holds concessions not only in Iraq, but also in Qatar and other principalities of the Arabian littoral. Moreover, its four parent organizations hold directly or indirectly concessionary rights not only in certain adjoining Middle Eastern countries but also in different parts of the world. Similarly, British Petroleum Company (formerly Anglo-Iranian Oil Company), has held and continues to hold not only concessionary rights in Iran, but also in Kuwait; at the same time it is spreading its refining operations to Aden and England. The Arabian American Oil Company, though attached exclusively to Saudi Arabia so far as production is concerned, is owned by four corporations whose holdings spread far beyond the Saudi kingdom. Even the small independent organizations that have entered the Middle East do not depend completely on their Middle Eastern sources, but rather treat them as an addition to their principal operations at home.

Although the companies have fulfilled the condition of geographical diversification of their activities sufficiently to feel fairly secure in case of crisis, in certain respects they still have weak spots in their armor. This is especially true of the transit problem. This matter will be discussed more fully later, but it may be stated here that Iraq Petroleum Company's dependence on Syrian transit has exceeded the point of reasonable safety. And though Aramco's position is less vulnerable, the fact that about 35 per cent of its exports depend on pipelines through Jordan, Syria, and Lebanon makes its situation, as well as that of Tapline, somewhat precarious.

As for the third condition of safety in case of crisis—the benevolent attitude of major competitors—it is of the utmost importance that competition among companies remain within the orbit of fair commercial practice. Should any company attempt to benefit directly from the juridicopolitical difficulties experienced by one of its com-

LEGAL AND POLITICAL SAFEGUARDS

petitors, it would invite chaos and the likelihood that sooner or later it would become the victim of similar circumstances. Thus far the record of mutual fairness has been rather satisfactory, despite the fierce struggle often developing among the competing companies in the initial phase of concession hunting. There has been considerable unhappiness in British oil circles over the fact that Aramco agreed upon a fifty-fifty profit-sharing formula with Saudi Arabia without informing other companies in advance, especially in view of the delicate revision negotiations between the Anglo-Iranian Oil Company and the Iranian government at that very time. Some critics even went so far as to assert that Aramco's decision had been the immediate impetus to the Iranian nationalization movement. This may have been true, although it would probably be safer to say that this decision was a contributory factor in a movement that had independent origins and that had assumed by the end of 1950 truly major proportions.

Regardless of what the whole truth in this case may be, neither Aramco, nor its parent companies, nor any other major Western corporations attempted to take advantage of the Anglo-Iranian difficulties either by offering a better deal to Iran or by purchasing nationalized Iranian oil. This solidarity of the major Western corporations has, to be sure, contributed to the charge that the international oil industry is operating as a cartel. Lack of solidarity in the Iranian affair would probably not have lessened these charges as they were often made before the dispute occurred. Disloyalty to each other in this case would have weakened one of their strongest defenses against arbitrary violations of concessions by the sovereign states of the Middle East.

This solidarity could be attained largely because the major companies are based in Western countries that are linked by many bonds of culture, business ethics, and politico-strategic interests. The "club" is fairly small in number: only the United States, Britain, Holland, and France, all members of the North Atlantic community, are represented in it, and the dominant interests are American and British. This situation is due partly to free historical development and partly to World War II, which eliminated, at least for a decade, three other major industrial nations—Germany, Japan, and Italy

—as active elements in international oil developments. With the gradual resumption, however, of advanced industrial status by these three nations, the hitherto exclusive position of the dominant group may become open to challenge. Should this challenge be played according to the rules thus far honored by the "club," i.e., should it be restricted to competition for new concessions, no major dislocations will be likely to result. Although there is no guarantee that these potential newcomers will not resort to methods unacceptable to the major Western companies, there might be political reasons militating against such a course of action, inasmuch as these former Axis states are now associated with the leading powers of the West in defensive alliances.

SELF-PROTECTION IN NORMAL TIMES

The remarks above apply, as we have said before, to the time of crisis. Yet despite the fact that crises of greater or lesser magnitude have been occurring from time to time, it is to normal times that companies have to gear their principal policies and operations. Barring a hostile attitude on the part of the host government for political reasons, the government-company relationship should be harmonious if company policies generate confidence in company actions. Such confidence must be based, first, on a scrupulous observance of the concession agreement by the company and, second, on the conviction of the host government that the company is not trying to take unfair advantage of gaps or uncertainties in the concession document. The time is past when it was possible and advisable to base one's concession on a brief and fairly general document. Although such a document might have been commendable on account of its brevity and seeming simplicity, it was also apt to leave too many matters to interpretation and thus carry within itself the germ of future disagreements. This is the reason that the concession documents have tended to become more and more exhaustive as time went on. The apogee of this trend has probably been reached by the Iranian Consortium agreement, which, though exasperatingly long, meticulously deals with a variety of details—doubtless because past experience showed that such details have

LEGAL AND POLITICAL SAFEGUARDS

been a frequent source of disagreements and mutual recrimination. The most striking evidence of this new approach is the first two articles of the agreement, which contain no less than twenty-four definitions of words and expressions to be used in the remainder of the document.

No matter how well the concession document is drafted, its implementation can never be automatic. Consequently, apart from the technical and commercial operations of the company, continuous contact and consultation between the company and the host government are needed. This is being accomplished, on the part of the governments, by special offices competent to deal with petroleum affairs and constituting a part of one of the ministries, such as Finance (Saudi Arabia), Public Works (Syria and Lebanon), or National Economy (Iraq).

The office is usually headed by an official with the rank of director or director-general, directly subordinate to the minister or his deputy. In some cases the importance of oil affairs has dictated a special setup, outside the usual ministerial structure. Such has been true in Iran, where a mixed board of five senators and five deputies was established for the purpose of enforcing the nationalization law and drawing up the statute of the National Iranian Oil Company, the latter to be responsible for the actual oil operations. Having accomplished its task, the board dissolved; its place was taken by NIOC, which henceforth began to represent the Iranian government in its dealings with the Consortium. In Lebanon petroleum affairs were entrusted in 1956 to a Minister of State whose special task was to negotiate a revision of existing pipeline and refinery concessions. In the same year negotiations with the concessionaire company prompted Syria to establish a special cabinet committee for oil affairs, composed of the Ministers of Finance, Public Works, and National Economy. The importance of petroleum affairs in the economies of the host countries has also been responsible for the fact that not only during the negotiations for a new or a revised concession, but also in the day-to-day relations between the company and the government, the latter is frequently represented by a cabinet minister in whose department the petroleum office is located.

Legislative branches (in the countries where they exist) have also often adapted their procedures and organization to the need of attending to oil affairs. In most of the Middle Eastern parliaments permanent oil commissions have been created, and seats on them have been assigned in proportion to party strength (where parties exist) in the parliaments. Membership in oil commissions is much coveted by deputies, and generally the members are recruited from the most influential leaders. Debates in these commissions have sometimes served as good barometers of the political currents in the parliaments as a whole.

It is with public bodies such as these that the companies have to work from day to day. Many matters of common interest to the companies and the host governments can undoubtedly be placed in a routine category, but nothing is completely automatic and virtually every action requires wisdom, knowledge, and tact. Moreover, even in normal times perplexing problems of law and economics are likely to arise in the company-government relations. For instance, an oil company conducting operations in a host country's border area will have to ascertain the exact course of boundaries if it wants to avoid complications. Similarly, if it expands its operations into submarine areas, it will want to know the limits of the territorial waters as defined by the host country and its neighbors. If the company decides to transport its oil by pipeline, it will have to agree on a mutually acceptable formula for transit rights and fees. These and other questions have to be attended to if continuity of operation is to be maintained. By handling them with skill and fairness, the companies protect their positions without recourse to outside assistance.

In the following chapters we will examine the methods and organization of the companies' bureaus of government relations as well as special problems in these relations.

CHAPTER VI

Handling of Government Relations by the Companies

THE growth in complexity of company relations with host governments has led to the appointment of special executives, recruited from the top echelons of the hierarchy, to act as permanent representatives vis-à-vis the host countries. These executives usually have large staffs, constituting vital and firmly established departments in the company organizations. It may be useful, by way of example, to present the organizational arrangements for the conduct of government relations of two major companies, Aramco and IPC.

ARAMCO'S GOVERNMENT RELATIONS ORGANIZATION

Aramco's table of organization [1] provides for a "Vice President, Government Relations and Finance." The government relations organization he presides over is expected to maintain contact with the government of Saudi Arabia both at the highest decision-making levels and at local levels. Directly under him stands the general manager of government relations, who is aided by an assistant general manager. These three officials are assisted by a "policy and planning staff" composed of four men selected for their executive experience and knowledge of Arab affairs. Immediately below this level comes a division of responsibility: one group of executives is charged

[1] For 1958; made available to the author by courtesy of the management.

with the conduct of basic political contacts with the Saudi government, while the other, a larger group, is responsible for local government relations. The first group is composed of three "company representatives" in the two national capitals of the kingdom, Riyadh and Jeddah, and the provincial capital, Dammam. These representatives serve as envoys and attend to the fundamental problems in government-company relationships, foremost of which are matters pertaining to the concession agreement. Prior to the establishment of this organization the company had only one "representative to the Saudi Arab government" in Jeddah and none in Riyadh. With the transfer of many government functions to Riyadh, a need arose to fill this gap—hence the provision for a new representative. It was also felt that there should be a senior company official with easy access to the governor of the Eastern Province, one who would transact matters transcending the daily routine work but on the local level. This explains the appointment of a third "company representative" for Dammam.

The second group of officials is entrusted with the conduct of local government relations and is headed by a manager, equal in rank to the company representatives. The manager is assisted by local representatives in the three main centers of the company's operations, Dhahran, Abqaiq, and Ras Tanura, as well as by heads of five divisions, namely, those of Surface Rights, Administrative Services, Saudi Arab Government Services, Arabian Affairs, and Translation. The Division of Saudi Arab Government Services handles what may be broadly called the company's technical assistance, both voluntary and by contract, to the government. It will be described in a later section of this chapter. The Arabian Affairs Division does basic research work in fundamental problems of Saudi society and government. It produces studies on its history, geography, economy, and social mores. It employs a staff of research-oriented specialists in each of these fields. Their studies and reports have frequently been of material value to the Saudi government, which is not adequately equipped to pursue such studies itself.

In addition to this basic organization, the company in the past has maintained local relations representatives at the pumping stations of Tapline located in Saudi territory, Qaisuma, Rafha, Badanah, and Turaif. More recently, however, these representatives have been subordinated directly to Tapline management, although the personnel has remained interchangeable as before.

The following list of activities of the Local Government Relations Department in a single month in 1955 illustrates the variety of problems to which it has to attend:

1. Permits. In co-operation with Saudi authorities the department had to process 1,200 applications for new or renewed drivers' licences for its personnel and to settle issues arising from the issuance by the Saudi passport office of re-entry visas to foreign contract employees.

2. Labor affairs. In a single month company representatives had to discuss sixty-seven new labor cases with the Saudi Arabian Labor Office in Dammam. Department representatives were also instructed to secure from the Saudi authorities copies of new labor regulations recently published by the government. They also conducted negotiations with the Saudi Labor Office following the latter's request that the company provide it with lists of non-American foreign contract employees which would include job descriptions and salaries. They also had to attend to the repatriation by Saudi authorities of foreign contract employees dismissed by the company for disciplinary reasons.

3. Police affairs. The department had to handle the deportation of a foreign contract employee guilty of violating the laws of the kingdom and the incarceration by the Saudi authorities of another foreign contract employee for gambling. It also co-operated with the local police in the installation of international traffic signs in the Eastern Province.

4. Educational affairs. Because the company provides the local population with school buildings, department representatives signed the transfer papers on a new school built by the company for the sons of its Moslem and Arab employees.

5. Medical affairs. Conversations were held with the Saudi authorities and plans were prepared in connection with the projected construction of a new government hospital in the Eastern Province.

Meetings were also held with the Saudi representatives to discuss the inauguration of a trachoma research program partly sponsored by the company.

6. Public relations. Certain matters basically pertaining to the company's Public Relations Department had, nevertheless, to be attended to by representatives of the Local Government Relations Department because they involved negotiations with Saudi authorities. Among these was the obtaining of official permission to publish an Arabic language review for company employees.

7. Customs matters. Department representatives had to attend to a number of routine and nonroutine matters connected with Saudi customs regulations. Among these were clearance for a shipment of Angostura Bitters for the use of company employees and exemption from duty of materials imported for the construction of recreational facilities.

8. Military escort. The department discussed with Saudi authorities the problem of a soldier escort for a company exploration party in a border region.

9. Construction and real property affairs. Numerous problems arising from the company's construction program had to be discussed with the Saudi authorities. They included consultations with the *qadhi* (judge) of Dhahran concerning the erection of a new mosque for company employees and with the governor of the province about the construction of new shops for the use of the company personnel.

10. Recreational matters. Department representatives discussed with Saudi coast guard authorities the problems connected with the use of certain beaches on the Persian Gulf by company employees. They also negotiated the matter of registration of boats owned by company personnel.

11. Surface rights. Expansion of Aramco's technical installations, especially the laying of new local pipelines, required negotiations with the Saudi authorities to ascertain property rights and to secure the right of way.

12. Pipeline operation. A number of problems connected with the operation of the pipeline pumping stations in the desert had to be attended to by department representatives. Among them was a temporary shortage of water at the Turaif station, which was likely to affect not only company personnel, but also Saudi officials, their families, and neighboring tribes.

13. Financial affairs. Negotiations with the Saudi authorities were necessary to obtain a stamp tax exemption and payment for services rendered the government by the company.

14. Technical assistance to the Saudi Arabian government. The department was involved in many matters connected with the technical assistance that the company is rendering the Saudi government. These included supplying the king's motor caravan with a number of items such as tires, spare parts, and air-conditioning equipment, helping in the installation of government-owned water wells, assisting in the recruitment of an engineer to supervise certain phases of the construction of a pier at Jeddah, installing various items in the royal garage in Riyadh, and furnishing asphalt to the Al-Khobar municipality.

The above list shows both the number and the variety of matters requiring the daily attention of the Local Government Relations Department. The department has built up its own personnel from a few men into a hierarchical organization, which by 1957 had 185 employees.[2] Service in Government Relations has become a career in itself and it attracts personnel with special skills, frequently different from those required by the usual divisions of an oil company. Young men with degrees in history or the social sciences and the necessary linguistic equipment find increasing opportunities for interesting work and advancement in the service of the department.

IPC'S GOVERNMENT RELATIONS

IPC's organization has been somewhat different, although it has likewise given recognition to the increasingly important sphere of company-government relations. In the case of IPC we are dealing with a group of three companies operating in Iraq itself (IPC proper, Mosul Petroleum Company, and Basrah Petroleum Company), but also, at one time or another, in four other Middle Eastern countries

---

[2] This figure also includes the personnel of the Arabian Research Division which forms part of the department.

(Jordan, Palestine, Syria, and Lebanon). Since the Palestinian War of 1948, which resulted in the stoppage of the flow of oil to Haifa, IPC's operations have been restricted to Iraq, Syria, and Lebanon. IPC's group organization provides for a differentiation between field operations and dealings with governments. Thus in Iraq field operations are under the charge of three general managers, one for each of the three companies of the IPC group. Their relations with the Iraqi government are looked after by a chief representative who represents all three companies. The general managers are directly responsible to the headquarters in London and so is the chief representative. There is thus a dual line of authority, independent of each other and receiving its instructions from London. This is in contrast to Aramco, where all authority is concentrated in the headquarters located in the host country and where co-ordination between government relations and field operations may easily be obtained on the spot. The chief representative thus plays a role comparable to the combined roles of the Vice President, Government Relations and Finance, and the representatives to the Saudi Arabian government in Aramco's organizational chart. But while Aramco's people confine themselves chiefly to the safeguarding of the concession, IPC's chief representative attends also to public relations and, to some extent, to local government relations. In his office there is a division of public relations, which has local representatives in Kirkuk, Ain Zalah (Mosul), and Basra.

A similar pattern prevails in other countries where IPC or its affiliates conduct operations. In Syria there are a general manager, with headquarters in Homs, in charge of pipeline and other field operations and a chief representative in Damascus. The latter, by the way, is a Syrian citizen, in conformity with Syrian legislation (see p. 93 above). In Lebanon also IPC has its general manager of the Tripoli refinery and a chief representative in Beirut. A similar organization exists in IPC's affiliate, the Qatar Petroleum Company. The IPC group is, moreover, represented by a chief local representative on the Trucial Coast. In Iraq itself company personnel engaged in government relations work in 1957 comprised some sixty employees. This figure did not include the personnel of the public relations division, which employed an additional thirty-two persons.

HANDLING GOVERNMENT RELATIONS

The small number as compared to Aramco's force may be explained partly by American organizational habits and partly by the character of the governments with which the companies work. It might at first appear that the more primitive the country the less elaborate an organization would be required. In practically every respect Iraq is more developed than Saudi Arabia, and yet fewer people seem to be needed to handle governmental relations. The explanation of this seeming paradox is probably to be found in the very special position occupied by Aramco as the first and foremost industrial enterprise in a country that until recently not only did not have any of the modern amenities of life, but also possessed a merely rudimentary government organization. With an increase in revenue from oil, the need for modern amenities has increased and the government has grown in size. Consequently there has been a proliferation of new government agencies to deal with the multiple new problems of a rapidly growing state and a steadily increasing demand by the royal court and by various agencies for technical advice, services, and supplies, which the company is in a position to provide. The new agencies and departments have often been hastily staffed by such personnel, native or foreign Arab, as was available. New laws and regulations have had to be passed to provide guidance, which in turn necessitated both knowledge of the law as well as ability to interpret it. For most of the government officials this has been a new experience. Moreover only in rare cases have they possessed the administrative training and skills requisite for efficient office operation. At the same time these employees have often been anxious to stress the sovereign character of their official acts and decisions, thus sometimes magnifying minor matters which Western states would consider routine. As a result the company has been obliged to maintain contact with an ever-growing number of Saudi government agencies and to attend to matters both normal and extraordinary, the latter arising from the company's role as a purveyor of special services to a fast-developing state. This, of course, was reflected in the increase in number of people serving in the company's government relations units.

SPECIAL PROBLEMS: TECHNICAL ASSISTANCE

As mentioned in the preceding section, an oil company, by virtue of its uniqueness in an underdeveloped country, may be called upon to perform many services in addititon to its normal operations. These services may be especially significant if they take the form of technical assistance to the host government. Practically every company operating in the Middle East has extended technical assistance to the host governments to a greater or lesser degree. The principal forms of this assistance are as follows: (*a*) assistance resulting from the operations necessary for the development of the oil industry itself, e.g., geographical and topographical surveying of the country, location of water resources, and construction of roads and utilities; (*b*) services requested and paid for by the host government; and (*c*) technical assistance furnished as a public service by the company.

The various companies' policies of assistance have not been uniform. They have been conditioned partly by the degree of development and the needs of the host country and partly by the will of the companies themselves to assist. There is no doubt that certain companies, notably Aramco, have consistently been doing more than others in this respect. In fact the degree to which the companies have been willing to extend technical assistance has frequently been taken as an indication of their progressiveness and friendly interest in the welfare of the host country.

Although this may be true in part, it is not the whole story. The host governments vary in their desire to be helped by the oil companies. A good contrast in this respect is Saudi Arabia and Iran since the conclusion of the Consortium agreement. In Saudi Arabia the government has relied heavily on Aramco. Medical services both preventive and curative and medical training and research, industrial training, construction of public schools, campaign against illiteracy, development of water resources, construction of roads and air strips, preparation of maps, agricultural development in the Al-Kharj oasis, construction of ports and of the Dammam-Riyadh railroad, rental of equipment, assistance in community and industrial development—all these make a long but not exhaustive list of technical services rendered the government by Aramco.³ Only a small part of these services were connected with the development of the oil industry itself; by far the greater part consists of projects not deriving from oil operations whether initiated by the company and accepted by the government or initiated by the government. Obviously the Saudi government desired the company to perform these services. By contrast Iran since 1954 has tried to be self-reliant in all those sectors in which it was believed that the government could do the job. This has been best expressed by the taking over by the National Iranian Oil Company of the so-called nonbasic functions in the operations of the oil industry (see p. 73 above). Iran has been following a deliberate policy of restricting the company to its essential task and reserving to itself those fields of activity in which the general development of the country is at stake.

SPECIAL PROBLEMS: TRIBAL PROTECTION

Among the special problems in company-government relations is that of tribal protection. This has led to a good deal of confusion and misunderstanding. The facts of the matter are fairly simple. A good part of the population in most of the Middle Eastern countries is still organized along tribal lines. In the past these tribes wielded considerable influence in certain desert regions where oil was being extracted or through which pipelines were laid. The strength of the tribes was in inverse proportion to the strength of the state. In Iran at the turn of the century or in Iraq after World War I conditions were such that the central government, despite its authoritarian form, was not fully in control in many outlying areas and the power of the tribes was paramount. It was precisely those periods in which the oil companies began their operations in the two countries. In Iran the company quickly sensed the reality of the situation and adapted itself to existing conditions by entering into a special relationship with two elements of local power, the sheikh of Mohammerah and the Bakhtiari khans. The sheikh of Mohammerah was a hereditary Arab ruler of a sizable territory on the

³ From a report made available to the author by courtesy of the management.

eastern bank of the Shatt al-Arab River. Although formally a subject of the Persian Empire, he was virtually autonomous, and the company soon found out that it was wiser to cultivate his friendship than to ignore him. A mutuality of interests based on the facts of life, so to speak, was established. The company benefited from co-operation with a local potentate and naturally preferred to see this harmonious relationship undisturbed by interference on the part of the distant central government. The sheikh, aware of the close link between the company and the British government, viewed his association with the former as a guarantee of freedom from the dictates of the Persian authorities. Thus what began as a practical adaptation to existing conditions ended as a vested interest of both parties in the preservation of a relationship that was not fully compatible with the unity and sovereignty of the Persian state.

The matter took a turn for the worse when on May 6, 1909, the British government, acting through its Resident in the Persian Gulf, Major Percy Cox, officially pledged itself to support the sheikh's autonomy as against the claims of the central government. No Persian government could accept such foreign interference in its domestic affairs with equanimity. Consequently, as soon as a strong regime emerged in Iran after World War I, it took stern measures to suppress the sheikh's autonomy. Although the British government realistically accepted the new dispensation, abandoning the sheikh to the tender mercies of Iran's new ruler, Reza Shah, the company suffered a serious setback because, by its somewhat shortsighted policy, it had allowed itself to be too closely identified with those centrifugal forces in Iranian politics that opposed the unification and strengthening of the country.

The company's relations with the Bakhtiari khans were also based on the fact that the latter were in effective control of certain areas deemed important to the company. What was chiefly involved was the security of wells and pipelines. In order to obtain the good will of the Bakhtiaris, a special arrangement was made as early as 1905 whereby they were to receive a subsidy of £3,000 a year as well as 3 per cent of the shares of any company that operated in their territory. A separate corporation, the Bakhtiari Oil Company, was founded in April 1909 with a capital of £400,000. It was to exploit oil in areas other than those worked by the First Exploitation Company (a concern created by the original concessionaire, William Knox D'Arcy). Twelve thousand pounds' worth of shares or 3 per cent of the total capital was issued to the Bakhtiari khans, who reciprocated by protecting the installations erected in their territory.

During World War I this arrangement with the Bakhtiaris underwent many vicissitudes, largely on account of intense German activity among the tribes of southwestern Iran. On a few occasions pipelines were punctured and production suspended. Although the attitude of certain Bakhtiari khans, from the British point of view, was dubious or traitorous, this fact underlined the need for vigilant cultivation of the tribesmen's good will. As in the case of the sheikh of Mohammerah, the British—both the company and the consular service—soon became deeply involved in local politics. It was, indeed, a combination of compelling circumstances and deliberate policy. The Bakhtiari khans were numerous and divided into factions. The British discovered before long that it was easier to deal with some representative major chieftain who could act in the name of the tribe as a whole than to negotiate with individual petty khans. Consequently they tried to back up certain individuals who sought to become *ilkhani* (paramount chief) of the Bakhtiaris. This naturally led them deeper and deeper into tribal feuds and politics and exposed them to rivalry with the central government of Iran. Furthermore, the Bakhtiaris and the sheikh of Mohammerah were at odds with each other, their mutual dislike occasionally flaring up into a violent outburst. Anxious to keep peace in the area of their operations, the British had to act as mediators and pacifiers, thus encroaching on the sovereign rights of the host country.

The rise of Reza Shah and his energetic policy of unification and centralization gave a blow to this British-tribal relationship. The tribes in Khuzistan and elsewhere were largely disarmed, and national security forces took over the task of maintaining law and order. With the occupation of Iran by British and Soviet forces during World War II, a relapse occurred. The weakening of the central government enabled the tribes to reassert their autonomy, and the British in their eagerness to maintain security and stability in the southern part of Iran again reverted to tribal diplomacy.

Although such a policy could be justified in the abnormal conditions of war and its immediate aftermath, it carried within itself a seed of danger, inasmuch as Iran, after two decades of nationalism under Reza, was much more conscious of its sovereign rights than it had been previously. Consequently delay on the part of the company or of British political representatives (they worked hand in hand) in adjusting themselves to the new sensitivity after the war was bound to have adverse effects on the company's status in Iran. This was amply proved in 1951, when the company's dealings with the tribal chieftains were frequently cited by the Iranian government as proof of interference in the internal affairs of the country.[4]

The Iraq Petroleum Company likewise resorted to tribal protection of its installations in the northern part of Iraq. The chief of the Shammar tribe contracted to assume this responsibility. Fortunately these security arrangements did not involve the company in any major quarrels with the government of Iraq. In fact, the latter fully approved of the arrangements as long as they were in operation. The contract with the Shammar came to an end in 1950–1951, after which date the camps and pipelines were protected by the specially established government oil police. Despite the nomadic character of the Shammar, who frequently migrate across the state borders in the deserts of northern Arabia, they were never made responsible for the IPC pipeline in Syria and Jordan, where security was provided by company-paid watchmen.[5]

No arrangements for tribal protection were ever made in Saudi Arabia. It may appear surprising that this method was not resorted to in a country where a majority of the people still live in tribes and where tribal allegiances and affiliations play an important role in domestic politics. The answer to this seeming paradox is to be found in the character of the central government. Iranian governments, with the exception of that of Reza Shah, were weak and ineffectual, and Iraq was under a British mandate for a long time, but Saudi Arabia was subject to the absolute rule of its strong-willed king, Abdul-Aziz ibn Saud. Although Ibn Saud's internal

[4] See "Iran Presents Its Case," *Oil Forum*, March, 1952, p. 81.
[5] On tribal protection, see S. H. Longrigg, *Oil in the Middle East: Its Discovery and Development* (London, 1954), pp. 81, 120.

HANDLING GOVERNMENT RELATIONS

policy often dictated conciliation and alliances with certain tribes, it tolerated no independent or autonomous groups in matters of public security. On this point the king was adamant. As a result it was he and his government alone that assumed responsibility for the security of Aramco's installations. It was not so much the actual physical protection exercised by his security forces as the general pacification of the country by his dedicated warriors which assured peace and tranquillity to the company.

To summarize: With the passage of time the company relations with host governments have tended to encompass an ever-widening territory. The growing automation of technical operations has not been accompanied by a corresponding process in government relations. On the contrary, the relatively simple man-to-man relationships of the pioneer days have given place to complex procedures. Two hierarchical structures—those of the company and the government—are facing each other, with the task of assuring the smooth functioning of operations and maximizing mutual benefits. The areas of contact are many, ranging from interpretation of concession agreement to the issuance of drivers' licenses and residence permits. In addition to their essential functions, the companies are frequently called upon to provide multifarious other services for the host governments. All of this requires special skills, as a result of which the government relations departments in the companies tend to become specialized agencies, employing full-time personnel and offering separate career opportunities. This is undoubtedly a welcome trend, inasmuch as in the delicate play of human relations—deeply permeated by politics—competence, concentration, and dedication are likely to produce better results than improvisation and amateurism.

◇◇ CHAPTER VII ◇◇

Territorial Claims: Submarine Areas

WITH the network of concessions covering large tracts of oil-bearing land in the countries around the Persian Gulf, it was inevitable that attention should be paid by both companies and governments to the expansion of oil operations in the submarine areas. Interest in sea-bed potentialities is not peculiar to the Middle East. On September 15, 1945, President Truman issued a proclamation laying claim on behalf of the federal government to the natural resources of the subsoil and sea bed on the continental shelf of the United States. The proclamation neither claimed sovereignty over the shelf nor asserted any rights to the high seas above it; it merely established United States jurisdiction over the natural resources. Moreover, it took into account the possibility that the shelf might be shared with another state, in which case it called for delimitation of boundaries by mutual agreement. Although the proclamation itself heralded a new doctrine in international law, the agreements that it envisaged had a precedent in the treaty signed in 1942 by Britain and Venezuela, whereby the two parties delimited their respective zones in the submarine areas of the Gulf of Paria between Trinidad and Venezuela.

CONTINENTAL SHELF PROCLAMATIONS

In the Middle East the first to follow the American example was Saudi Arabia. A royal decree of May 28, 1949, defined the territorial waters of the kingdom as six miles wide, and a pronouncement of the same date laid claim to the "subsoil and sea bed of those areas of the Persian Gulf seaward from the coastal sea of Saudi Arabia, but contiguous to its coasts" as appertaining to the kingdom and subject to its jurisdiction and control.[1] The decree applied to all coasts of Saudi Arabia, but the pronouncement dealt specifically with the Persian Gulf. The decree stipulated that where the kingdom's territorial waters overlapped the waters of another state boundaries would be determined by the government "in agreement with the State concerned in accordance with equitable principles" (Art. 8). An almost identical formula was used by the pronouncement with regard to the sea bed and subsoil of the extraterritorial sea.

Although following in its general concept the Continental Shelf Doctrine enunciated by President Truman in 1945, the Saudi proclamation on submarine areas did not explicitly mention the continental shelf, in all probability because the Persian Gulf, a shallow basin less than 100 fathoms deep, does not possess the typical shelf formation. It merely spoke of the areas "contiguous" to Saudi territorial waters, without determining how far seaward these areas were to go. Moreover, instead of limiting its claim to natural resources it asserted jurisdiction over the sea bed and its subsoil in general, thus exceeding the American claim and posing such potential problems as control of submarine cables or pipelines laid on the sea bed of the offshore areas.

The Saudi initiative was promptly followed by the sheikhdoms of the Persian Gulf. On June 5, 1949, the sheikh of Bahrein proclaimed that the submarine areas adjacent to the territorial waters of Bahrein were subject to his "absolute authority and jurisdiction." The outer limit of these areas was not defined and was left to determination after consultation with neighboring governments "in accordance with the principles of justice, when occasion so requires." [2] In quick succession other Gulf sheikhdoms issued similar proclamations, almost identical in form. Their sequence was as follows:

[1] Royal Decree no. 6/5/4/3711 issued on 1 Sha'ban 1368 (May 28, 1949), in *Umm al-Qura*, suppl. no. 1263, 2 Sha'ban 1368 (May 29, 1949).

[2] Proclamation no. 37/1368, June 5, 1949. An English translation can be found in the supplement to the *American Journal of International Law (AJIL)*, 43 (1949), 185-186.

Qatar	June 8, 1949
Kuwait	June 12, 1949
Trucial sheikhdoms:	
Abu Dhabi	June 10, 1949
Dubai	June 14, 1949
Sharjah	June 16, 1949
Ajman	June 20, 1949
Umm al-Qaiwain	June 20, 1949
Ras al-Khaimah	(date not available)

Spurred by this example, Iran also proceeded to enact legislation with regard to the offshore areas. A bill to this effect, first introduced in Parliament on May 19, 1949, was promulgated on June 19, 1955. It declared that "the areas as well as the natural riches of the sea bed and subsoil up to the limits of the continental shelf, which extends from the coasts of Iran and the Iranian islands in the Persian Gulf and the Sea of Oman, belong to the Iranian government and are under its sovereignty" (Art. 2). The law excluded from these provisions the Caspian Sea, to which the "rules of international law concerning the closed seas" were to apply. In case dissension occurred over the limit of the continental shelf extending to the shores of another state or shared with a neighboring state, it was to be "resolved in conformity with equity," and the government was to take "the necessary diplomatic measures" (Art. 3).[3] The bill confirmed the law of 24 Tir 1313 (July 15, 1934) defining the limits of territorial waters and the maritime zone under Iranian control.[4] Unlike the proclamations of Arab states of the Persian Gulf, the Iranian law used the term "continental shelf." The new Petroleum Act of July 29, 1957, likewise referred to the continental shelf, as did the subsequent proclamations on November 2 and December 2 declaring certain zones open for oil operations. The latter comprised certain areas of dry land as well as the offshore areas adjacent to Khuzistan, Makran, and Baluchistan.

The last to assert its claims was the government of Iraq, which on November 27, 1957, issued a statement defining its rights to waters

---

[3] *Iran-Presse* no. 9, Jan. 10 and 11, 1958.

[4] According to that law, Iran claimed a 6-mile limit for its territorial sea. However, a draft law submitted to the Senate on Dec. 20, 1958, extended the limit of Iran's territorial waters to 12 miles from the lowest tide line (*Iran-Presse*, no. 280, Dec. 21, 1958).

"contiguous to Iraqi territorial waters" which closely resembled the proclamations made earlier by Saudi Arabia and the Persian Gulf sheikhdoms.[5] There is no doubt that its timing was directly related to the increased Iranian interest in the offshore zones, as manifested by the above-mentioned proclamations, the grant of a contract to the Agip Mineraria of Italy, and the reassertion of Iranian claims to Bahrein in the fall of 1957.[6] Actual study by Iran of bids presented in the spring of 1958 by various oil companies with regard to these "open" zones prompted Iraq to issue a new statement on April 9, 1958, in which the government reiterated that "operations or installations made or to be made in this area of contiguous waters are subject to Iraqi sovereignty, and are not permitted to be carried out except by Iraqi authorities." The statement further announced the government's "adherence to the principles of international dealings in this regard and to the principles of equal distances which guarantee to Iraq the freedom of passage to or from the high seas." It ended with a warning that the government "does not recognize any statement, announcement, legislation, or demarcation related to territorial waters or to contiguous waters issued by any neighboring country which might be contrary to this statement." [7]

In terms of international law the offshore proclamations were actually superfluous. A state can claim control of contiguous submarine areas by virtue of the theory of "Appurtenance," meaning an automatic extension of its jurisdiction without the requirement of proclamation or occupation, real or fictitious, similar to the case of the territorial waters.[8]

UNCERTAINTIES AND ARBITRATIONS

While laying down an important principle, the Continental Shelf Doctrine left a number of questions unanswered. First of all, it left open the definition of the continental shelf. Neither the Truman proclamation of 1945 nor the subsequent acts of other states have

[5] The text is in *Al-Waqai' al-'Iraqiyah* (Official Gazette), no. 4069 of Nov. 27, 1957.
[6] See p. 134 below.   [7] *Az-Zaman* (Baghdad), April 10, 1958.
[8] For a more detailed discussion, see Richard Young, "Legal Status of Submarine Areas beneath the High Seas," *AJIL*, 45 (1951), 225 ff.

dealt uniformly with this question. There has been a tendency, at least in the Western Hemisphere, to consider as continental shelf a sea bed not exceeding, at its outer limits, a depth of 100 fathoms or 200 meters (approximately 600 feet). Such at least was the definition given unofficially in the United States and officially in Mexico. The Canadian government favored the actual edge of the shelf as the outer limit except in cases where the edge might be poorly defined or where there is no shelf in a geographical sense. In these instances it was inclined to set the boundary "at such a depth as might satisfy foreseeable practical prospects of exploitation of the natural resources of the seabed adjacent to a particular state." [9] Certain American states, such as Nicaragua, Chile, and Costa Rica, gave more extravagant definitions, either disregarding the depth of the shelf or claiming as many as 200 marine miles seaward.[10] In its report presented in 1956 to the General Assembly, the United Nations Law Commission defined the term "continental shelf" as the "seabed and subsoil of the submarine areas adjacent to the coast but outside of the territorial sea to a depth of 200 metres (approximately 100 fathoms) or, beyond that limit, to where the depth of the superjacent waters admits of the exploitation of the natural resources of the said areas." [11] This report, together with the records of debates at the Eleventh Session of the General Assembly where it was discussed, served as a basis for discussion at the International Conference on the Law of the Sea, which met in Geneva between February 24 and April 24, 1958, on the initiative of the General Assembly, to codify the rules of maritime law. The result of its deliberations was embodied in the "Convention on the Continental Shelf," Article 1 of which defined the continental shelf as "the seabed and subsoil of the submarine areas adjacent to the coast but outside the area of the territorial sea, to a depth of 200 metres or, beyond that limit, to where the depth of the superjacent waters ad-

[9] "International Conference on the Law of the Sea," *External Affairs* (Ottawa), Jan., 1958.

[10] See Richard Young, "Recent Developments with Respect to the Continental Shelf," *AJIL*, 42 (1948), 849 ff.

[11] Art. 67 of the report. The full text can be found in *Official Records of the U.N. General Assembly, 11th Session,* suppl. no. 9 (A/3159).

mits of the exploitation of the natural resources of the said areas." [12]

The second issue not solved by the offshore proclamations was whether concessions for the exploitation of mineral resources in national territories extended to the submarine areas and, if so, how far. This question has acquired special importance whenever a country has granted concessions to different companies for its national territory and for the offshore submarine areas. The matter was brought to a head when the rulers of Qatar and Abu Dhabi, whose territories were already covered by agreements with the IPC group, decided to grant separate concessions for the offshore areas to other companies. In Qatar the Petroleum Development (Qatar), Ltd., which in 1935 had acquired a 75-year lease on the entire territory, challenged the validity of the concession subsequently granted to Superior Oil Company and the Central Mining and Investment Corporation of London to operate jointly in a 12-mile offshore belt around the Qatar Peninsula. In 1949 the dispute was submitted to arbitration by the ruler of Qatar and Petroleum Development (Qatar), Ltd. The arbitrators, under the chairmanship of Lord Radcliffe, held hearings in Doha in February 1950 and in April of that year issued an award which contained the following points:

1. Petroleum Development's concession includes the islands over which the sheikh ruled at the date of the concession whether or not they are shown on the map attached to the concession.

2. The concession includes the bed and subsoil of all the inland or national waters of the mainland of the state of Qatar and its islands.

3. The concession includes the sea bed and subsoil beneath the territorial waters of the mainland and islands of the state of Qatar.

4. The concession does not include the sea bed or subsoil beneath the high seas of the Persian Gulf contiguous to Qatar's territorial waters.[13]

The dispute between Petroleum Development (Trucial Coast), Ltd., holder since 1939 of a 75-year concession covering the territory

[12] *AJIL*, 52 (1958), 858 ff.
[13] "Qatar Arbitration Award," *Petroleum Times*, June 2, 1950, p. 391; see also Sir Hersch Lauterpacht, ed., *International Law Reports, 1951* (London, 1957).

of Abu Dhabi, and the latter's ruler over his 1950 grant of a concession to Superior Oil Company in the offshore areas also ended in arbitration. In a judgment rendered in Paris on August 28, 1951, the umpire, Lord Asquith of Bishopstone, stated that Petroleum Development's concession extends to the subsoil of the territorial waters (including the territorial waters of the islands) but not to the subsoil of the shelf, which he defined as the submarine area contiguous to Abu Dhabi outside its territorial zone. The zone was defined in the judgment as three miles wide.[14]

The Doha and Paris arbitration awards were important inasmuch as they established precedents clarifying a hitherto obscure question of the outer limit to be placed on a land-based concession. Yet they left an important point untouched, namely, the way in which the boundary should be drawn between states whose shorelines are adjacent to each other. In a case where the land boundary between two such states runs in a straight line toward the coast, touching the latter at a right angle, the matter will not present great difficulty, inasmuch as by projecting the line into the water space one can determine the continuing boundary on a reasonable basis of equity and logic. But what should be the solution if the land boundary approaches the coast line at a sharp angle? Should it be projected at the same angle into the water or should the water boundary follow a line drawn at a right angle from the coast? The difference between the two formulas might mean different ownership of some rich submarine oil pool in the wedge of the sea bed situated between the two possible lines under discussion. This is just one example of the multiple problems likely to arise in adjacent seas and national territories. Although not peculiar to the Persian Gulf, they are acquiring increasing importance there, in some cases bordering on urgency,[15] on account of the proved or nearly proved oil deposits in the offshore areas. Unfortunately no unified approach to such issues has yet been agreed upon, and international practice abounds in widely divergent solutions and formulas. Some formulas have been

[14] The text of the award is in the *International and Comparative Law Quarterly*, 1, pt. 2 (1952), 247–261.

[15] The problem of exact delimitation of such water boundaries is expected to arise in connection with the recently granted offshore concessions in the Neutral Zone of Saudi Arabia and Kuwait.

SUBMARINE AREAS

adopted by mutual agreement, but others, unilaterally devised to reflect national claims, are still awaiting recognition.[16]

In view of the gaps in international law, special importance must be attached to legislation and agreements among the states directly concerned. Fortunately both the original Saudi Arabian pronouncements as well as the subsequent proclamations of the Gulf sheikhdoms were couched in terms allowing mutual agreements among the riparian states in case of overlapping claims. Moreover, the views of the sheikhdoms were notably uniform, probably the result of British guidance. Saudi Arabia's attitude apparently underwent a modification, and on February 16, 1958, it claimed sovereignty over the 12-mile belt of sea contiguous to the shore instead of the 6-mile belt specified in the previous law.[17]

Although its timing suggested that the decree was primarily designed to assert Saudi (and general Arab) claims over the controversial Gulf of Aqaba at the impending United Nations conference on maritime law in Geneva, its immediate impact was on relations with Bahrein and Qatar, whose territorial waters overlap those of Saudi Arabia. A week after the issuance of the decree the ruler of Bahrein, Sheikh Salman ibn Hamad al-Khalifah, journeyed to Riyadh, where, on February 22, 1958, he and King Saud signed an agreement defining the maritime boundaries between the two countries. Pointing to the fact that "the regional waters between the Kingdom of Saudi Arabia and the Government of Bahrein meet together in many places overlooked by their respective coasts" and referring to the Saudi and Bahrein offshore proclamations of 1949, the agreement traced the boundary which would divide not only the overlapping territorial waters but also the contiguous maritime zones located north of Bahrein and east of Saudi Arabia. Starting at a point situated south of Bahrein, the boundary line was to envelop the island on its western side at a mid-distance between the coasts of

[16] For a detailed discussion of this problem, see S. Whittemore Boggs, "Delimitation of Seaward Areas under National Jurisdiction," *AJIL*, 45 (1951), 240 ff., and, by the same author, "National Claims in Adjacent Seas," *Geographical Review*, 41 (1951), 185 ff.

[17] Royal Decree no. 33, 27 Rajab 1377 (Feb. 16, 1958), in *Umm al-Qura* no. 1706 3 Sha'ban 1377 (Feb. 21, 1958). It repeals Decree no. 6/5/4/3711 of 1 Sha'ban 1368 (May 28, 1949).

the two countries and, upon reaching a point north of Bahrein, was to extend straight forward in a northeasterly direction toward the central area of the Persian Gulf. The agreement declared that "everything that is situated to the left of the above-mentioned line . . . belongs to the Kingdom of Saudi Arabia and everything to the right of that line to the Government of Bahrein," with the reservation, however, that an irregular hexagonal area (approximately double the size of Bahrein) located on the high seas to the left of the dividing line was to be subject to an equal division of profits should oil be exploited there. The king of Saudi Arabia was to determine the way in which exploitation of the oil resources would be carried out. The clause in question was not to be construed as infringing upon the rights of Saudi Arabia with regard to sovereignty and administration over the above-mentioned area.[18]

THE BAHREIN CONTROVERSY

The Saudi-Bahrein agreement constituted an important step forward in settling the controversial boundary issues by mutual consent. Yet despite its peaceful intent and the clarity of its text it had international ramifications. In the Exclusive Agreements concluded in 1880 and 1892 between Britain and the ruler of Bahrein, the latter had promised "to abstain from entering into negotiations or making treaties of any kind with any state or Government other than the British without the consent of the British Government, and to refuse permission to any Government other than the British to establish diplomatic or consular agencies or coaling depots in [Bahrein] territory, unless with the consent of the British Government." [19] In practice this meant that Britain assumed the actual conduct of Bahrein's foreign relations through her Political Agent stationed in Manama. Consequently the Ruler's personal negotiation in Riyadh constituted a major departure from established practice, the significance of which was not greatly diminished by a subsequent statement of the British government that it approved of the agreement.

Another ramification, or, to be more precise, complication,

[18] *Umm al-Qura*, no. 1708, 17 Sha'ban 1377 (March 7, 1958).

[19] C. U. Aitchison, *A Collection of Treaties, Engagements and Sanads Relating to India and the Neighboring Countries* (5th ed.; Calcutta, 1933), XI, 237–238 (hereafter referred to as Aitchison).

stemmed from the protest that the government of Iran lodged both with Britain and Saudi Arabia as soon as it learned of the Saudi-Bahrein agreement. The Iranian protest concerned not so much the tenor of the agreement as the fact that it had been concluded by the ruler of Bahrein, whom the government of Iran considers its subject. A discussion of the merits of the Iranian claim to sovereignty over Bahrein is not germane to this study, but during the past four decades Iran has repeatedly asserted its claim to the island principality, invariably without effect on the *status quo* in view of British power in the Persian Gulf and the Arab character of the ruling Khalifah dynasty in Bahrein.[20] No matter how slender are the chances for Iran to recover its dominion over the island and how impractical is the raising of the claim against the background of the more fundamental orientation of Iranian policy as between the East and the West, the issue of Bahrein has become a matter of *amour-propre* for Iranian cabinets, which rarely miss an opportunity to include it on their agenda. The most recent—and most radical—manifestation of this attitude occurred on November 11, 1957, when, following the instructions of the shah, the Iranian cabinet approved a draft law designating Bahrein as the fourteenth province of Iran in the new administrative redivision of the country. Soon afterward Dr. Ali Gholi Ardalan published a long article in which he set forth in considerable detail all the reasons why Iran claimed the island.[21] In an outburst of nationalism somewhat reminiscent of the Mossadegh era (and with the blessing of the authorities) the Iranian press began speculating about future parliamentary representation of Bahrein in the Majlis and the possible appointment of an Iranian governor of the island.

The British and the Saudi reactions to the Iranian claims were understandably negative. For Britain the matter had been settled once for all with the conclusion of the above-mentioned Exclusive Agreements, which subjected the rulers of Bahrein to a virtual British protectorate. For Saudi Arabia the fundamental question was the ethnic character of the bulk of Bahrein's population, which it claimed to be Arab. In fact, Saudi Arabia was the first to protest when Iran

[20] For a detailed treatment of this dispute, see Fereydoun Adamiyat, *Bahrein Islands: A Legal and Diplomatic Study of the British-Iranian Controversy* (New York, 1955).
[21] *Dunia* (Teheran), Nov. 23, 1957.

drafted the law including Bahrein in its administrative structure. "Bahrein," said an official Saudi statement at that time,

> is a natural extension of the Arabian Peninsula and is an integral part of it. It is a country which has its own foreign status. Its people are striving for freedom and independence so that they may participate with their Arab brothers and neighbors in a united Arab world. To doubt this principle would be to doubt the most elementary geographical facts. We can believe anything but for Bahrein to be claimed as an Iranian province or as an indivisible part of Iran.[22]

When the Iranian government protested against the Saudi-Bahrein maritime boundary agreement a few months later, Riyadh's position remained unchanged and the kingdom's official mouthpieces reiterated practically *in toto* their earlier statements refuting Iran's claims to the island.[23]

Although the Iranian-British-Saudi dispute over Bahrein is not likely to have immediate adverse repercussions upon the status and security of such oil operations as are now being conducted in Bahrein or as are contemplated in the waters between the island and the Arabian mainland, it nevertheless introduces, at least from the legal point of view, a complicating factor into the state-oil relationships of the Persian Gulf area. It also makes one wonder about the significance of references to the "continental shelf" in the Iranian law of 1955, which omitted any mention of agreements to be negotiated and instead spoke vaguely of "diplomatic measures" to be taken should jurisdictional disputes arise.

In conclusion, while progress has undoubtedly been achieved in clarifying the riparian states' claims to submarine areas, in setting precedents for differentiation between territorial and nonterritorial waters in the granting of concessions, and in promoting peaceful solutions through agreements, enough legal and political uncertainty remains to dictate considerable caution to the oil companies in their steps toward exploitation of sea-bed resources.

---

[22] News commentary broadcast by Radio Mecca on Nov. 30, 1957; subsequently published in *Umm al-Qura*, no. 1695, 14 Jamada I, 1377 (Dec. 6, 1957).

[23] "Radio Mecca Commentary on the Offshore Agreement between Saudi Arabia and Bahrein and on Iranian Claims to Bahrein," *Umm al-Qura*, no. 1708, 17 Sha'ban 1377 (March 7, 1958).

◇◇ CHAPTER VIII ◇◇

Territorial Claims: Desert Borders

EQUAL to the maritime problems in complexity are the uncertainties concerning land boundaries in the Arabian Peninsula. The idea of a fixed boundary line in the desert is alien to the bedouin mind. Since time immemorial tribesmen have wandered in the wastes of the peninsula in search of water and grazing grounds. Claims to ownership were usually limited to a coastal town, an oasis, or a water well. The desert in between could be likened to a high sea, to which no one could justifiably lay exclusive claims of control. And as long as the Arabian Peninsula was wholly or largely under the Ottomans, there was no need to trace inland boundaries between the various administrative divisions.

THE BOUNDARY AGREEMENTS

Two events, however, demonstrated the need for exact delimitation of desert borders. These were the growth of British influence in the coastal states of the Persian Gulf, in Oman, Aden, and the Hadhramaut and the discovery of oil. The first major attempt to settle the boundary question occurred on July 29, 1913, when Britain and the Ottoman Empire concluded a convention regarding the Persian Gulf area. Article 5 of the convention dealt with the boundaries of Kuwait, and Article 11 provided for a boundary between the Ottoman Sanjak of Nejd and the Qatar Peninsula. The latter boundary, designated as the "Blue Line," was to follow a straight north-south course from the Gulf of Zakhnuniya, west of the base of the Qatar Peninsula, to the Rub al-Khali desert.[1] Although never ratified, the convention of 1913 was not unimportant inasmuch as two of its provisions—those concerning Kuwait and the boundary of Nejd—were subsequently invoked in later documents and negotiations. The first of these was the Anglo-Turkish Convention of March 9, 1914, whereby Britain and Turkey agreed on the boundary dividing the Aden Protectorate from Ottoman territory in Arabia. This was ratified. Its Article III defined the boundary as a line (thereafter known as the "Violet Line") running at a 45-degree angle from Lekemet al-Shoub northeast toward the Rub al-Khali, where, at latitude 20 degrees north, it was to join the Blue Line traced by Article 11 of the 1913 convention. Thus shortly before the outbreak of World War I the combined Blue and Violet Lines became the official boundary between Ottoman possessions and British-protected territories of southern and southeastern Arabia. Although the war and the subsequent collapse of the Ottoman Empire brought about a radical change in the political status of the Arabian Peninsula, this boundary was repeatedly invoked by Britain in her later territorial disputes with Saudi Arabia and could not be ignored by oil companies claiming concessions in the disputed areas.

The rise after World War I of the independent kingdom of Saudi Arabia, which in due time became heir to the former Ottoman provinces of Nejd, Hasa, Hejaz, and Asir, caused a number of new boundary problems. The latter were rendered more acute by the fact that the kingdom's territory was adjacent to the British-protected Arab states in the general area where oil deposits were proved or expected to exist. Consequently the delimitation of King Ibn Saud's possessions became a major item on the postwar agenda of British-Saudi relations. Great Britain, it should be noted, appeared not only as a traditional "protector" of the Persian Gulf states and other territories in southern Arabia, but also as a power exercising mandatory control north of Saudi Arabia, in Iraq and Transjordan. It was, in fact, this northern boundary that claimed first attention in British-Saudi relations. At a conference at Oqair (December 2, 1922)

[1] Hurewitz, **I,** 269 ff.

British and Saudi representatives agreed on the boundary separating the sultanate of Nejd from Kuwait and Iraq. At the same conference the unratified 1913 convention was referred to. Regarding the southern boundary of Kuwait, the Oqair convention said:

> The frontier between Najd and Kuwait begins in the west from the junction of the Wadi al Aujah with the Batin, leaving Raq'i to Najd; from this point it continues in a straight line until it joins latitude 29° and the red semi-circle referred to in Article 5 of the Anglo-Turkish Agreement of 29 July 1913. The line then follows the side of the red semi-circle until it reaches a point terminating on the coast south of Ras al-Qali'ah and this is the indubitable southern frontier of Kuwait territory.[2]

The red semicircle referred to in this passage had been described as follows in Article 5 of the 1913 draft convention:

> The autonomy of the Shaykh of Kuwayt is exercised by him in the territories the limit of which forms a semi-circle with the town of Kuwayt in the center, the Khur al-Zubayr at the northern extremity and al-Qurayyin at the southern extremity.[3]

The difficulty of this definition has been its lack of precision. Al-Qurayyin is a name given both to a hill and to a group of wells located between two and three miles apart. The town of Kuwait covers an area of several square miles, the center of which is a matter of considerable controversy. Consequently, despite the outwardly complete description of the boundary, there are enough ambiguities to make firm claims to immediately adjacent lands virtually impossible. Similarly lacking in precision is the way in which the convention describes the limits of the Neutral Zone to be located south of Kuwait:

> The portion of territory bounded on the north by this line [i.e., the southern boundary of Kuwait] and which is bounded on the west by a low mountainous ridge called Shaq (Esh Shakk) and on the east by the sea and on the south by a line passing from west to east from Shaq (Esh Shakk) to 'Ain al 'Abd (Ain el Abd) and thence to the coast north of Ras al Mish'ab (Ras Mishaab), in this territory the Government of Najd and Kuwait will share equal rights until through the good offices of the Government of Great Britain a further agreement is made between Najd and Kuwait concerning it.[4]

[2] Aitchison, XI, 213-214.   [3] Hurewitz, I, 270.   [4] Aitchison, XI, 213-214.

The Oqair conference also produced a protocol defining the boundaries between Iraq and Nejd and providing for another Neutral Zone to be established west of Kuwait. Here again ambiguities crept in. "The looseness of the wording," wrote one of the participants subsequently, "was to lead to trouble five years later."[5] Inasmuch as the main objective of the Oqair agreements was to delimit the northern boundaries of Nejd, Kuwait's northern borders were not discussed. Eventually they were settled in 1932 by an agreement between Kuwait and Iraq, following the grant of independence to the latter.[6]

The remaining portion of the Saudi northern boundary, i.e., the boundary with Transjordan, was the subject of the so-called "Hadda Agreement" of November 2, 1925,[7] followed by the Treaty of Jeddah of May 20, 1927. In both cases Britain, as the mandatory power, represented Transjordan. The Hadda Agreement described the boundary in exact terms using meridians and parallels, thus eliminating the possibility of serious mistakes. It, however, was incomplete, inasmuch as it dealt only with the eastern boundary of Transjordan, leaving the southern undetermined. This incompleteness was due to rival claims to the Aqaba and Maan districts, which, though originally part of Hejaz, had been occupied by the British and annexed to Transjordan in 1925. A subsequent (1927) attempt by the British to settle this issue resulted in partial success only. The main text of the Treaty of Jeddah contained no reference to boundaries. In the appended exchange of notes, however, the British government gave its own definition of the frontier between Hejaz and Transjordan. The Saudi government refused to accept this definition but for the sake of "cordial relations" expressed its "willingness to maintain the

[5] H. R. P. Dickson, *Kuwait and Her Neighbours* (London, 1956), p. 276. The book contains an interesting eyewitness account of the Oqair conference, pp. 270-280.

[6] Sir Rupert Hay, "The Persian Gulf States and Their Boundary Problems," *Geographical Journal*, Dec. 1954, pp. 433-445.

[7] The Hadda Agreement was concluded at Bahra. It should not be confused with the Bahra Agreement, signed the preceding day, Nov. 1, 1925, which dealt with tribal migrations and aggressions on the border between Nejd and Iraq. The text of both agreements may be found in *British and Foreign State Papers* (London, 1929), CXXI (pt. I), 1925.

status quo in the Maan-Aqaba district" pending a final settlement.[8]

The Saudi-Transjordanian boundary has been free of controversy and sensational incidents since the conclusion of the Treaty of Jeddah. Because the border area east of Aqaba was not seriously presumed to contain oil, no attempts at drilling were made by the oil companies holding concessions on either side of the boundary. Both Aramco and the Transjordan Petroleum Company, Ltd., had at one time held concessionary rights to the areas directly adjacent to the boundary line, and should oil be discovered in this region the boundary issue would probably be revived.

Saudi Arabia's boundary with Yemen was partly settled by the Treaty of Taif of May 20, 1934, which traced its line for a distance of about four hundred miles inland from the Red Sea coast between Muassam and Maidi to the edge of the Rub al-Khali.[9] To avoid misunderstandings a line of some 240 pillars was erected. No attempt was made to demarcate the boundary in the Rub al-Khali itself. The Yemen-Saudi boundary has remained free of disputes ever since the conclusion of the treaty.

CONTROVERSY OVER SOUTHEASTERN BOUNDARIES

Thus, however imperfectly, Saudi Arabia's frontiers have been delimited in the areas adjacent to Kuwait, Iraq, Jordan, and Yemen. The boundaries with Qatar, the Trucial Coast, Muscat-Oman, and the Aden Protectorate have remained undefined, and the only valid international document on which border claims can conceivably be based is the earlier-mentioned Anglo-Turkish Convention of 1914. By the mid-1930's it became evident that there was considerable disparity in the views of Saudi Arabia and Britain as to the outer limits of their spheres in this general area. While this disparity had been latent ever since the rise of the Saudi house as a unifier and conqueror of various Arabian provinces, in 1934 it broke into the open following the grant of a concession by the Saudi government to

---

[8] *Ibid.* (London, 1936), CXXXIV, 1931.
[9] For details see H. St. John Philby, *Arabian Jubilee* (London, 1952), p. 187.

Aramco's predecessor, Standard Oil of California. A royal proclamation promulgating the concession agreement stated:

The Standard Oil Company of California is permitted to exploit petroleum and its extracts *in the eastern portion of our Saudi Arab Kingdom, within its frontiers,* in accordance with the conditions and regulations laid down in the agreement signed by Our Minister of Finance and the representative of the said Company at Jeddah on the 4th day of Safar of the year 1352 [May 29, 1933].[10]

Thus the boundary problem was revived. The subsequent grants of concessions to IPC's affiliates by the rulers of Qatar (1935) and Muscat-Oman (1937) and by the sheikhs of the Trucial Coast (1937 and 1939) added new importance to this matter. The first step to clarify the boundaries was taken by the United States government in 1934 when it made inquiries on this subject in Turkey and Britain. In reply the British pointed to their 1913 and 1914 conventions with the Ottoman Empire, asserting that the Blue and Violet Lines provided therein should be considered as valid boundaries. The British soon afterward informed the Saudi government of their view. The Saudi reaction, expressed in a note, dated May 13, 1934, by Acting Foreign Minister Fuad Bey Hamza to the British ambassador in Jeddah, Sir Andrew Ryan, was to deny the validity of these lines while declaring readiness to enter a discussion to define the frontiers.

The negotiations which ensued were conducted intermittently in Jeddah, Riyadh, and London between 1934 and 1938, to be interrupted for a period of ten years by World War II. The basic Saudi Arabian position was expressed in a memorandum by Fuad Bey Hamza, dated April 3, 1935, defining Saudi Arabia's boundaries with Qatar, the Trucial Coast, the sultanate of Muscat and Oman, and the territories in southern Arabia. In the memorandum the Saudi minister claimed the following boundaries (thereafter referred to as the Fuad Line) for his country:

1. With Qatar: A line starting on the west coast of Qatar, some fifteen miles up the bay of Dauhat as-Salwah, intersecting the Qatar Peninsula in a general southeastern direction, and reaching its east coast some seven miles north of Khaur al-Udaid. This proposed

---

[10] Italics mine. The text is in *Umm al-Qura,* July 14, 1933.

boundary would not leave the Qatar Peninsula a single entity under the sovereignty of one ruler but would cut off its southern part, representing approximately one-sixth of the total area, in favor of Saudi Arabia. In fact, the proposed line would divide the Qatar Peninsula in the way that Jutland Peninsula is divided between Denmark and Germany.

2. *With Abu Dhabi and other states of southeastern Arabia:* A line starting some sixteen miles south of Khaur al-Udaid and embracing in a wide irregular arc most of the Rub al-Khali desert as well as a sizable area in the hinterland of the Trucial Coast. This line was to meet at latitude 17 degrees north the Violet Line of the Anglo-Turkish Convention of 1914 and follow the latter's course westward.

Notable in Fuad Bey's proposal were three features: (*a*) that it claimed for Saudi Arabia about twenty-three miles of the coast line southeast of Qatar, thus separating the latter from the nearest Trucial Coast sheikhdom of Abu Dhabi; (*b*) that, although claiming a substantial part of the Trucial Coast hinterland, it left its two major oases—Liwa and Buraimi—outside Saudi territory; and (*c*) that with respect to the Aden Protectorate it accepted some 250 miles of the Violet Line as established in the Anglo-Turkish Convention of 1914.

Britain's position was formulated in a memorandum of November 25, 1935, which offered a new definition of the frontier, thereafter to be known as the "Riyadh Line." The latter was broadly parallel to the Fuad Line, i.e., it also left the bulk of the Rub al-Khali desert within the confines of Saudi Arabia. In contrast to the Fuad Line, however, it cut off from Saudi Arabia the southernmost part of the Qatar Peninsula, providing for a junction of the Qatar and Abu Dhabi territories and depriving the Saudis of access to the seacoast east of Qatar. Moreover it claimed more territory for the Trucial Coast sheikhdoms in the hinterland and assigned somewhat vaster areas to Muscat-Oman along the outer edges of the Rub al-Khali. According to the British government, the Riyadh Line was based on "an exhaustive reappraisal of the tribal situation in the disputed areas." In other words, Britain was willing to take into consideration the principle often advanced by the Saudis in negotiations, namely, that boundaries should be based on the historical and actual allegiance of various borderland tribes to one or another party in the boundary dispute. Noteworthy in this connection was Britain's recognition that the Blue Line of the 1913 convention, drawn up arbitrarily along the 50th meridian, no longer served its purpose as a boundary and that most of the Rub al-Khali should be considered as part of Saudi Arabia.

After World War II the controversy was resumed in 1949 in connection with the activities of Aramco's exploration parties in borderland areas. British Political Officers in the Trucial Coast strongly objected to the appearance of Aramco's personnel (accompanied by Saudi escorts) in what they claimed to be Abu Dhabi territory. This brought protest notes from Saudi Arabia. As a result negotiations between the two governments were resumed, and on October 4, 1949, Saudi Arabia presented a new definition of the boundary. The Saudi government repeated its earlier claim to a belt of territory roughly fifteen miles wide at the base of the Qatar Peninsula. The Saudis also claimed some 175 miles of the coast east of Qatar between Khaur al-Udaid and Mirfa, hitherto considered by the British to belong to Abu Dhabi. From Mirfa the proposed Saudi boundary was to follow a southeastern course for a brief stretch and then to turn east toward the oasis of Buraimi. The boundary as thus defined would reduce the Trucial sheikhdoms to a narrow belt of land along the coast and incorporate into the Saudi territory most of their hinterland including the oases of Liwa and Buraimi. This was in strong contrast with the British view that Liwa constituted an integral part of Abu Dhabi. Buraimi, an oasis of fifteen square miles composed of eight villages, was claimed by the sheikh of Abu Dhabi (six villages) and by the sultan of Muscat (two villages). Its total population was estimated at 10,000.

Believing the Saudi claim to be extravagant, the British countered on November 30, 1949, with the statement that under the circumstances the Blue Line of the Anglo-Turkish Convention of 1913 should be regarded as the only valid boundary. Instead of coming closer to each other's point of view, the two parties were drawing further apart.

In an attempt to break the deadlock Emir Faisal, Saudi Arabia's Foreign Minister, visited London in 1951. His conversations in the British capital led to an agreement to call a round-table conference to be attended by Saudi Arabians on one hand and the rulers of Qatar and the sheikhdoms and a representative of Muscat—the latter group under British chairmanship—on the other. Furthermore, it was agreed that until the conclusion of the contemplated conference the activities of the oil companies on both sides and also the movements of the Trucial Oman levies (a British-officered force in Abu Dhabi) would be restricted to areas outside the disputed territory, without prejudice to the ultimate rights of both parties.

The round-table conference was held in Dammam in January and February 1952. It was attended by the rulers of Qatar and Abu Dhabi, by Emir Faisal, and by Sir Rupert Hay, British Resident in the Persian Gulf. Much of the discussion centered around the historical allegiance of various tribes in the borderland area between Abu Dhabi, Qatar, and Saudi Arabia, especially the Bani Yas and Manasir. The collection of *zakat,* i.e., religious tithe, by one or another ruler in the area was frequently invoked by the disputing parties as evidence in support of their respective claims to sovereignty. The British side insisted on taking the 1935 Saudi proposal—the so-called Fuad Line—as an authoritative expression of Saudi views, which would thus preclude consideration of more extensive Saudi claims in the postwar period. On their part the Saudis denied the validity of their 1935 proposal. The latter, they asserted, had been suggested only as a compromise and, having been rejected by the British, became a "dead letter proposal" and could not be considered as binding on Saudi Arabia. By mid-February the conference had adjourned without an agreement. The stage was thus set for unilateral actions.

THE BURAIMI DISPUTE

Desirous of asserting its sovereignty in the disputed area, the Saudi government in August 1952 sent a party headed by Emir Turki ibn Abdullah ibn Ataishan to the Buraimi oasis to act as governor. According to the Saudi version,[11] "he arrived with a civilian staff of some 40 clerks, attendants, technicians, and policemen." The British termed it an act of aggression, asserting that Turki's party comprised eighty men, of whom fifty were armed troops. According to the Saudis, within a month after Turki's arrival fifty-nine tribal chiefs of the Buraimi region had reaffirmed their allegiance to King Ibn Saud, to be followed by others at later dates. The British, while not disputing these facts, called it a policy of "blandishments and bribes" and countered Turki's activities by dispatching detachments of the Trucial Coast and Muscat forces to the neighboring districts. In addition, the Sharja-based airplanes of the RAF carried out frequent flights over Buraimi, dropping anti-Saudi messages.

It looked for a time as if an armed clash was inevitable. The Saudis, however, sought the good offices of the United States. As a result, on October 26, 1952, a "standstill agreement" suggested by the American ambassador in Jeddah, Raymond Hare, was signed by British and Saudi representatives. Saudi Arabia then proposed a plebiscite in the disputed area. Rejecting the plebiscite, the British proposed arbitration. After considerable bickering over the presence of British-led troops around Buraimi—referred to as a blockade by the Saudis—an arbitration agreement was concluded in Jeddah on July 30, 1954. It was also agreed that all forces should be withdrawn from Buraimi except a 15-man police detachment to be maintained by each party in the oasis. British oil operations were to be authorized in the northern part of the contested region and Saudi operations in the south and the west. (An IPC subsidiary had actually begun operations at Fahud, south of Buraimi in the hinterland of Muscat-Oman, early in 1954.)

Arbitration proceedings began in Nice on January 22, 1955. The tribunal was composed of Dr. Charles De Visscher (Belgium), chairman, Dr. Ernesto Dihigo (Cuba), Dr. Mahmoud Hasan (Pakistan), Sir Reader Bullard (Great Britain), and Sheikh Yusuf Yasin (Saudi Arabia). In the course of the proceedings both parties presented

[11] Saudi Arabia, Permanent Delegation to the United Nations, *The Buraimi Dispute, A Summary of Facts Regarding British Seizure of Territory in Southeastern Saudi Arabia after Terminating Arbitration Proceedings for a Peaceful Settlement* (New York, n.d.).

long memorials stating their cases. Toward the end of August the British agent informed the tribunal that at its next September meeting he would submit five complaints on Saudi violations of the arbitration agreement. These were: (*a*) that the Saudi police contingent exceeded by four to six men the maximum of fifteen; (*b*) that passengers were carried in airplanes supplying the police post; (*c*) that attempts were made to send arms to the disputed territory; (*d*) that the Saudis prevented British relief supplies from reaching fire victims in Hamasa, a village in the Buraimi group; and (*e*) that Saudi officials had offered bribes to certain Buraimi chiefs. The latter was by far the most serious charge. The British asserted that a bribe of £30,000,000 ($85,000,000) was offered in one cash payment to the brother of the ruler of Abu Dhabi if he, a resident in the area, would cast in his lot with Saudi Arabia. At the session beginning on September 11 the tribunal began hearings on the British charges. Its verdict—to censure or exonerate Saudi Arabia—was expected within a week. On September 16, however, the British member of the tribunal, Sir Reader Bullard, suddenly presented his resignation and despite the pleas of the chairman walked out. This was promptly followed by the resignation of the chairman and two other members of the tribunal. Arbitration proceedings thus came to an end.

A few weeks later, on October 26, the British-officered forces of the sheikh of Abu Dhabi and the sultan of Muscat entered Buraimi, overpowered and ejected the Saudi police contingent, and occupied the whole of the oasis. Simultaneously in a note handed to the Saudi government Britain proclaimed a new boundary line, which in general followed the Riyadh Line of 1935, and warned against violating it. Saudi protests, followed by condemnation of the British action and a call by the Arab League's Political Committee on November 14, 1955, for neutral supervision of the disputed area, were of no avail. British-officered troops remained in occupation of Buraimi, and in a general movement strengthening Britain's position in southeast Arabia the forces of the sultan of Muscat in December seized the town of Nizwa, seat of the imam of Oman, religious and secular leader of Muscat-Oman's hinterland area.

Thus one phase of the dispute was concluded rather abruptly by the use of force and Britain's refusal to continue arbitration. Militarily Saudi Arabia was obviously in a weak position and could not engage in open hostilities. Politically, however, she had quite a few trump cards. In the first place, Britain's withdrawal from arbitration, together with repeated refusals to allow neutral supervision of the disputed area, was likely to arouse sympathy for Saudi Arabia and throw doubt on Britain's arguments in support of her boundary claims. In the second place, Saudi Arabia retained considerable ability to manipulate the tribes in the contested region and could take advantage of disputes which might erupt between the sultan of Muscat and his unruly subjects in the vast hinterland stretching westward toward the Rub al-Khali. In this area the second act of the desert drama was soon to be played.

THE QUESTION OF INNER OMAN

The sultanate of Muscat-Oman has been facing more than boundary problems. It has also been beset by difficulties stemming from the fact that its hinterland, i.e., inner Oman, has had its own traditional leaders who only reluctantly accepted the sultan's suzerainty and not infrequently are in a state of open rebellion. The sultan's effective authority is generally limited to the narrow coastal plain north and south of the town of Muscat. Inner Oman is separated from the coastal strip by the high and barren Al-Hajar range, of which Jabal al-Akhdar (Green Mountain) constitutes the most prominent feature. On the western slopes of Green Mountain the strongholds of the Ibadi sect such as Nizwa and Ibri are situated. The sect, predominant in inner Oman, is headed by elected chieftains, who combine religious with temporal authority under the title of the imams of Oman. One of the Ibadi chiefs acting for the imam on September 25, 1920, signed a "treaty" with the sultan of Muscat at Sib, whereby the internal autonomy of Oman was recognized.[12] Following this agreement, the forces and agents of the

---

[12] The text may be found in *The Status of Oman and the British Omanite Dispute*, published by the Arab Information Center (New York, 1957). References to this agreement may be found in Aitchison, XI, 284; Wilfred P. Thesiger, "Desert Borderlands of Oman," *Geographical Journal*, CXVI (Oct.–Dec., 1950), 162; G. J. Eccles, "The Sultanate of Muscat and Oman, with a Description of a

sultan of Muscat stayed clear of the imam's territory. In fact, the latest incumbent of Muscat's throne, Sultan Said bin Taimur, never ventured beyond the limits of his coastal area until December 1955. In that month, however, acting in concert with the British, he undertook his first expedition into inner Oman. Starting from Dhofar on the coast of the Arabian Sea, he personally led a motorized caravan across the desert northeastward to Nizwa, the imam's capital. There he was met by the main body of his British-officered army, which had approached Nizwa from the east. Both columns entered Nizwa on December 24, 1955, forcing the imam to flee to the mountains. The sultan's flag was hoisted in other towns of inner Oman as well, and tribal chieftains hastened to pledge allegiance to the conqueror. After this show of force the sultan returned to Muscat, leaving his officials and some token forces to administer the newly subjected territory.[13]

Peace did not last long. The fugitive imam, Ghaleb ibn Ali, did not accept defeat and from his hiding place tried to rally to his cause not only the temporarily subdued tribes in Oman, but also the independent Arab governments. Not content with mere reassertion of autonomy, he began to insist on complete independence and international recognition of it. By maintaining a permanent delegation in Cairo, this center of militant Arab nationalism, and another in Riyadh, he kept in touch with the outside powers that on political and ideological grounds were likely to abet him. The degree to which either of these states has aided him has not been fully ascertained. British sources maintain that an Omani volunteer force numbering some 500 men underwent training near Dammam in Saudi Arabia under Saudi and Egyptian instructors. The Omanis have obtained some military equipment from outside, including land mines, but the immediate origin of these weapons has not been disclosed. Thus trained and armed, the imam's men rose in rebellion

---

Journey into the Interior Undertaken in 1925," *Journal of the Central Asian Society*, XIV (1927), 23–24; R. Vadala, "Mascate," *Asie française*, XXIII (May 1923), 135; and Bertram Thomas, *Arab Rule under the Al Bu Said Dynasty of Oman, 1741–1937* (London, 1938), p. 26.

[13] For a vivid description of this expedition, see James Morris, *Sultan in Oman: Venture into the Middle East* (New York, 1957).

149

against the sultan of Muscat in July 1957. In quick succession the rebellious tribes reoccupied Nizwa and Ibri, restoring their control over large areas of inner Oman west of Green Mountain and temporarily cutting off the usual road of access from Buraimi to IPC's oil operations in Fahud.

In contrast to previous rebellions, this one received considerable publicity abroad, partly because of indefatigable Cairo propaganda and partly because of genuine alarm in the British Parliament. It was obvious that the sultan of Muscat had suffered serious reverses and that if left alone to face the imam he was too weak to regain his authority in the interior. In fact, his ability to maintain himself in his coastal strip should a determined rebel attack be launched against it was dubious. The implications of such a rebel success would be serious inasmuch as it would present a serious challenge to the whole political structure in southeast Arabia and Britain's position therein.

Moreover, from the British point of view the problem was somewhat complicated by the lack of any definite treaty commitment to defend the sultan if attacked.[14] Before the rebellion Muscat's meager forces, partly officered by the British, had been sufficient to maintain a precarious balance between the sultan's authority and the centers of Ibadi strength in the interior. But by the summer of 1957 it was clear that only substantial British reinforcements could restore that balance. Eventually, after what appeared like a few weeks of indecision, such reinforcements were flown in from Kenya in early August and an expeditionary force commanded by a British general crossed into inner Oman, determined to crush the rebellion. The imam's tribesmen offered resistance, but within a few days the sultan's force reoccupied most of their strongholds. With the fall of Nizwa on August 13 the open phase of the rebellion came to an end, and the rebels sought refuge in the neighboring mountains. But neither the imam nor his well-wishers in Cairo and Riyadh were reconciled to defeat. The imam's delegates in these two capitals were

[14] The British-Muscati Treaty of Friendship, Commerce and Navigation of Dec. 20, 1951, nowhere mentions either a protectorate relationship or an obligation to defend Muscat in case of an attack. The text is in *Treaty Series*, no. 44 (Cmd. 8633; London, 1952).

maintained, eager to inform the outside world of guerrilla activities carried out against the sultan in Oman's interior in 1957 and 1958.

CONCLUDING REMARKS

The Buraimi dispute and the Oman war have provided eloquent testimony to the complications that are likely to arise in connection with undefined desert borders. Moreover, what looks like a technical boundary issue may have far-reaching ramifications. It is worth noting that both in the Buraimi and in the Oman crises the contestants were not only prone to accuse each other of greed based on the supposed existence of oil in the disputed regions, but were inclined to implicate the oil companies in the controversy. Thus a Saudi Arabian pamphlet on the subject of Buraimi quoted the London *Times* of December 20, 1955, as stating that the Muscat field force was "financed partly by the Sultan and partly by a subsidiary of the Iraq Petroleum Company." [15] And the British memorial to the Buraimi arbitration tribunal tried to connect Aramco with the Saudi territorial claims by referring to its readiness in the postwar period "to embark upon new Arabian ventures" and accusing it of serious "incursions . . . into Abu Dhabi territory." [16]

Hovering above these accusations was the more fundamental problem of British rule in the Persian Gulf versus the Saudi dream of unity for the whole of the Arabian Peninsula. The Saudis were not reluctant to state their position in blunt terms when presenting their case to the Buraimi arbitration tribunal:

In taking up the considerations of other than a legal character which in its view are relevant to the questions before the Tribunal, Saudi Arabia would avert first to a fundamental point: the anomalous character today of the political system in the Persian Gulf built up during the nineteenth century by the British Empire. By this Saudi Arabia does not intend to question the purely legal authority of the United Kingdom to act in the present arbitration on behalf of the Ruler of Abu Dhabi and the Sultan of Muscat; but it does suggest that this surviving monument of the im-

[15] *The Buraimi Dispute*, p. 4.
[16] *Arbitration concerning Buraimi and the Common Frontier between Abu Dhabi and Saudi Arabia: Memorial Submitted by the Government of the United Kingdom of Great Britain and Northern Ireland* (London, 1955), I, 92.

perial age is not a regime in harmony with modern needs and aspirations in the Gulf region. . . . It is the firm opinion of Saudi Arabia that, were it not for the British, there would be no serious difficulties regarding these boundaries.[17]

[17] *Arbitration for the Settlement of the Territorial Dispute between Muscat and Abu Dhabi on One Side and Saudi Arabia on the Other: Memorial of the Government of Saudi Arabia* (11 Dhu al-Hijjah 1374—31 July 1955), I, 519, 523.

◆◇ CHAPTER IX ◇◆

Pipelines: In Search of a Formula

At first glance it may appear that the complexity of relations between the companies and the governments in the producing countries could hardly be matched in any other aspect of oil operations. This may have been true during the first few decades of the industry's activities in the Middle East. But from 1948 onward the seemingly simple problem of transportation began claiming an ever-increasing share of attention, emerging in the later fifties as a real "problem child" in company-government relations.

Middle Eastern pipeline operations have gone through several stages, corresponding to definite economic and legal formulas or situations.

The first stage coincided with the Franco-British mandatory regime in Syria, Lebanon, Palestine, and Transjordan. The Iraq Petroleum Company was the main actor on the scene, asking for and obtaining the right of transit for its oil to Mediterranean terminals in Lebanon and Palestine. The procedure was simple: a Western-owned corporation sought a right of way for its pipelines in Western-controlled territory. The negotiations were not complicated by the demands of a sovereign native state. The mandatory powers did not think it proper to burden the company with special obligations. The company's privilege was to be their privilege as well. Consequently, when on March 25, 1931, IPC concluded its first transit conventions with Syria and Lebanon, it was expressly exempted from all transit fees, import duties, and taxation while gaining the right to construct, maintain, and operate the pipelines for seventy years, with the privilege of renewal. The company's conventions concluded with the British mandatory governments for Palestine and Transjordan closely resembled the Syrian and Lebanese conventions and likewise exempted the company from any payments.

EXIT MANDATES—ENTER ARAMCO

This situation prevailed until the period immediately following World War II. The second phase in the history of pipelines was marked by two significant developments. One was the achievement of independence by the mandated territories of Syria, Lebanon, and Transjordan, coupled with the establishment of a Jewish state in Palestine. The other was the appearance of another company—wholly American-owned Aramco in Saudi Arabia—which also developed an interest in pumping its oil through a pipeline to the Mediterranean coast. For purposes of negotiation and operation Aramco's parent companies established a separate nonprofit subsidiary, the Trans-Arabian Pipe Line Company (Tapline). Tapline's first task was to conclude transit agreements with the four countries that its pipeline was to traverse: Saudi Arabia, Jordan, Syria, and Lebanon. In contrast to IPC in the 1930's, the company had to deal with fully sovereign Middle Eastern states. Of the four, Saudi Arabia was distinctive inasmuch as it was the producing country. Consequently its interests closely paralleled those of the company. Its convention with Tapline, dated July 11, 1947, provided for no transit fees during the first fifteen years. The company undertook, however, to pay for "all reasonable and necessary expenses" incurred by the government at the ports and pumping stations for protection, administration, customs, health and municipal works, and tax and land-use formalities. After fifteen years the company was to pay the Saudi government a transit fee commensurate with the highest fees paid by Tapline or any other pipeline company in the Middle East, the length of the line, and the quantity of oil transported. The company also obligated itself to establish schools for children of Saudi employees and government officials as well as hospitals for employees, travelers, and residents in the area of the pipeline stations.

PIPELINES

Tapline's conventions with Jordan and Lebanon were signed on August 8 and 10, 1946, respectively. They were followed by a convention with Syria on September 1, 1947. All three conventions contained many identical provisions, the differences among them being due to such special circumstances as the anticipated construction of terminal facilities in Lebanon. Each convention was to be valid for seventy years, with the privilege of renewal, and each provided for a transit fee of £1.5 per 1,000 tons transported, with a minimum payment of £20,000 a year (£1 to be valued at $4.03). The conventions stipulated, in addition, a protection fee not less than £40,000 a year for Jordan and Syria and £25,000 for Lebanon. A subsequent (1950) Sidon agreement with Lebanon provided for a loading fee of 2.88 pence per ton, with a minimum of 2,600,000 piastres a year for ten years, together with a fee of 650,000 piastres a year for the right to operate a ship-to-shore radio service. Syria was to receive 50 per cent of the above-mentioned terminal fees under a separate agreement between the two countries concluded on June 10, 1947. The basic formula of all three conventions treated the pipeline as a transportation company. The company was expected to pay a transit fee without regard either to the value of the goods transported or to the profits realized by the operation.

Tapline and Syria: Drama of Ratification

Although the transit conventions with Syria, Jordan, and Lebanon were almost identical, the political circumstances of the Syrian convention set it apart from the others. Barely three months after the signing of the convention the United Nations on November 29, 1947, adopted a resolution to partition Palestine. The reaction of the Arab states, leading to the war with Israel in 1948, is too well known to be recapitulated here. In the oil industry the impact of the war was particularly felt in the delayed ratification of Tapline's convention with Syria. All over the Arab East voices were demanding the use of "the oil weapon" as a means of pressure on the West. But inasmuch as all-out action—nationalization of oil in the producing countries—did not materialize because of the cautious policy followed by Saudi Arabia and Iraq, refusal to permit the transit of oil was the most likely form of pressure available. Of the three transit countries, it was natural that Syria—a veritable heartland of Arabism—should feel and act more strongly than the other two. Jordan, as a Hashimite kingdom in alliance with Britain, was certain to behave with moderation. Lebanon's old Phoenician traditions precluded an extremist policy likely to result in serious economic losses. In fact, 1948 witnessed a unique psychological duel between Lebanon and Syria over the Tapline convention. Sensitive to a highly emotional public opinion, the Syrian government and parliament delayed ratification despite their desire to benefit from such pipelines as might be constructed through their territory. After the conclusion of active hostilities in Palestine in the latter half of 1948 Lebanon grew more and more insistent on early Syrian ratification.

The Lebanese press played a not inconsiderable role in this duel, materially helping its government in attempts to secure Syria's agreement. Its editorials dwelt at length on the benefits both countries would derive from the pipeline operation. Particular publicity was given to a story that Aramco and other oil companies intended to carry out a miniature Marshall Plan in the Middle East by spending within the next five years up to $2,000,000,000 on expansion of the industry.[1] The story was based on public statements by W. F. Moore and B. E. Hull, presidents of Aramco and Tapline, respectively. The Lebanese also played rather skillfully on Syrian fears lest the company abandon its project in the Levant countries and divert the line to Egypt. Thus the Beirut press gave a big display to the news that, as a result of the delays in Syria, Tapline had decided to liquidate the greater part of its initial organization in the Levant, dismiss such local help as it had hired, stop new recruitment, and suspend further shipments of supplies. The Lebanese eagerly seized upon any hint that the threat of diverting the pipeline to Egypt might actually materialize. Such hints were not lacking. Despite the highly inflammable tone of the Egyptian press about using oil as a weapon in the Palestinian crisis, there were unofficial indications that once the effervescence subsided the Egyptian government might not be averse to considering the passage of the pipeline through the Sinai Peninsula. With an obvious eye to the Syrian audience one of the Beirut papers spoke of a visit by ten Aramco directors to Egypt and of an Egyptian mission about to go to the United States with a view to clarifying

[1] *Ad-Diyar, Az-Zaman,* and *Al-Amal* (all of Beirut), Nov. 16, 1948.

the pipeline question. The paper loudly criticized Egypt and Syria for their attempts to deprive Lebanon of benefits from the projected pipeline.[2] Pursuing the same line, another Lebanese daily quoted a spokesman of the Saudi Arabian legation in Cairo to the effect that his government would be compelled to agree with Tapline in its search for another outlet in case Syria refused to ratify the convention.[3]

This combination of abuse and goading, of appeals to the solidarity of two sister nations and threats of dire economic consequences should the pipeline project be abandoned, did not fail to impress the Syrians. Syrian newspapers gradually began to pick up the themes suggested by the Lebanese press, thus modifying their previous intransigent attitude. Toward the end of November 1948 one of the Damascus dailies carried a statement, attributed to a "high Syrian official," which said: "Consideration by the company of the possibility of constructing its pipeline across Egypt or Palestine has caused us to undertake a serious study of this question." The statement added: "Syria welcomes the passage of Tapline through its territory and would be glad to view the pipeline as a step toward happiness and comfort so long as this does not affect the sovereignty of the country."[4] The Syrian press was still far from united in its attitude, and some papers were relentless in their hostility toward ratification. Notable in this respect was a series of articles published in January 1949 by the former Minister of Defense, Ahmad Sharabati, in which he strongly opposed ratification as detrimental to Syria's interests. Accusing the company of starting false rumors about rerouting the pipeline to Egypt, he stressed that Syria's original plan had been to have the terminal located in its own territory and he deprecated the size of the royalties Tapline was offering. In conclusion, he advised persistence in refusing ratification so long as the Palestinian issue had not been solved in favor of the Arabs.[5]

These negative attitudes were in no small measure due to the influence of certain political groups and parties. Foremost among the

[2] *Al-Hadith* (Beirut), Nov. 25, 1948.
[3] *Beirut al-Massa* (Beirut), Dec. 6, 1948.
[4] *An-Nidal* (Damascus), Nov. 30, 1948.
[5] From the articles published in *Assa al-Jinnah* (Damascus), Jan. 14, 21, 1949, and *An-Nasr* (Damascus), Jan. 18, 1949.

convention's opponents were the Socialist Party of Akram Hourani, the Arab Renaissance Party led by Michel Aflaq and Salah al-Bitar, and the People's Party headed by Rushdi Kikhya and Nazem al-Qudsi, both prominent citizens of Aleppo. The first two represented progressive and left-wing tendencies, and the third was basically conservative, but at this particular juncture, albeit for different reasons, they were united in opposing ratification. Of the three, only the People's Party had sizable representation in the Parliament. Its thirty-five deputies accounted for the bulk of the group of forty which early in 1949 was reputed to oppose Tapline's project. Taken together this was not a negligible force, and it was one prepared to resort to extraparliamentary pressure to achieve its ends. An example of its tactics was supplied on February 26 when at a meeting at the Syrian University students staged a demonstration calling for a general strike against Tapline's convention.

As for the government, its counsels were divided. A strong group in the cabinet favored speedy ratification. Its leading member was Majdeddin Jabri, Minister of Public Works. On February 18 he issued a statement pointing to three important advantages likely to be derived from the Tapline convention: (*a*) influx of hard currency, (*b*) delivery to Syria of 200,000 tons of oil per year at a low price, and (*c*) availability of the company's equipment if oil was discovered in Syria.[6] Jabri was not alone in his attitude. Apart from his colleagues in the cabinet he could count on the support of important business circles, which could exert considerable influence on Parliament. Early in March, Aref al-Lahham, secretary of the Merchants' Association in Damascus, published an article urging speedy ratification of the convention.[7] This was promptly followed by release of a study prepared under the auspices of the Syrian Chambers of Commerce, which viewed the projected pipeline as a positive development.[8]

By the end of March, Parliament was reported ready to consider ratification, following a government motion to this effect, and it was expected that a majority would vote in favor. Then, suddenly, on March 30 Colonel Husni az-Zaim overthrew the government in

[6] *Al-Balad* (Damascus), Feb. 18, 1949.
[7] *At-Tusha* (Damascus), March 5, 1949.   [8] Syrian press, March 12, 1949.

a bloodless coup, causing new doubts as to the fate of the convention. Eleven days later, however, he received Tapline's representative and on May 16 in the absence of the dissolved Parliament he and his Council of Ministers, by a "legislative decree," ratified the convention, together with two subsidiary Syro-Lebanese agreements pertaining to the operation of and profits from the pipeline. Thus the last legal obstacle in the path of the projected pipeline was removed. The promptness with which the dictator acted gave rise in the ever-suspicious East to a number of rumors, each purporting to explain the reasons for his coup or the alleged connection between the coup and ratification. One version was that the coup had been engineered by the British, who were trying to prevent the ratification of Tapline's convention by placing in power a man ready to do their bidding.[9] Another was that, failing to secure ratification under Syria's parliamentary regime, the Americans had not hesitated to overthrow the government in favor of a dictator, who promptly paid his debt of gratitude. Neither version could provide convincing documentary evidence. Zaim's rapprochement with Egypt and simultaneous estrangement from Iraq disproves the theory that he was a British puppet. The accusation against Tapline had an even shakier foundation. The fact was that Syria's Parliament had been definitely prepared to approve ratification toward the end of March, i.e., just prior to the coup, and that both Tapline and Syrian negotiators were taken by surprise and annoyed at a new complication when the coup occurred.

Construction of the pipeline began without much delay. It was completed on December 2, 1950. The pipeline was 1,058 miles long and was equipped with four pumping stations, Qaisumah, Rafha, Badanah, and Turaif, all located in Saudi Arabian territory. Its terminal was at Sidon, and its throughput capacity was 16 million tons a year (325,000 barrels per day). Of the total length, 754 miles of the line between Qaisumah and Sidon were owned and operated by Tapline; the remainder was the property of Aramco and subject to its management.

[9] This version was accepted by the French author, Louis Charentenay, in his article "Où va la Syrie?" *L'Afrique et l'Asie,* 3rd trimester, 1949, and by the Lebanese daily, *Ad-Diyar* (Beirut), April 21, 1949.

The First Revisions

Despite their initial enthusiasm about the pipeline, the Lebanese were soon disappointed. Once the pipeline had been completed, the sizable labor force employed during the construction was dismissed, and only about 500 persons remained on Tapline's payroll in Lebanon. The financial benefits accruing to the Lebanese treasury fell short of expectations also, amounting to about $250,000 a year. Consequently voices were soon raised demanding upward revision of the payments. Syria also asked for a change in the financial clauses. The company complied. Agreements signed on May 21, 1952, with Syria and on July 17, 1952, with Lebanon provided for the payment of $600,000 a year to each country, in addition to the original transit and security fees. Moreover both agreements (especially that with Lebanon) stipulated a number of "fringe" benefits.[10]

The conclusion and subsequent revision of Tapline's conventions were bound to affect IPC's agreements also. Revisions were incorporated in a number of instruments signed by IPC and Syria and Lebanon between 1947 and 1950. The basis of calculation in each case was the amount of oil transported. Like the Tapline conventions, both original and revised, the new IPC agreements adopted the "transportation company" formula, i.e., the company was to pay transit fees, but its earnings were to be of no concern to the transit countries. Thus was completed the third, or revision, phase of the pipeline story.

A Coup in Lebanon: Profit Sharing or Taxes?

Tapline's 1952 agreement with Lebanon was not destined to enjoy smooth implementation. In September of that year a *coup d'état*

---

[10] The original convention with Lebanon was actually modified by exchanges of letters and agreements in 1949, 1950, and 1952. The stipulated payments according to the latest (1952) version included six varieties: (a) a transit fee of £1.5 per 1,000 tons, with a minimum £20,000 a year; (b) a security payment of £25,000 a year plus any excess cost the government might incur by mutual agreement; (c) tax commutation payments of Lebanese pounds 60,000 annually to the Beirut and Sidon municipalities; (d) a loading fee of 2.88 pence per ton, with a minimum of 2,600,000 piastres a year for 10 years, together with a fee of 650,000 piastres a year for the right to operate ship-to-shore radio service; (e) a compensation of £75,000 annually for road deterioration caused by company vehicles; (f) a payment of $600,000 a year, one-third of which could be paid in pounds sterling at the company's option.

brought to power a government which promptly demanded an upward revision of the freshly concluded agreement, thus opening a new, fourth phase in the history of the pipelines. Supported by a press campaign and various nationalist organizations, the new cabinet challenged the validity of the agreement, claiming that it had been concluded by a government which did not represent the will of the people. Lebanon's nationalist circles asserted that the country's revenues from pipeline operations should be calculated on one of two bases: either the companies should share their profits with the transit countries according to the fifty-fifty formula prevailing in the producing countries or they should be subjected to an income tax on corporations.

With regard to the first basis, the Lebanese asserted that "the geographical position of the countries through which oil passes in transit represents a natural resource for this region exactly as the production of oil does for the producing countries." [11] Thanks to the pipelines the companies, it was claimed, effected considerable savings stemming from the difference between the posted price in the Persian Gulf ($1.75 a barrel in 1952) and that at the Mediterranean pipeline terminal ($2.41 a barrel). From these savings (66 cents a barrel) Tapline should be permitted to deduct its costs of amortization and administration, the balance representitng the net profit subject to equal sharing with the transit countries. As for IPC, it should not be entitled to deductions because, without the right of transit through Syria and Lebanon, it would have had to build a pipeline from Kirkuk to the Persian Gulf, i.e., a distance similar to that between Kirkuk and the Mediterranean. One Lebanese economist calculated that the profits effected by IPC on 8 million tons transiting toward the Mediterranean amounted to $35,000,000, while those made by Tapline on 16 million tons in transit were $50,000,000.[12] IPC's profits should be divided in equal shares between the company on the one hand and Syria and Lebanon on the other. Half of Tapline's profits should be divided between Lebanon, Syria, and Jordan.

[11] Said Himadeh, "Need to Reconsider the Proposal of Revision of the Oil Agreements," *An-Nahar* and *Beirut* (Beirut), Nov. 30, 1952. See also *Le Commerce du Levant* (Beirut), Dec. 6, 1952.
[12] *Ibid.*

If such a settlement should be inacceptable to the companies, they should be subjected, the Lebanese claimed, to an income tax, which would be 42 per cent of corporation profits. Together with other taxes, such as harbor fees and import duties, this would amount to 45 per cent.

The companies refused to accept either basis for calculation. Tapline's argument was, first, that its freshly concluded agreement, on which "the ink had barely dried," was valid and did not require revision. Furthermore, the company claimed that the transit of oil should be treated like the transit of any other merchandise, i.e., it should be subject to transit fees but not to the sharing of profits by a company conducting its business. Tapline, said one of the pamphlets published by its public relations office, "was not an oil company but a transportation company, transporting goods for others in the same manner as a railroad company, a trucking company or a steamship line." [13] A deadlock ensued and for more than three years neither party would give up its position.

SPOTLIGHT ON IPC: NEW PIPELINES AND AGREEMENTS, 1953–1958

While Tapline and Lebanon remained deadlocked, IPC displayed considerable activity in three directions. First, in 1952 it inaugurated a new 30-inch pipeline linking Kirkuk with Banias on the Syrian coast. Secondly, armed with the fact that it was within its power to limit its transit and terminal operations to Syria's territory exclusively, it concluded on January 6, 1955, a new agreement with Lebanon, whereby the latter's revenue was considerably raised. The agreement was a diplomatic victory for the company inasmuch as it was still based on a transit-fee formula.[14] Thirdly, the company on

[13] From a company pamphlet, *Sanatan min at-Taqaddum* (*Tapline: Two Years of Progress*) (Beirut, March 15, 1953), p. 2.

[14] Its terms provided for five kinds of payment: (*a*) for services and assistance, £65,000 annually, retroactive to 1953; (*b*) as a transit fee, £2.16 per 1,000 tons as of May 15, 1952; (*c*) as tax commutation, £L (Lebanese pounds) 35,000 annually, retroactive to include 1952; (*d*) for road maintenance, £L30,000 for 1952, £L25,000 for 1953 and 1954 each, and £L25,000 annually thereafter; (*e*) £207,500 annually as of May 15, 1952, in lieu of supplying crude at a reduced price. A simultaneously amended loading agreement called for the payment of 2.88 pence per ton loaded on ships as of May 15, 1952, with a minimum of £4,400 annually.

November 29, 1955, concluded a revised convention with Syria, which brought about a radical change in the pattern of payments, making Syria's revenue from pipeline operation four times what it had been the preceding year.[15]

By the end of 1955 it seemed that the company had brought to a successful conclusion an important phase of its relations with the transit countries. This success was, however, shortlived. Spurred by the advantageous terms of the Syrian convention, the Lebanese government promptly demanded revision of its own 1955 agreement, pointing to the considerable difference in payments to the two countries. The company resisted this demand, stressing that the ton-mile formula applicable to such agreements naturally favored Syria over Lebanon. The government [16] refused to accept the ton-mile formula; Lebanon's importance, it claimed, should be measured in terms of the availability of its territory for transit purposes rather than in terms of its size. The company countered with an intimation that exaggerated demands on the part of Lebanon might lead it to consider the skirting of Lebanese territory in any future expansion plans. In the spring of 1956 the government issued a virtual ultimatum that unless the company agreed to a radical upward revision of payment, it would be subjected to taxation. On June 29, 1956, the threat was carried out when the Parliament voted to apply the existing income tax law to the oil companies. The measure was primarily directed at IPC, as Tapline had recently come forward with an offer potentially more advantageous to the Lebanese than the terms of the tax law. The law itself was not immediately implemented, as Parliament postponed its application pending further negotiations with the companies.

In the spring of 1958 serious consideration was being given to a

[15] Ratified on Dec. 6, 1955, by the Syrian Parliament, the new convention provided for a transit fee of 1s. 4d. per 100 ton-miles of crude transported; a terminal fee of 1s. 1d. per ton loaded on ships at terminals in Syria; £250,000 annually for protection and other services; and £8,500,000 in settlement of all outstanding past obligations. In addition to the pipeline convention proper, an amended oil-loading agreement stipulated payments of 3 pence per ton, with a minimum payment of £20,000 a year, to be guaranteed for 10 years, and a refined products agreement making available to Syria the products of the Tripoli (Lebanon) refinery at advantageous prices.

[16] Headed by an outspoken Moslem nationalist, Abdullah al-Yafi, with an ambitious ex-premier, Saeb Salam, as Minister of State in charge of Petroleum Affairs.

proposal whereby Lebanon's income from IPC's operations would be tripled, reaching £1,066,000 a year if the existing throughput of 7.5 million tons of crude continued through Lebanon's territory. This proposal was a compromise solution following neither the ton-mile formula nor the tax-law provisions, and for political reasons Camille Chamoun, then president of Lebanon, was anxious to submit it to Parliament before the summer recess. The civil war in the summer of 1958, however, caused the suspension of all pipeline negotiations, and they were not resumed until the end of the year. At the present time (April, 1959) IPC's dispute with Lebanon is still awaiting settlement.

TAPLINE'S PROFIT-SHARING FORMULA

The prolonged deadlock in Tapline's relations with Lebanon was broken on May 1, 1956, when the company in a rather dramatic gesture offered to divide its profits equally between itself on one side and Lebanon, Syria, Jordan, and Saudi Arabia on the other.[17] Outwardly it looked like acceptance of the Lebanese theory. In reality it meant the inclusion of a new element, Saudi Arabia, which had hitherto been omitted from Lebanese calculations. Tapline's offer was made on the assumption that the four countries would divide among themselves their joint half of the profits. Lebanon promptly accepted the formula, and Syria and Jordan followed soon thereafter. From Tapline's point of view the situation presented the advantage that, though constituting a major concession, it transferred further negotiations from the company-government level to an intergovernmental level, permitting the company to stand aside until agreement was reached.

Before long, however, two new complications arose. One stemmed from Lebanese passage of an income tax on oil companies. Although, as pointed out above, the law was primarily aimed at IPC, Tapline was formally affected also, despite Lebanon's earlier accept-

[17] Inasmuch as Tapline is a nonprofit organization set up to effect savings in transportation for its parent companies, it was in reality an offer to share those savings rather than its own nonexistent profits with the four countries in question.

ance of the company's basic offer to the four countries. Tapline lodged a protest and received informal assurances that, once the agreement among the Arab governments was reached, means would be devised to exempt it from the provisions of the law. In the meantime application of the law to Tapline was postponed.

The other complication was due to a difference of views among the four Arab governments as to the division of the profits. Lebanon, Syria, and Jordan subscribed to a formula that would divide the Arab joint share into four equal parts, but Saudi Arabia claimed that account should be taken of the length of the line located in her territory, i.e., 71.5 per cent of the total.[18] Appeals to Saudi Arabia by the three other countries to accept their formula proved vain. In fact, the Saudi government soon channeled the discussion into an altogether new direction by asking that the profits obtained by pipeline operation be divided into equal parts between Aramco and itself after deduction of the payments to Jordan, Syria, and Lebanon as costs of operation. Although the Saudi claim was essentially a matter to be settled between Saudi Arabia and Aramco, it profoundly affected the relations between the three transit countries and Tapline inasmuch as it left open the question of what these "costs of operation," i.e., payments for transit should be.

Independently of the outcome of the Saudi-Aramco dispute [19] and without abandoning its basic fifty-fifty principle, Tapline decided to reopen negotiations with Lebanon, Syria, and Jordan by proposing a new formula. The latter was to be based on two elements: the investment of the company in its pipeline and terminal facilities and a ton-mile principle. The investment was divided as follows: 11.5 per cent in the Sidon (Lebanon) terminal and 88.5 per cent in the pipeline. The latter figure had to be divided among the four countries according to the length of the line located in each. Although Saudi Arabia's share in the total length was 71.5 per cent, her share in the total investment (i.e., including the terminal), was 63 per cent. Consequently Tapline proposed that the remaining

[18] The distribution of mileage from Qaisumah to Sidon was as follows: Saudi Arabia, 538.909 miles; Jordan, 110.317; Syria, 79.068; and Lebanon, 25.834—a total of 754.128 miles.
[19] For details, see p. 80 above.

37 per cent should be distributed among Lebanon, Syria, and Jordan as follows: Lebanon 14.5 per cent, Jordan 13 per cent, and Syria 9.5 per cent of the total offered to the four countries.

The transit countries' response was not discouraging, and it was confidently expected that an agreement would be concluded in 1958 along the lines proposed by Tapline. In the spring of that year, however, the negotiations with Syria had to be adjusted to the merger of Syria and Egypt into a single state—the United Arab Republic. It was not clear what the effect of this merger upon the pipeline negotiations would be. Initially it seemed that these negotiations would be continued by Syrian representatives acting on behalf of, and guided by the interests of, the Syrian region of the United Arab Republic. Following the Lebanese civil war and the Iraqi revolution of the summer of 1958, the general situation in the Arab East underwent a considerable change, which also affected the mutual relationships of the Syrian and the Egyptian regions of the newly unified state. This was reflected in a greater centralization of power in Cairo, replacing the hitherto loosely co-ordinated policies of regional authorities. A presidential decree of September 1958 ordered the establishment of a General Petroleum Authority for the United Arab Republic. Its task is to formulate general oil policies. Consequently it is likely that all Syrian pipeline negotiations, including those pending with Tapline, will be reviewed in the light of the interests of the united republic and that the ultimate decisions will be reached by the newly centralized cabinet in Cairo rather than by the provincial authorities in Damascus.

◈◇ CHAPTER X ◇◈

International Control: From Big Power Pacts to Pan-Arab Plans

RATHER early in the history of their expansion in the Middle East the oil companies began to face the possibility of some form of international control over the area's petroleum and their own activities. The Berenger-Long Agreement, which was linked to the San Remo peace settlement of 1920, was the first example of intergovernmental action in this respect. It provided, as we know, for division of Iraq's oil resources between British and French interests, in a 75:25 per cent ratio.[1] The exclusion of American interests prompted the latter to demand and, with the aid of the United States government, to obtain a share in the development of Iraq's oil resources. A result of these negotiations was the intercompany Red Line Agreement of 1928, in which private and governmental undertakings were closely interwoven.

The *modus vivendi* established by the Red Line Agreement seemed sufficient in the interwar period. The revolutionary changes brought about in the Middle East and the world at large by World War II were bound to subject the agreement to strains and stresses. A new Middle East began to emerge from its semicolonial status; Britain's position as the paramount power in the area was challenged; and the world saw the simultaneous rise of three potent new factors in Middle Eastern politics: the rebirth of Soviet ambitions, the eco-

[1] See p. 15 above.

nomic and political power of America, and the explosion of turbulent Arab nationalism.

Against this broad political background certain indisputable facts pertaining specifically to Middle Eastern economics were bound to attract increasing attention in the postwar period. These were: (a) the concentration of substantial petroleum reserves in the Middle East; (b) the development of these resources by a restricted group of major American, British, Dutch, and French companies; (c) the presence of proved oil reserves in some states and not in others; and (d) the general economic underdevelopment of the area, leading to striking contrasts between the wealth of the few and the poverty of the masses.

Considerable thought began to be given to various problems likely to arise on the basis of these facts. Governments and private interests began to consider the general problem of accessibility to these resources and of safeguards against artificial exclusion. Ways and means to avoid international rivalries and dangers of war were discussed. Consumers' interests found their advocates. Because of the poverty of the Middle East increasing attention began to be paid to the utilization of oil revenues to develop the region as a whole rather than piecemeal. Linked with this was the rising clamor of the have-not countries to share in the bounty of the haves. And some people were beginning to demand that the nations in whose territory oil was found should assume decisive control over their own resources —either by nationalization pure and simple or by participation in the concessionaires' decision-making processes and their marketing organizations.

Various motivations were at work, and ideas could not be divorced from the sources from which they emanated. Special interests clashed with crusading idealism, and much wishful thinking was practiced. Certain proposed "solutions"—assuming that any solution was needed —were obviously based on misinformation or on distortion of facts.

Ultimately four patterns of thought, each implying a solution which in some way would subject the Middle Eastern oil industry to national or supranational control, emerged from the welter of ideas. These patterns were as follows: (a) elimination of international rivalry through big-power agreement; (b) protection of consumers' interests through international control under United Nations auspices; (c) assurance of peace and stability through internationally supported regional development schemes; and (d) control of oil resources by the Middle Eastern states themselves.

ATTEMPTED REGULATION THROUGH ANGLO-AMERICAN AGREEMENTS

On August 8, 1944, the United States and Britain signed a petroleum agreement, the ostensible aim of which was to promote an orderly development of the international oil trade, to assure equal opportunity in the quest for new concessions, and to ensure adequate supplies of oil to all countries. The agreement did not mention the Middle East specifically. Nevertheless the real motive, so far as the United States government was concerned, was to assure for American interests a fair share in Middle Eastern oil exploitation and to obtain from Britain—hitherto regarded as the paramount power in the area—recognition of the legal and political validity of existing American concessions, especially in Saudi Arabia.

While the motives of the United States, a newcomer in the Middle East, were understandable, it was less clear what Britain hoped to accomplish through the agreement. A perusal of the history of the negotiations shows that initially the British were somewhat reluctant to negotiate at all. This was due partly to the earlier—and by then abandoned—attempt of the United States government to enter the Middle Eastern oil business directly by constructing a government-owned pipeline between the Persian Gulf and the Mediterranean and partly to general British reluctance to admit the legitimacy of any interests but their own in an area which London still regarded as falling within its exclusive influence. The British decision to negotiate an agreement could be explained only in terms of some *quid pro quo* that Britain hoped to obtain from it. Such a compensatory advantage could conceivably be of a triple nature. (a) While reassuring actual and potential American interests in the Middle East, a pledge of noninterference and fair play could be obtained from the United States with regard to existing British interests. This would be an intangible advantage only. But the British could not be oblivious to the fact that their strength as a great power was seriously sapped by the war and that they would soon have to contend with extremely dynamic American business enterprises which might disregard various legal niceties in the finely constructed edifice of Britain's imperial influence in the Persian Gulf and the mandated areas. (*b*) By recognizing the legitimacy of American interests in the Middle East, Britain could obtain from the United States a promise of some regulation of international trade that would safeguard certain markets, and possibly open new ones to British oil interests. Britain, like many a less dynamic European competitor, might be motivated by fear of a sweeping conquest of world markets by an efficient and expanding American business organization. (*c*) Moreover, once the principle of mutual consultation and agreement was established, it would be easier for Britain to maintain her entrenched position in certain areas and to put a check to any undue expansionist tendencies as exemplified by the earlier project of a government-sponsored pipeline.

The agreement was subject to ratification in both Britain and the United States. When its terms were released in Washington, it encountered considerable opposition in the Senate and by the oil industry. Apart from the merits of the agreement, the industry objected because it had not been consulted during the negotiations. As a result, the government withdrew the agreement. A new agreement with Britain was signed on September 24, 1945. Despite some changes in the text of the original document, the new agreement differed little from its predecessor. Its main provisions were as follows:

1. The signatories recognized "that the prosperity and security of all nations require the efficient and orderly development of the international petroleum trade," which "can be best promoted by international agreement."

2. Adequate supplies of petroleum should be accessible "to the nationals of all countries on a competitive and nondiscriminatory basis."

3. The principle of equal opportunity should apply to the acquisition of oil concessions, but "all valid concession contracts and lawfully acquired rights" were to be respected and there should be "no interference directly or indirectly with such contracts or rights."

INTERNATIONAL CONTROL

4. Pending the conclusion of a multilateral petroleum agreement which would establish a permanent International Petroleum Council, the signatories agreed to establish an Anglo-American "International" Petroleum Commission, the task of which would be to study the problems of the international oil trade and to make recommendations to both governments.

5. Nothing in this agreement was to be construed as affecting any existing or future legislation relating to the importation of oil into the country or territories under the jurisdiction of either government or as applying to the operation of the domestic petroleum industry of the signatories.[2]

The hearings on the new agreement in the Senate Foreign Relations Committee were delayed until June 1947, but on July 1 the committee reported favorably on it, with a few minor amendments. Despite the committee's approval no action was taken by the Senate.

The defeat of the agreement was largely due to the opposition of the oil industry in the United States, which, with a few exceptions such as Jersey Standard and the American Petroleum Institute, regarded it with misgiving and suspicion. The phrasing of the text was so general and in some instances so vague as to permit a variety of interpretations. It was these implied and potential dangers rather than any overt provisions of the agreement which produced fears, especially among independent producers. The latter objected on many grounds, foremost among which was alarm lest the Anglo-American agreement lead to undue expansion of Middle Eastern oil production, which would result in increased imports to the United States and the flooding of domestic markets by "cheap" Arab oil. Clauses referring to "orderly" international petroleum trade had an ominous ring, implying as they did government-imposed regulation, generally resented by spokesmen for free enterprise. An international oil commission, however emasculated its powers, was opposed as leading to government interference. It was particularly feared that the commission would make recommendations on the allocation of markets and thus set a dangerous precedent. The inde-

[2] U.S. Senate, *Petroleum Agreement with Great Britain and Northern Ireland: Hearings before the Committee on Foreign Relations* (80th Cong., 1st Sess., June 2–25, 1947; Washington, 1947).

pendents were fearful, moreover, that the agreement might result in the creation or semiofficial recognition of an international petroleum cartel and felt that the principle of equal accessibility to foreign oil resources was too general to give them a guarantee that independent interests would be considered on an equal basis with those of the major companies.[3]

The demise of the Anglo-American agreement marked the end of a definite phase in attempts to introduce some sort of international regulation into the Middle Eastern oil industry. This phase was characterized by the dual assumption that oil developments in this strategic area required intergovernmental understandings and that the latter should be spearheaded by an agreement between the two big powers with a major stake in Middle Eastern oil. It is perhaps idle to speculate on what would have been the consequences had the agreement been ratified. It probably would not have had much impact upon the domestic oil industry in America. Such an impact might have been beneficial from the narrow point of view of independent interests, because the only practical way in which the agreement could have affected the course of events in the Middle East would have been to delay, for legal reasons, the spectacular expansion of Aramco's production in Saudi Arabia, thus decreasing the threat of major imports which the independents have always feared. Article II of the agreement, which spoke of respect for "lawfully acquired rights," might have strengthened—through the support of the British government—the claims of the British, French, and Dutch interests that the restrictive clauses of the 1928 Red Line Agreement limited the freedom of the American partners in the Iraq Petroleum Company to expand into Saudi Arabia. Had this occurred, the settlement out of court which Jersey Standard and Socony reached with other IPC partners late in 1947 would perhaps have been more difficult to achieve. But even granting such obstacles, it is doubtful whether they would have definitely checked the basic trends. If not Jersey and Socony, some other major American companies might have entered the picture, perhaps not with identical

[3] *Ibid.* For a detailed treatment of the whole episode, see Benjamin Shwadran, *The Middle East, Oil, and the Great Powers* (New York, 1955), pp. 325 ff.

results but nevertheless with the fundamental purpose of maximizing the opportunities afforded by the great resources of Arabia.

A CRUSADE FOR UNITED NATIONS CONTROL

In the second phase—beginning about 1946—emphasis shifted radically. The new thesis, promoted by the International Co-operative Alliance, was that Middle Eastern oil should be viewed primarily not as a matter of concern to the governments and the industry, but as a subject of interest to consumers in the world at large. Consequently it was the consumers' view that should prevail. As an exponent of a definite social and economic philosophy, the Alliance believed that consumers' interests were best served by co-operatives; hence the production and, especially, the marketing of oil should be subjected to co-operative endeavor. The Alliance gave concrete expression to these ideas at its Sixteenth Congress held in Zurich on October 7–10, 1946, by passing a resolution on the control of world oil resources. The resolution contained the following main points:

1. Invoking the authority of the Atlantic Charter, it stressed the principle of equal access to natural resources.

2. It warned of the dangers to peace stemming from international rivalry, "either on the part of predatory private monopolists or on the part of imperialistic governments," for "control of oil resources of the Middle East."

3. It expressed belief in the efficacy of consumer-owned co-operatives in checkmating "monopolistic concentrations."

4. It emphasized "in strongest terms the immediate need of placing control and administration of the oil resources of the world under an authority of the United Nations, and, as a first step in that direction, the oil resources of the Middle East, by and with the consent of the States involved, these resources to be administered in such a way that Co-operative Organizations can be assured of receiving an equitable share." [4]

[4] International Co-operative Alliance, *Report of the Proceedings of the Sixteenth Congress of the International Co-operative Alliance* (London, n.d.), pp. 184, 185.

Inasmuch as the International Co-operative Alliance had been recognized by the United Nations as a nongovernmental organization of Category A, with consultative status in the Economic and Social Council, it promptly availed itself of the privilege granted to such organizations to request that its resolution be placed on the agenda of the next session of the Council. In an accompanying memorandum the Alliance suggested that "the control might best be technically implemented by means of an international convention to be drafted by the Council, which would then invite the countries concerned to give their ratification and bring it into effect." [5] The Council considered the matter at its Fifth Session in August 1947. Although a majority of its members expressed understanding of, and sympathy with, the ideas of the Alliance, the Council was reluctant to act on the resolution and, instead, adopted a British motion "to do no more than take note of the proposals at the present session." [6] The Alliance found an enthusiastic backer of its resolution in another nongovernmental organization, the World Federation of Trade Unions (WFTU), which went on record as unreservedly supporting the Alliance's plea for world oil control.

The Council's failure to act on the Alliance's resolution did not deter the latter from attempting to bring the matter before the Council again. With this in view the Alliance, at its congress at Prague in 1948, adopted a second resolution calling for international control of oil. In essence this was a repetition of the first resolution. It did not, however, specifically mention the Middle East, and in one paragraph attention was drawn to the world shortage of oil, which was blamed on restrictive practices of monopolistic combinations.

During the discussion on the proposed British draft the Soviet delegate moved an amendment which would have added the adjective "capitalistic" before the words "monopolistic combinations, cartels and trusts"; the latter were being castigated for excess profits, restrictive practices, and domination of markets. Moreover, he proposed that the concluding paragraph of the resolution should make

[5] International Co-operative Alliance, *Report of the Seventeenth Congress at Prague, 27th to 30th September, 1948* (London, n.d.), p. 32.
[6] *Ibid.*

INTERNATIONAL CONTROL

it clear that the United Nations was being asked "to undertake control of the distribution of oil by capitalistic monopolies, trusts, and cartels" only. These amendments were opposed by the British delegate, who pointed out that monopolistic control of oil was not limited to capitalistic cartels but was also exercised by certain governments or by a combination of state and private capital. Consequently, to quote the British delegate, "it would be a tremendous mistake if this Congress were to agree that where oil is owned by the State there shall be no enquiry; that such States shall be free to do exactly what they like, and that the old rivalry between States over oil shall go on." [7] On these grounds the British delegate opposed Soviet amendments. His views were shared by the delegate representing the International Co-operative Petroleum Association (of which more will be said later), who moved that both "private and State capitalistic" monopolies should be expressly mentioned in the text. In the ensuing vote the British-sponsored resolution was adopted, with the amendment introduced by the Petroleum Association, by 626 votes against 353 that favored the Soviet version. The substantial number of votes cast for the Soviet amendments was indicative of the extent to which the thinking of many delegates was conditioned by anticapitalistic spirit and of their apparent readiness to follow the Soviet lead in this respect.

Like its predecessor, the Prague oil resolution was submitted to the United Nations for inclusion in the provisional agenda of the Economic and Social Council, which after some delay considered it at its Twelfth Session held at Santiago in February 1951.[8] There it encountered strong resistance on the part of the British delegation. The latter proposed the removal of the resolution from the agenda, arguing that it had already been submitted to the Fifth Session in 1947 and that its principal motivation, the need to cope with the world shortage of oil, had outdated it under the new conditions of abundance. To the chagrin of the Alliance's representative, who according to the rules was not allowed to take part in the procedural

[7] *Ibid.*, p. 94. The text of the resolution is on p. 92.

[8] The accompanying memorandum questioned "the principle and practice of the concession system, both from the point of view of international law and the promotion of peaceful relations between countries." See "Study of World Oil Resources," *Cartel*, Jan., 1951, pp. 101 ff.

debate, the Council deferred to the British objection by 8 votes to 4, with 6 delegates abstaining.

Despite the frustration of its efforts in the United Nations the Alliance did not cease to show a lively interest in the world petroleum industry. Its house organ, *Cartel,* frequently carried articles or editorial comments on various aspects of the oil trade, publishing studies on the structure of the Royal Dutch–Shell group and the Iranian oil Consortium as well as devoting space to the report of the United States Federal Trade Commission on international oil cartels.[9]

The attempt to get international control of world petroleum resources did not exhaust the Alliance's endeavors so far as oil was concerned. In a parallel action the Alliance sponsored the creation of an offshoot organization, the International Co-operative Petroleum Association. The decision to establish the association was reached in October 1946, when fundamental rules for it were adopted. The latter provided for an authorized capital of $15,000,000 and the launching of the association when central co-operative petroleum organizations had subscribed $500,000 of capital stock. Subsequently, central organizations in about twenty countries expressed interest in forming the association and pledged initial capital in excess of $1,000,000. The association came into being officially on April 15, 1947. Its headquarters was set up in New York, and its first efforts centered on the distribution of lubricants. Its avowed aims included the acquisition of producing facilities, refineries, and tankers and the gaining of access to Middle Eastern oil reserves. "The I.C.P.A.," stated one of its early reports, "should look forward to acquiring crude oil sources with strategically and economically located refineries and pipe lines to assure its members a plentiful supply of petroleum products at reasonable prices." [10]

In connection with the association's interest in Middle Eastern oil, an interesting episode took place in 1951. In July of that year, at the height of the Iranian oil crisis, Howard A. Cowden, secretary of the ICPA, submitted to President Truman a memorandum proposing that Iranian oil properties be operated by a board consisting of nine members: three nominated by the British, three by

[9] *Cartel,* Oct., 1950; Oct., 1954; and Oct., 1952.
[10] International Co-operative Alliance, *Report of 17th Congress,* p. 200.

INTERNATIONAL CONTROL

Iranians, and three by "world consumer interests," the latter to be represented by his association. Anxious to have his ideas accepted, Mr. Cowden made a trip to Iran during which he discussed them with Premier Mossadegh. He also held talks on this subject with Washington officials but apparently did not secure any commitments except an expression of "interest" in his plan.[11]

Despite its opposition to big business the association in its report of 1954 averred that "a readjustment of many phases of international economy has brought about a resurgence of competition" in the petroleum trade and that great oil corporations were gradually expanding their power "through a system of control and through ability to effect economies by mass production and mass distribution."[12] This seemed like a tacit admission that the association's opportunities were somewhat limited in the face of the superior performance of privately owned corporations.

OIL-SUPPORTED REGIONAL DEVELOPMENT

The shelving by the Social and Economic Council of the proposals to subject world oil resources to United Nations control coincided with the first major political crisis in Middle Eastern oil affairs—the nationalization of the oil industry in Iran. As the crisis deepened, it became increasingly evident that the eruption of Iranian nationalism should be viewed not in the narrow framework of government-company relations but against the broader background of Anglo-Iranian relations and the social and economic conditions prevailing in the Middle East. The latter provided an obvious breeding ground for social conflict, and political relations between Britain and Iran suffered from a residue of imperial notions, ill adapted to the new nationalist era in Asia. With the settlement of the Iranian dispute by the creation of the Consortium, the spotlight turned to an area of almost chronic turbulence, the Arab world. Here the traditional woes of the Middle East, socioeconomic imbalance and the clash of Western influence with native nationalism, were accentu-

---

[11] *Petroleum Times*, Nov. 2, 1951, p. 965.
[12] International Co-operative Alliance, *Report of the Nineteenth Congress at Paris, 6th to 9th September, 1954* (London, n.d.), pp. 252 ff.

ated by a third complication, the Arab-Zionist conflict. Consequently the sociopolitical climate in this zone was potentially and actually even more explosive than that in Iran. Although the Western-owned oil companies had thus far emerged relatively intact from the repeated crises in which the Arabs and the West were involved, only a very short-sighted person could ignore the danger which certain of the companies had faced at times.

The interests of the oil industry were not the only interests that mattered. There was the general interest of the area itself in achieving peace and stability based on solid economic foundations. There was also the interest of the West as a whole to maintain its economic and political relations with the Middle East undisturbed by tensions and conflicts. And there was the newly created state of Israel, whose chances of survival depended on its ability to achieve a *modus vivendi* with the surrounding Arab world. To determine the conditions upon which Arab-Israeli coexistence could be based was one of the most baffling issues of mid-twentieth-century diplomacy. There was a strong belief in many quarters, especially in Israel itself, that the channeling of Arab energies into constructive programs of regional development and reform would free Israel from its unenviable position as a continuous target of enmity. Thus all other interests were reinforced by the desirability of obtaining a permanent Arab-Israeli settlement.

There was nothing specifically new in the plans of many Middle Eastern governments or unofficial groups to launch comprehensive programs of development for their countries. We have already discussed such programs with special reference to Iran, Iraq, and Kuwait. Nor was there any great novelty in the hope of linking these programs with expanding oil revenues. In fact the two were closely connected. What was new was the growing conviction, beginning in the mid-1950's, that the region should be treated as a whole and that its already existing difficulties should not be compounded by making the have countries richer and thus allowing the contrast between them and the have-nots to grow beyond the limits of safety.

There is no point in attempting to list all the plans that have been proposed with regard to regional development of the Middle East. Consequently we will limit ourselves to summary description

INTERNATIONAL CONTROL

of a selected few typical of certain trends of thought or important because of the source from which they emanated. For the sake of clarity suggested plans may be divided into three broad categories. First, some plans specifically mentioned the oil revenues as the source of financing. Secondly, some provided for financing by the Arab states but refrained from assessing the oil revenues directly. Thirdly, others called for outside—mostly Western—contributions but admitted, and sometimes expected, simultaneous financing from Middle Eastern sources. A fourth and special category consisted of plans possessing the elements of some of the first three types but differing from them in motivation, which was to assure a political settlement between Israel and the Arab states.

The forerunner of many plans in the first category was probably a British proposal submitted to the United States in 1945 in connection with the projected Anglo-American oil agreement,[13] of which reportedly it was to be a part. It provided for the establishment of a bank for regional economic development to which oil companies would contribute a proportion of their profits.[14] The failure of the Senate to ratify the treaty buried the proposal, which was not revived until some twelve years later and then under different auspices.

In 1957 two similar proposals were made emanating from different sources and having different motivations. The author of one of them was Emile Bustani, a deputy in the Parliament of Lebanon, a former cabinet member, and a well-known engineering contractor. Bustani first advanced his proposal in an article published in Beirut in 1953,[15] but during his visit to Britain and the United States in 1957 he updated it and made it more specific. Many tensions in the Middle East, according to Bustani, were due to the discrepancy in income between the oil-producing countries and the nonproducing ones. To eliminate these, surplus funds of the producing group should be made available for development purposes to the nonproducing states. This could be achieved by the creation of an Arab oil investment bank. The capital of this bank would be created by

[13] For a summary of the agreement, see p. 170 above.
[14] See Doreen Warriner, *Land and Poverty in the Middle East* (London-New York, 1948), p. 139.
[15] *Al-Kulliyah* (the magazine of the Alumni Association of the American University of Beirut), May, 1953.

annual contributions from the producing countries and the oil companies, each of which would give at least 5 per cent of its profits to the bank. At the current production rate in the Middle East this should amount to about $100,000,000 a year.

The nonproducing countries, Lebanon, Syria, Jordan, and Egypt, could borrow from the bank for development purposes. The rate of interest would be fixed at 3 per cent a year. An additional advantage to be derived from such a scheme would be the real stake that the nonproducing countries—also serving as transit territories—would acquire in the continuity of oil operations. These transit countries could then be induced to sign treaties with the producing states, guaranteeing against sabotage or the shutting off of the pipelines. This might lead to expansion of the pipeline network in the area.[16]

A second plan, in certain respects very like Bustani's, was advanced by a group whose primary motive was not so much the equalization of prosperity between the have and the have-not countries of the Middle East as a desire for peace, mostly in terms of an early Arab-Israel settlement. In 1957 the Public Affairs Institute of Washington, D.C., published a study containing detailed suggestions of how funds should be contributed to a proposed Middle East Regional Resources Development Agency. The latter, according to the study, should be capitalized at $1,000,000,000. To this capital fund the United States should contribute $450,000,000. The remainder was to be assured from three sources in the Middle East in the following proportions: (a) The Arab oil-producing countries (Kuwait, Saudi Arabia, Iraq, Qatar, and Bahrein) would pay 10 per cent of their royalties. The first payment, based on 1956 oil royalties, would amount to $82,000,000. This sum would be renewable annually for five years. (b) The oil companies would also pay 10 per cent of their profits. They were asked to advance $450,000,000, the anticipated profits of five years' earnings. (c) Other Middle Eastern countries were to contribute the remaining $18,000,000 "in order to qualify for membership." In contrast to Bustani's plan, which dealt with the Arab countries only, the Institute plan proposed that the De-

---

[16] *New York Times*, Oct. 23, 1957. See also Emile Bustani, *Doubts and Dynamite in the Middle East* (London, 1958), pp. 139 ff.

INTERNATIONAL CONTROL

velopment Agency should serve the needs of the region as a whole, "including the Arab States and Israel." It devoted considerable space to background political information, and "Nasserism" was described as an evil not to be appeased. It also contained a strong plea for the resettlement of Palestinian Arab refugees, who could be easily absorbed, it claimed, by the Arab states.[17]

It may be in order, perhaps, to ask about the chances of either of these plans being accepted. The plan presented by the Public Affairs Institute would founder on Arab intransigency toward Israel. Arab states would never agree to become partners with Israel in a common economic endeavor no matter what the financial incentive might be. This uncompromising hostility to Zionism has led the Arab states in the past decade to adopt many policies that could hardly be justified on purely economic grounds. The prevailing trends of Arab nationalism infallibly subordinate economic advantages to political considerations.

But even if Israel is excluded from the program, would other non-Arab states participate? The attitude of Iran, the only non-Arab major oil-producing country in the Middle East, would obviously be crucial, and its attitude is far from encouraging. Its production in 1957 reached the pre-Mossadegh level, with a corresponding rise in oil revenues. But Iran has to heal the self-inflicted wounds of the Mossadegh era; its domestic needs must have priority in the allocation of funds. Moreover, as a non-Arab country it has little reason to help the Arabs, with whom it has a few unsettled political accounts such as the conflicting claims to Bahrein, the differences over navigation in the Shatt el-Arab, and its reservations about unilateral Egyptian moves regarding the Suez Canal. It differs from the Arab world in its general policy orientation, which is expressed by its participation in the Baghdad Pact as opposed to prevailing Arab neutralism. For all these reasons Iran is more than reluctant to consider any regional collaboration with the Arab states for the benefit of the latter. This reluctance was given forceful expression when the Iranian government in the spring of 1958 publicly dis-

[17] Public Affairs Institute, *Regional Development for Regional Peace. A New Policy and Program to Counter the Soviet Menace in the Middle East* (Washington, D.C., n.d.).

avowed and recalled its ambassador to the United States, Dr. Ali Amini, for suggesting in a Washington address that Middle Eastern countries should pool their oil incomes for development purposes.[18] Although the reasons for the ambassador's recall were broader than those publicly stated, the latter were believed sufficient to provide an official excuse.

What about the attitude of the Arab producing states? Would not a call for solidarity with less fortunate Arab nations make enough impression to cause them to part with a portion of their revenues? Alas, self-interest and inter-Arab rivalries would be a strong deterrent to any concrete action aiming at sharing wealth. The general attitude of the producing states, though carefully concealed behind verbal assurances of Arab solidarity, is one of reluctance to make any firm and long-range commitments affecting even a small portion of their oil revenues. This is proved by their behavior and can be ascribed to economic as well as political reasons. The ruler of tiny Kuwait with oil revenues well over $300,000,000 a year is known to invest most of the surplus funds in securities in the London stock market. The government of Saudi Arabia is chronically beset with budgetary difficulties and if it makes any major transfers of funds to other Arab countries it usually does so as loans and outright grants for political services and attitudes. Although frequently criticized abroad, this manner of acting is understandable in view of the rivalries plaguing the Arab East and as a countermeasure to the willful and sometimes ruthless actions of the Saudi king's enemies. As for the third major producer, Iraq, its prolonged political isolation from the rest of the Arab world and the apparent results of the revolution of July 1958 make unlikely an offer to subsidize its adversaries, including Nasser's Egypt.

In spite of the enmities and jealousies in the Arab camp strong unifying forces are at work. These forces have led the Arab League to plan and sponsor many projects, political, cultural, and economic, which link the Arab states together through institutional arrangements. Such a project was adopted by the Arab League in the early

[18] The ambassador's speech may be found in the supplement to the *Iran Review*, Jan.–Feb., 1958. For highlights of his proposals, see also the *New York Times*, Feb. 1, 1958.

INTERNATIONAL CONTROL

summer of 1957 when it decided to establish an Arab Financial Institution for Economic Development.

On June 4, 1957, all nine members of the Arab League [19] signed an agreement containing the detailed rules of the proposed institution.[20] Its membership was to be open to members of the Arab League as well as to "any Arab state or other Arab country" which was admitted by the board of governors of the institution. This provision left the door open for the admission of such semidependent territories as Kuwait, which as a nonsovereign entity was prevented from joining the League. The initial capital of the institution was to be £E20,000,000 ($56,000,000). The institution was to begin operations with the subscription of 75 per cent of this amount. The founding members were to subscribe in the ratio of their contributions to the Arab League budget. The percentages would be: United Arab Republic, 50.29; Iraq, 15.98; Saudi Arabia, 14.57; Sudan, 6; Lebanon, 5.64; Jordan, 2.82; Yemen, 2.82; and Libya, 1.88. This was a compromise between the reluctance of the producing countries to participate in any common development endeavor and the desire of the have-nots to share in the former's bounty. Egypt, a have-not with respect to oil, was to be the largest contributor, thus to some extent defeating the whole idea of sensible sharing. But at least a beginning was made. Henceforth the members of the Arab League were enjoined to think in terms of economic collaboration. With political strife rampant in the Arab world in 1957–1959, they still had a long way to go to implement their agreement. Despite its moderate assessment Iraq refused to join and even went so far as to declare its inability to pay its regular installments to the League's budget. This was due partly to the hostility between Iraq and Egypt, which the 1958 Iraqi revolution did not abate, and partly to the substantial losses which Iraq suffered as a result of the sabotage of IPC's pipelines in Syria following the war over Suez. Lebanon was also reluctant to join, largely because of the worsening of its rela-

[19] Jordan, Sudan, Syria, Iraq, Saudi Arabia, Lebanon, Libya, Egypt, and Yemen (in the Arabic alphabetic order in which they were listed in the agreement).

[20] *Ittifaqiyat bi-Insha al-Muasasat al-Maliyah al-Arabiyah lil-Inma al-Iqtisadi* (Agreement for the Establishment of the Arab Financial Institution for Economic Development), mimeographed text by courtesy of the Economic Committee of the League of Arab States.

183

tions with Egypt during the same period. Kuwait, an obvious candidate among the outsiders, remained noncommittal. Despite these difficulties, a number of original signatories gradually pledged their subscriptions. Libya's signature on January 12, 1959, brought the total subscription above the required $42,000,000, and on the same day the Arab League Economic Council, then holding its fifth session in Cairo, proclaimed the establishment of the institution. The latter, commonly referred to as the Arab Development Bank, was promptly (on January 16) assured of co-operation in the form of technical advice and service by Eugene R. Black, president of the International Bank for Reconstruction and Development.

The mounting crisis in the Middle East from 1957 onward acted as a stimulant to more intensive thinking in the West about the right solution to the area's problems. Many statesmen thought that the West should take the initiative before the situation deteriorated beyond repair. On December 6, 1957, Italian Foreign Minister Giuseppe Pella submitted to Secretary of State John Foster Dulles a proposal that the United States and Western Europe create a financial pool to underwrite a program for economic development of the Middle East. Under this proposal 20 per cent of European repayments on Marshall Plan loans, scheduled to begin in 1958, would be transferred to a Middle Eastern fund to be set up by the Organization for European Economic Cooperation (OEEC), and European countries without Marshall Plan aid would contribute proportionately.[21] The State Department's reaction was negative. It feared adverse criticism in the Arab East, where such a fund would be viewed as an extension of NATO and decried as a new manifestation of imperialism. It also objected to putting the main burden of financing on the United States while depriving it, as a nonmember of the OEEC, of a voice in supervision. Furthermore, under the terms of the Marshall Plan repayments were to be made in European currencies, thus favoring fund purchases in European markets.

A modified version of the plan was presented in mid-summer 1958 by Italy's new premier, Amintore Fanfani, during a visit to Washington. The Fanfani plan proposed that the fund should be estab-

[21] For Pella's plan, see the *New York Times*, Dec. 7, 1957. See also *Oriente Moderno*, Jan. 1958, p. 31.

INTERNATIONAL CONTROL

lished under United Nations auspices and that the contributions of the OEEC members should be larger than those originally suggested.[22] It is difficult to say what effect these Italian initiatives had on subsequent developments, inasmuch as when the Fanfani plan was presented the White House was preparing a presidential declaration on the Middle East. This was in connection with the momentous events taking place in Lebanon, Jordan, and Iraq in July 1958, the landing of the American Marines in Beirut, and subsequent calls of Soviet Premier Khrushchev for a summit meeting. On August 13 President Eisenhower personally presented to the United Nations General Assembly his own six-point program for peace in the Middle East. This included a proposal for the establishment of an Arab development institution on a regional basis. The relevant passages of the President's speech were as follows:

> To help the Arab countries fulfill their aspirations, here is what I propose:
> First, that consultations be immediately undertaken by the Secretary General with the Arab nations of the Near East to ascertain whether an agreement can be reached to establish an Arab development institution on a regional basis.
> Second, that these consultations consider the composition and the possible functions of a regional Arab development institution, whose task would be to accelerate progress in such fields as industry, agriculture, water supply, health and education, among others.
> Third, other nations and private organizations which might be prepared to support this institution should also be consulted at an appropriate time.
> Should the Arab states agree on the usefulness of such a soundly organized regional institution, and *should they be prepared to support it with their own resources,* the United States would also be prepared to support it.
> The institution would be set up to provide loans to the Arab states as well as the technical assistance required in the formulation of development projects.
> The institution should be governed by the Arab states themselves.
> This proposal for a regional Arab development institution can, I believe, be realized on a basis which would attract international capital, both public and private.
> I also believe that the best and quickest way to achieve the most desirable

[22] *New York Times,* Aug. 1, 1958; also *Oriente Moderno,* Aug., 1958.

result would be for the Secretary General to make two parallel approaches. First, to consult with the Arab states of the Near East to determine an area of agreement. Then, to invite the International Bank for Reconstruction and Development, which has vast experience in this field, to make available its facilities for the planning of the organization and operational techniques needed to establish the institution on its progressive course.

I hope it is clear that I am not suggesting a position of leadership for my own country in the work of creating such an institution. If this institution is to be a success, the function of leadership must belong to the Arab states themselves.[23]

If subjected to close scrutiny, President Eisenhower's proposal reveals certain new and interesting features. First, it recognizes Arab nationalism, its aspirations and sensibilities. In contrast to many earlier official pronouncements in which references to the Middle East were usually phrased in the formula "Israel and the Arab states," this statement was silent as to Israel. The proposed development institution was to be for the Arab countries only, thus excluding not only Israel but also other non-Arab states of the Middle East. The statement advocated Arab leadership of the institution and called for collaboration with such international bodies as the United Nations and the World Bank, thus emphasizing United States disinterestedness. Furthermore, although the President offered American financial aid, he made it conditional upon the willingness of the Arab states to support the institution with their own resources. The latter provision implied that there was no lack of financial resources, at least in some Arab states, and that they should be utilized for constructive purposes. This could be interpreted as a hint that better use could and should be made of the oil revenues on a regional basis. No direct reference was made to the oil companies, and the only suggestions that they might play a role in the proposed program occurred in the passages that spoke of consultations with private organizations and the attraction of private capital.

In some ways the President's statement represented a new departure in official American thinking about the Middle East. This was mainly in its recognition of the exclusivity of the Arab com-

[23] *New York Times*, Aug. 14, 1958. Italics mine.

munity and the admission of the need to consult it before any program was adopted.

Although the President's program for peace, of which the economic development proposals were only a part, was not formally accepted by the General Assembly, some of its fundamental points found their way into the resolution finally adopted, which was sponsored by the Arab states themselves. In it the General Assembly invited the Secretary General "to consult as appropriate with the Arab countries of the Near East with a view to possible assistance regarding an Arab development institution designed to further economic growth in these countries" and requested "member states to cooperate fully in carrying out this resolution." [24]

The Assembly resolution had an effect in Arab countries despite a generally negative attitude to the President's peace proposals as a whole. In fact, barely a week later Abdul Khalek Hassouna, secretary general of the Arab League, declared that the League would reconsider its initial plan to set up a development institution in the light of the newly adopted resolution. He also took this opportunity to express the hope that Iraq would now join the institution and that Kuwait might do so as well.[25]

Although the Eisenhower declaration and the later Arab resolution did not mention oil profits, it is fairly certain that, once a regional development institution begins functioning, oil revenues will be prominent in the considerations of all interested parties.

PAN-ARAB OIL ACTIVITIES

At an early date the Arab League developed an interest in oil affairs of both a negative and a positive character. The negative interest was the idea that denial of oil to unfriendly states could be used as a weapon in the Arab political struggle. An attempt at implementation of this idea occurred at the time of the Palestinian

[24] *Ibid.*, Aug. 19, 1958. It should be pointed out that the creation of an Arab Regional Development Institution had also been proposed by Secretary General Dag Hammarskjöld in his statement to the emergency special session of the Assembly on Aug. 8, 1958.
[25] *Ibid.*, Aug. 26, 1958.

crisis of 1947–1948. Some of the events in this connection have already been related, but a brief recapitulation may be in order at this juncture.

In anticipation of hostilities in Palestine and foreseeing Western support of the Zionist cause, the Arab League at its Bludan meeting in June 1946 passed a set of resolutions, one of which called for denial of Arabian oil to the West. This decision was not implemented when actual hostilities took place in Palestine in 1948, despite urgings by Iraqi leaders. This was due largely to opposition by Saudi Arabia, which believed that a commercial oil operation should be divorced from political considerations. Notwithstanding this failure, the interests of the oil companies were affected by the Arab-Jewish conflict in three ways: (*a*) in an outburst of resentment against the American policy Syria delayed ratification of the Tapline convention about twenty months; (*b*) in a gesture of defiance toward Israel and out of solidarity with other Arab states Iraq stopped the movement of oil by pipeline to the Israeli-held Haifa terminal and caused construction of the parallel line between Kirkuk and Haifa to cease; and (*c*) boycott measures against Israel by the Arab League gradually affected the transactions of a number of oil companies with Israel. The most tangible example of this kind was supplied by the virtual ultimatum that the Arab League Economic Council early in 1956 addressed to Shell and Socony to desist from further activities in Israel. As a result of this pressure two British companies, Shell and British Petroleum, announced on July 23, 1957, that they had decided to dispose of their marketing interests in Israel. The official reason was that in the last few years their operations in Israel had "not been commercially attractive." Their decision was to affect their marketing operations. The holdings of both companies in the Haifa refinery (then operating at 30 per cent of capacity) were to be continued, at least until further notice.[26]

On a more positive side was the Arab League's realization of the importance of oil for Arab economics and concern that this natural resource had, according to their view, served primarily the interests of foreign companies and foreign countries. More particularly the

[26] "Two British Companies to Sell Oil Marketing Agencies in Israel," *ibid.*, July 24, 1957.

INTERNATIONAL CONTROL

Arab League complained that its member states had no voice in deciding the level of production and prices and in determining the direction of exports. It thought that the existing concessions allowed too great latitude to the companies and prevented adequate supervision of their activities by the host governments.[27] These considerations led the League to make some fundamental decisions in regard to oil matters of both an organizational and a political nature. Oil affairs were the subject of deliberation in three of the principal organs of the League: the Council of the League, the Political Committee, and the High Economic Council, the latter composed of the Ministers of Finance or of National Economy of the member states. At its meeting of August–September 1951 the Political Committee decided to establish an Oil Experts Committee, which was promptly set up. It became a central agency of the League for the formulation of oil policies. From 1953 onward the committee, as well as the Arab League as a whole, paid considerable attention to oil problems. In the course of that year the committee presented three recommendations, which were subsequently adopted by the Arab League Council at its twentieth regular session held in January 1954 in Cairo. As amended by the Political Committee the recommendations made "with a view to formulating a general Arab oil policy" were as follows:

(a) a permanent Petroleum Office in the Arab League secretariat should be established; (b) member states should provide for the exchange of statistics and information concerning oil among themselves pending the establishment of the Office; and (c) measures should be taken to construct new refineries in Arab countries, to enlarge the existing ones, and to set up national Arab companies for the distribution of oil products.

In conformity with these recommendations the Permanent Petroleum Office was set up by the secretary general of the League in 1955. An Iraqi engineer, Mohammed Salman, became its first director. From the outset he endeavored to awaken an "oil con-

[27] "Shuun al-Petrol" (Petroleum Affairs), a chapter in Jamiat ad-Duwwal al-Arabiyah, al-Amanat al-Ammah, *Jami'at ad-Duwwal al-Arabiyah wa al-Qadaya allati 'Alajatha, 1945–1957* (League of Arab States, Secretariat-General, *League of Arab States and the Matters It Has Handled, 1945–1957*), prepared by Taufiq Ahmed al-Makri and Ibrahim Shakrallah (Cairo, 1957), pp. 375 ff.

sciousness" among the Arab states, to gather statistical data and information, and to draft the principles of a general Arab oil policy. The creation of the Permanent Office marked the beginning of steady and systematic interest on the part of the Arab League in petroleum affairs. Before long Salman submitted to the Oil Experts Committee his "Plan for the Co-ordination of Petroleum Policy of Arab States." At its fourth meeting, in November 1955, the committee discussed this plan and approved most of its ideas. Subsequently it adopted its own set of eighteen recommendations, which were forwarded to the third session of the High Economic Council, which met in Cairo on January 21–26, 1956. Of the forty-five resolutions passed by the council, nineteen concerned oil and were based for the most part on the recommendations submitted by the committee. Among the most important resolutions were the following:

1. A more effective economic boycott of Israel should be assured:
   (a) By inducing Arab emirates that are not members of the League as well as companies operating in them not to supply oil to Israel.
   (b) By issuing a warning to Shell and Socony to desist from further activities in Israel within a certain time limit.
   (c) By exerting pressure on Italy to reconsider its trade agreement with Israel whereby the latter receives $1,500,000 worth of oil from Italy, most of it presumably of Arab origin.
2. Oil-producing states should help the transit states to obtain a fair share of the royalties for the oil transported through their territories.
3. Future concessions should contain a stipulation forbidding foreign governments to own stock in the concessionary companies.
4. "Arabization" of the oil companies should be secured by provisions forcing the companies to employ Arab labor and personnel and by assuring the training of Arab technical cadres by foreign experts.
5. Arab states not members of the League should be invited to attend the meetings of the Oil Experts Committee as observers.[28]

[28] *Ibid.*, pp. 376–377. According to this source, Kuwait, Bahrein, and Qatar subsequently consented to send delegates to the meetings.

INTERNATIONAL CONTROL

6. Arab oil-producing states should follow the example of Iraq in approaching the Iranian government with a view to reaching an agreement on a unified oil policy with regard to companies operating in both Iran and the Arab states.
7. A most-favored-nation clause should be included in future concession agreements with regard to housing facilities provided by companies for their employees.

It was significant that the Economic Council decided to pass over one important recommendation of the committee. This was one asking the secretary general to draw the attention of member states to the great benefit accruing to Israel from being supplied with a major part of its petroleum requirements from the Soviet Union in exchange for citrus fruit, so that the Arab states might take whatever action they deemed appropriate in this regard.

Between its formation in 1951 and the fall of 1957 the Oil Experts Committee held six meetings. Of greatest importance was undoubtedly its fifth meeting held at Cairo, April 15–25, 1957, during which sixteen recommendations concerning basic oil policy for the Arab states were adopted. These recommendations were subsequently submitted, through the Third Committee on Boycott, Oil, and Communications to the High Economic Council, which discussed them at its fourth regular session in Cairo, May 25–June 3, 1957. The Council eventually adopted fourteen resolutions on oil. Two of these were based on earlier recommendations of the Oil Experts Committee: one enjoining the Arab governments not to grant concessions to companies whose stock was owned, even in part, by foreign governments, the other resolving to send a final ultimatum to the Shell Company to cease its dealings with Israel under penalty of boycott in the Arab states. The other twelve resolutions reflected the sixteen 1957 recommendations of the Oil Experts Committee, some of which were repeated *verbatim* and others amended and consolidated. These resolutions are summed up below in the order in which they appeared in the original Arabic text:

1. In connection with the investigation of the charge that some oil tankers sailing between Sidon and Italy are unloading their oil in Israel, Arab states are urged to ascertain in each case the destina-

tion of the tanker and to make sure that oil is unloaded according to the logbook.

2. An Arab oil congress should be held in Cairo not later than February 1958.

3. An exchange program for Arab oil experts should be arranged with a view to acquainting them with various phases of oil operations.

4. Arab states should consult each other in order to develop a unified policy with regard to refineries and their production.

5. Arab states should take the necessary steps to construct refineries in a central area of the Arab world and strive toward the production of aviation gasoline.

6. An Arab tanker company should be established to carry oil to foreign outlets as well as to Arab states.

7. Arab governments are urged to oppose the creation of any international organization whose purpose would be to control the production and marketing of Arab oil.

8. In connection with the proposal of the Saudi Arabian delegate that an Arab pipeline company should be set up to carry oil from the sources of production to the port terminals through Arab lands and in view of the great importance of the matter, the Council recommmends that the subject be referred back to the Oil Experts Committee for further study, with a request that it be resubmitted to the next meeting of the Council.

9. Arab governments are enjoined to furnish the Permanent Petroleum Office with all necessary information concerning oil in their respective countries in order that the data may be published periodically.

10. In view of the danger that the proposed Israeli pipeline from the Gulf of Aqaba to Haifa presents to Arab national economy in general and the Suez Canal in particular, Arab governments are urged to exert their influence to persuade the potential users of such a line, notably Kuwait, Bahrein, Qatar, and Iran, to refrain from using it.

11. Arab states are urged to apply any grants given them by the oil companies to the promotion of higher education in oil affairs.

INTERNATIONAL CONTROL

12. Arab states are enjoined to put into effect the recommendations adopted by the preceding, third, session of the Economic Council.[29]

Resolution no. 2 calling for the convocation of an Arab oil congress contained a number of subsidiary points. All Arab states, i.e., even nonmembers of the League, were eligible to attend. The director of the Permanent Petroleum Office was instructed to contact some of the petroleum companies to enlist the aid of their experts in the preparation and delivery of certain lectures. Necessary steps were to be taken to invite non-Arab oil-producing countries (Iran and Venezuela were specifically mentioned) whose conditions were similar to those of the Arab states to attend the congress. An "oil brain trust" to discuss and review on the Cairo radio various aspects of the oil industry in the Arab world was to be formed. Adequate press and radio publicity was to be assured, and films dealing with the oil industry were to be shown in theaters in Arab countries during sessions of the congress. It was also decided to organize an exhibition in which the progress of the oil industry in each Arab country would be demonstrated. The agenda of the congress was to include the following subjects: oil legislation, prospecting, drilling, exploitation, transport, refining, and marketing; petrochemical industries; world and social conditions relating to the oil industry; the economics of oil; and the part played by oil in the economic development of Arab countries. To implement these instructions a committee of three was appointed.[30]

Of particular importance was the resolution calling for a further study of the proposal to establish an all-Arab pipeline to be owned by Arab states. The proposal originated, as the resolution pointed out, with the Saudi delegate. The latter had earlier submitted a

[29] The recommendations may be found in Jami'at ad-Duwwal al-Arabiyah, al-Amanat al-'Ammah, al-Idarat al-Iqtisadiyah, *Qararat al-Majlis al-Iqtisadi fi daur Inaqadihi al-'Adi al-Rabi'*, mlf 7/9/25 (League of Arab States, Secretariat-General, Economic Administration, *Decisions of the Economic Council during Its Fourth Ordinary Session;* Document mlf 7/9/25; Cairo, May 25–June 3, 1957), pp. 3–7.

[30] Composed of Mohammed Salman, director of the Permanent Petroleum Office, Dr. Mahmoud Abu Zeid of Egypt, and Zuhair Sabri of Syria.

memorandum in which he pointed to the profits realized by the oil companies through the operation of their own pipelines. Tapline was given as an example. According to the memorandum, the posted price of crude in the Persian Gulf was $1.97 a barrel, whereas the Sidon posted price of the same oil pumped through the pipeline was, until the closure of the Suez Canal in 1956, $2.46. From the 49-cent differential 21 cents should be deducted as the cost of operating the pipeline. The other 28 cents represented a profit. After the blocking of the Suez Canal the posted price at the Sidon terminal was increased to $2.69 a barrel, thus adding 23 cents a barrel and increasing the profit to 51 cents a barrel. This, according to the memorandum, was a substantial profit, in which the transit countries had shared only to a limited degree by receiving fixed transit fees. Since oil production was bound to increase and new transportation facilities would have to be provided, the Arab countries should aim at building their own system of pipelines by a joint effort.

Financing for such a venture could be obtained in various ways. One would be through an official agency, such as the United Nations or the World Bank, or by a private banking group. The pipeline would be owned jointly by the producing and transit countries in a proportion representing a combination of the percentage of crude oil produced and the length of the pipeline in the various participating countries. Another alternative would be to open the stock to private subscription by citizens and residents of the Arab countries. A third alternative would be to invite the oil companies already operating to participate with the Arab countries in the financing of the project, which, when completed, would serve the oil industry on a common carrier basis. A point in favor of this plan, added the memorandum, would be the desire of the participating Arab countries to assure a steady flow of oil through the projected pipeline, inasmuch as no Arab country would want to injure its own interests by cutting off the line. A collective agreement to protect the line could be concluded to strengthen this material guarantee.[31]

[31] The memorandum in question was drafted by the Petroleum Adviser to the Saudi government, James McPherson. It was submitted to the Arab Oil Experts Committee by Sheikh Abdullah Tariki, Director General of Petroleum and Mineral Affairs of Saudi Arabia. Though never officially released, it became known, first in broad outline and later in considerable detail, after the meeting

INTERNATIONAL CONTROL

Some of the council's recommendations were discussed further at the sixth meeting of the Oil Experts Committee which took place in Baghdad on November 2–7, 1957. The meeting was attended by the delegates of all member states of the League with the exception of the Sudan and Libya and by observers from Kuwait, Bahrein, and Qatar. It had two principal objectives: to make arrangements for the proposed Arab oil congress and to study certain subjects referred to it by the fourth session of the Economic Council. Up to this time (April 1959) the Arab League secretariat has not released any official documents pertaining to the meeting. According to various press and radio reports, its agenda included: (*a*) preparations for the Arab oil congress; (*b*) discussion of the Saudi plan for the creation of a Pan-Arab pipeline; (*c*) discussion of the council's recommendation to create an Arab tanker fleet; (*d*) discussion of an agreement between the producing and the transit countries aimed at guaranteeing the pipelines; and (*e*) tightening the boycott of Israel by preventing oil shipments from reaching the Haifa refinery.

With reference to the Pan-Arab pipeline plan, it was reported that the original Saudi proposal was modified by the inclusion of Iran as a potential participant and user of the projected line. The plan encountered opposition on the part of Iraq, whose delegate declared that his government would not be able, in the foreseeable future, to divert the large funds required for the financing of such a project. It was also reported that differences of opinion arose between the producing and the transit states with regard to the proposed guarantee of the safety of the pipelines, the transit states expecting a more tangible *quid pro quo* in terms of a larger share of the oil revenues. The committee was reported to have reached agreement eventually on three main points, namely, the arrangements for the Arab oil congress, a unified policy by producing countries toward oil companies, and the strengthening of the economic blockade of Israel by administrative and diplomatic means. No agreement was reached on the Saudi plan, and the problem of guaranteeing the pipelines was deferred for further study. No information was published concerning the committee's action on the Arab tanker fleet

---

of the League's Economic Council in May–June 1957. By the very nature of its proposals it was not intended to remain confidential long.

proposal.[32] The committee's recommendations were approved by the League's Economic Council at its fifth session held in Cairo on January 5–14, 1959.[33]

The Arab Oil Congress

Following several postponements, the projected Arab oil congress convened in Cairo on April 15–23, 1959, attended by about 420 delegates. Official members of the Congress represented the Arab League states with the exception of Iraq, which refused to participate owing, no doubt, to its current difficulties with the United Arab Republic. Unofficial members included observer delegations from Iran and Venezuela, from the oil-producing Persian Gulf principalities, as well as from various oil companies. The agenda covered legislative, economic, and technical matters. Of the many papers which were read and discussed, a few dealt with controversial issues, stirring up considerable interest among those in attendance. Frank Hendryx, an American legal adviser to the Saudi petroleum department, asserted that a government can unilaterally modify a concession agreement inasmuch as it "cannot by contract hamper its freedom of action in matters which concern the welfare of the state." It was not clear whether his remarks did or did not have prior approval from the chief Saudi delegate, Sheikh Abdullah Tariki. His views were challenged by certain Arab delegates during the ensuing discussion. They contrasted, moreover, with the statement made shortly afterward by Anis Qasem of the Libyan Petroleum Commission that "Libya believes in the sanctity of freely negotiated contracts and will honor them." [34]

Similarly controversial was the paper read by Emile Bustani, head of the Lebanese delegation, entitled "Sharing the Oil Wealth," in which he reiterated his plea, voiced earlier on several occasions,[35] that the producing countries and the companies should each lend 5 per cent of their annual profits to the Arab development bank at 2.5 per cent interest. He claimed to have the endorsement of the

[32] Information compiled from *Oriente Moderno*, Nov., 1957, p. 725; *Cahiers de l'Orient Contemporain*, XXXV (1957), 18; *Al-Hayat* (Beirut), Nov. 8, 1957; and *Al-Hurriyah* (Cairo), Nov. 8, 1957.

[33] Al-Jami'at ad-Duwwal al-Arabiyah, *Nashrah*, Jan. 14, 1959 (League of Arab States, *Bulletin*, Jan. 14, 1959).

[34] *New York Times*, April 20, 1959.   [35] See p. 179 above.

ruler of Kuwait for this proposal. The delegations of the oil-producing countries believed his proposal ill-timed, coming as it did shortly after the world-wide reduction of oil prices, which adversely affected oil-based revenues.

The congress was conducted in an atmosphere of dignity and soberness. Except for Hendryx' paper, the delegates' speeches were free of extremism. Those present obviously tried to avoid political overtones and the controlled press of the United Arab Republic handled the meetings with remarkable restraint and faithfulness to facts. It was understood that this was due to definite instructions from the Cairo government, which at this juncture was anxious to avoid excitement likely to create an unfavorable impression abroad. Apart from the official proceedings, a good deal of behind-the-scenes activity went on. It was reported that the Venezuelan delegation tried hard to convince the Arab delegations of the need for co-ordination of price and output policies among the world's principal producing countries. However, any attempt, formal or informal, to create a common front in this respect was bound to suffer a setback owing to the absence of Iraq. In fact, Saudi Arabia was the only major producing country among the official delegations, inasmuch as Kuwait as a nonmember of the Arab League and Iran as a non-Arab country were represented by observers only.

General moderation was also characteristic of the final twelve resolutions which partly conformed to the earlier recommendations of the High Economic Council and partly reflected new ideas. Foremost among the resolutions were: (*a*) a call for the "improvement of participation of oil producing countries [in the profits] on an equitable and reasonable basis"; (*b*) an expression of the desirability of integrating the petroleum industry; (*c*) a recommendation to set up national oil companies to operate side by side with existing private companies; (*d*) a call for consultation among governments with a view to co-ordinating and harmonizing their actions toward the steady and regular flow of oil to world markets (and, implicitly, the maintenance of stable oil prices); (*e*) approval, in general terms, of the Arab pipeline plan proposed by Abdullah Tariki of Saudi Arabia; (*f*) referral to "competent authorities in the Arab League" of Bustani's proposal for oil-based financing of the Arab development bank; and (g) a recommendation that the companies should consult with the host governments before making price changes. The remaining resolutions called for the setting up of machinery to co-ordinate, at the national level, the conservation, production, and export of petroleum; an increase in the refining capacity in the oil-producing countries; the setting up of petrochemical industries; the creation of institutes to develop national technicians; and the holding of Arab oil congresses annually.[36]

SUMMARY AND CONCLUSION

The contrast between the oil haves and have-nots has been conducive to international jealousies. The inequality has been accentuated by the fact that the Middle Eastern oil industry is controlled by a few powerful companies. This has led to complaints of monopoly in the dual sense of control by a few privileged states and by a restricted number of corporations. The general economic underdevelopment of the Middle East and the resulting political instability have stimulated much thinking about the proper utilization of the area's oil revenues to assure an orderly regional development. Both motives—jealousy and a desire for the socioeconomic improvement of the region as a whole—have led to many plans and proposals aiming at various forms of supranational control of Middle Eastern oil resources or pooling of oil revenues. These have ranged from draft agreements providing for close co-operation among the major outside powers, such as Britain and the United States, through plans aiming at the recognition of consumers' interests, exemplified by numerous memoranda of the International Co-operative Alliance, to various projects to establish regional development institutions.

While the outside world was busy planning and crusading, the Arab states themselves were attempting to formulate a unified oil policy. The Arab League served as a focal agency in this respect. Although unanimous on one important point—denying oil to Israel—the Arab states were divided on a number of issues, stemming

[36] Information compiled from a variety of sources including the Arab League *Bulletin*, April 24, 1959; *Al-Ahram* (Cairo), April 24, 1959; *Al-Jamahir* (Damascus) and *Al-Wahdah* (Damascus), both of April 24, 1959; and personal communications from participants in the congress.

largely from the dichotomy between the producing and the transit countries. Certain general schemes, such as the creation of an Arab tanker fleet or an Arab distributing company, seemed to find universal approval among the members of the League, but they were destined to remain in the realm of planning so long as adequate capital was unavailable for their financing.

Arab states are not motivated by "internationalist" considerations. Their own national interest—more and more interpreted in a Pan-Arab sense—dominates their thinking. Thus they are unlikely to give much comfort to such organizations as the International Co-operative Alliance. There is a certain paradox in Arab attitudes. Many Arabs are critical of the domination of world markets and transportation by a few major oil companies. And yet, if we assume that such a condition is contributing to the maintenance of stable world prices for oil, it is the Arab states which will gain by their close association with these companies.

As for the companies themselves, they can no longer hope either for anonymity in their actions or for an indefinite restriction of their relations with the host countries to an individual company-government relationship. Too many eyes are focused on their operations to permit them to remain oblivious to the numerous plans and projects involving them and their contributions to the welfare of the world or of the Middle East. And they cannot hope for long to remain immune to various manifestations of a unified Arab policy.

/ PART THREE

COMPANIES AND THE PUBLIC

IN THE HOST COUNTRIES

◈◈ CHAPTER XI ◈◈

Typical Attitudes: Nationalist Themes

IN Part Two we dealt with one aspect of the companies' position in the host countries, the relations between the companies and the governments. Now we propose to turn to another aspect, that of relations between the companies and the public. As mentioned earlier (see p. 4), it is virtually impossible to ignore the public. Under certain conditions it may be played down, but it is always there. It infuses the host country–company relationship with a truly political content, which, though not lacking in relations on the official level, is somewhat overshadowed by the formal legalistic considerations. Our discussion must therefore embrace a wide range of public attitudes and actions related directly or indirectly to the operations of the oil companies in the Middle East. First we will review the content of public opinion as manifested by its most typical expressions. We will follow this by a review of the ways and means whereby these attitudes are expressed. And finally we will try to show how the companies react to these manifestations of public opinion, taking account of both the contents of this reaction and the methods employed.

The importance of public opinion as a major factor in the shaping of relations between the company and the host country depends on the political development and stability of the latter. In countries possessing an absolute or dictatorial form of government public opinion is less important than it is in those with more developed constitutional forms and democratic practices. In the patriarchal states of the Arabian Peninsula the official relations between the company and the host government are undoubtedly paramount. They are closely followed, in order of importance, by the companies' industrial relations, which invariably take precedence over public relations. In countries which have reached a higher stage of development, public opinion may claim equality with the official relationship on the government level or even take precedence. The latter situation seems to have existed in Iran in the winter of 1950–1951, when the crusading activity of a small but articulate group of deputies led by Dr. Mossadegh virtually overshadowed the official government proceedings.

As for the relative importance of public relations versus industrial relations in the more developed countries, much depends on the character of the political regime in the country as well as on the kind of operation being conducted. In Lebanon, where the oil companies are restricted to transit and refining and where the personnel employed is small in numbers, industrial relations have generally followed a steady course, relatively free of crises and tensions. Public relations have had to be emphasized there in view of the prevailing political sophistication and the multiplicity of ways in which public opinion could be expressed. In Syria the situation has somewhat resembled that in Lebanon, but it has also differed in several respects. The Syrian public, especially in major towns, is politically awakened, and whenever it enjoys political freedom it finds many means to express its views and feelings. During periods of political tolerance Damascus and other urban centers have known a proliferation of newspapers and periodicals comparable to that in neighboring Lebanon. Given free rein, these organs of public opinion have occasionally gone to extremes of excitability, surpassing similar manifestations in Lebanon. But such periods have alternated with periods of dictatorship when the press was muzzled and the number of newspapers was restricted. In addition, Syria has known specific "twilight situations" when officially the press was free but unofficially it could emphasize only certain themes as a result of intimidation practiced by certain political parties in alliance with the military. As for priority of public relations over industrial relations, it doubtless existed both in the periods of complete political freedom and in those of "twilight," but it was never so pronounced as in Lebanon, inasmuch as industrial relations have not had an equally smooth sailing in Syria. In fact, from time to time industrial conflict has provided the main source of public excitement.

For the purpose of describing typical public attitudes it may be useful to distinguish between the phase prior to the granting of the concession and the phase following it.

The process of negotiating and granting the original concessions in the four major producing countries, Iran, Iraq, Saudi Arabia, and Kuwait, was rather uneventful so far as the role of public opinion was concerned. In Iran the concession was granted in the early years of the twentieth century, under the Qajar dynasty, before the introduction of the constitutional system. Negotiations were conducted in secret among a few men, and because of the novelty of petroleum as a useful commodity and the uncertainty of its existence in Iran's subsoil the matter could hardly lend itself to public debate or controversy, assuming that the latter were feasible under the existing political conditions. The original 1901 concession agreement was not published officially either in Iran or in Britain. It made its first public appearance in the *Official Journal* of the League of Nations in December 1932 at the time of the British-Iranian dispute over the unilateral repudiation of the concession by Reza Shah.

The circumstances under which the new concession of 1933 was granted were obviously different. By that time oil had been found and extracted, and the Iranians had had ample opportunity to learn about its value. The new document was signed after a prolonged dispute and after unilateral actions on the part of Iran and Britain, the latter employing naval demonstrations in the Persian Gulf. This time public opinion did play a part; yet it was controlled if not actually manufactured by the government. Consequently, although the few Iranian papers then in existence expressed themselves in terms which suited the government at the time, one can hardly speak of public opinion as a truly independent factor in the situation. But, however restricted, it was in existence. Such was undoubtedly the view of the Communist Party of Iran, which tried to exploit the Anglo-Iranian dispute not only to launch a campaign against the Anglo-Iranian Oil Company but to discredit Reza Shah.[1] The government abandoned its earlier secrecy and in May 1933 published the new concession in the Iranian *Official Gazette* and in the annual collection of "Acts and Regulations for the Year 1312" as Decree 2395 of the Ministry of Finance.

There was similarly little opportunity for the Iraqi public to air its views in the negotiating phase of the original concession and of its first major revision in 1931. Fundamental decisions such as the division of ownership among British, Dutch, and French interests and the subsequent admission of Americans, were made outside Iraq, without the participation of its government and with the Iraqi public unaware of the issues at stake. Moreover, the concession agreement was concluded at a time when Iraq was under British mandate and when it was possible to negotiate and sign with a minimum of publicity. Nevertheless no attempt was made to conceal the text of the revised agreement, which was duly published in the *Official Gazette* in its Arabic and English versions.[2]

The public had little or nothing to do with the original concession in Saudi Arabia. Both the primitive state of the country in the early 1930's and the absolutist character of its government determined the character of negotiations. The concession agreement was published in the official journal,[3] but its contents have never been given much publicity and according to a later Aramco statement, the Saudi Arabian government was opposed to its release to the public.[4]

In Kuwait also public opinion had no part in negotiations and signing. The concession agreement has never been published, remaining to this day a confidential document.

As for the pipeline conventions concluded in the interwar period between Iraq Petroleum Company and Syria, Lebanon, Transjordan, and Palestine, public opinion seems not to have influenced the

---

[1] L. Magyar, "The Fight for Persian Oil and the Tasks of the Communist Party of Iran," *International Press Correspondence*, XII (1932), 1239 ff.

[2] The Arabic version was published in *Al-Waqai' al-'Iraqiyah*, no. 982 of May 19, 1931, and no. 1135 of May 28, 1932, and the English version in the *Official Gazette*, no. 21a of May 21, 1931, and no. 27 of July 3, 1932.

[3] *Umm al-Qura*, July 14, 1933.

[4] *New York Times*, July 8, 1947.

proceedings. The conventions were signed in 1931 at a time when all four of these countries were under the mandatory administrations. Consequently, although the conventions were negotiated with the native governments of Lebanon, Syria, and Transjordan, they had to be approved by the French High Commissioner in the case of Syria and Lebanon and by the British High Commissioner in the case of Transjordan. In Palestine the British High Commissioner negotiated directly with the company, as there was no native government in existence. Furthermore, not only was conclusion of the conventions reserved for approval of the sovereign foreign authorities, but the latter were also made responsible for their implementation. Thus Article 29 of the Syro-Lebanese-IPC convention stated: "With regard to the execution of the present convention, relations between the government and the company will be established, for the duration of the mandate, through the intermediary of the High Commissioner of the French Republic in Syria and Lebanon." [5]

In accordance with the administrative standards set up by the mandatory authorities the transit conventions were duly published in the official journals of the countries in question.[6] The presence of foreign rule in these mandated territories was comparable in its effects to the existence of authoritarian governments in Iran or Saudi Arabia when the agreements were being concluded. In either case the people did not participate in the political processes. Although the countries of the Mediterranean seaboard were generally more advanced than areas located further east, public opinion there tended to focus on the major issues of independence or Zionist-Arab conflict rather than on the minor question of transit rights for negligible quantities of barely discovered Iraqi oil.[7]

The situation underwent a radical transformation in the period following World War II. Between 1947 and 1949 entirely new con-

[5] Al-Jumhuriyat as-Suriyah, Wizarat al-Maliyah, *Ittifaqat an-Naft, 1950* (Republic of Syria, Ministry of Finance, *Petroleum Agreements, 1950*; Damascus, n.d.), p. 19.

[6] In Lebanon in *Al-Jaridat ar-Rasmiyah*, no. 2509 of June 5, 1931; in Syria in *Bulletin officiel des actes administratifs du Haut-Commissariat*, no. 25, Dec. 31, 1931, p. 11.

[7] Production in Iraq began in 1927. In 1930 it amounted to only 120,000 tons and in 1931 to 110,000.

ventions had to be negotiated between Tapline and the transit countries for the construction of a pipeline. Some four years later the principal producing countries went through a phase of major revisions of the existing concessions. In the mid-1950's the transit countries again emerged as a problem area with their demands for a substantial revision of pipeline conventions. This time the existence and value of oil were known to the public. Political awakening had gone hand in hand with interest in, and alertness to, oil questions. No longer could the public be accused of innocence born of ignorance. If anything, it had become oversensitive in matters pertaining to oil, tending to develop the somewhat exaggerated notions characteristic of awakened but not yet mature organisms.

Because the spectacular expansion of the oil industry in the postwar period coincided with the process of political emancipation and social revolution in the Middle East, it was bound to be linked with and affected by this process. Consequently oil became a subject of great expectations and overoptimistic hopes, but it also became a symbol of foreign influence and one of the favored targets of national xenophobia.

Nothing could better illustrate the popular attitude of hopeful expectation than the reactions of the Lebanese and to some extent the Syrian public to the proposal to construct the Trans-Arabian pipeline, which was to link the Persian Gulf area with the eastern Mediterranean seaboard. An account of these negotiations has been presented earlier (Chapter IX). It is sufficient to say here that the story of the negotiation, signing, and ratification of the Tapline convention in Syria and Lebanon provides an excellent illustration of the power of Middle Eastern public opinion in the postwar period. More concretely this story reveals: (*a*) the power of public opinion in giving priority to political over purely economic considerations (as in the delay in ratifying due to the Palestinian affair vs. the obvious economic advantage of an early ratification); (*b*) the way in which a determined press campaign in favor of a particular solution can influence public and governmental attitudes (as in the energetic Lebanese press campaign aiming at early ratification of the convention by Syria and the resultant response of Syrian press and political circles); and (*c*) the proneness of the public to consider oil as an instrument in the political struggle (as in frequent references by political leaders and newspaper editors to oil as "the sharpest weapon" the Arabs have in their resistance to Western schemes in Palestine).

No matter what the point of departure was (approval or disapproval of the concession agreement), public attitudes invariably tended to evolve as time went on. This evolution was due in part to the change in the character and scope of employment opportunities which a new oil enterprise provided. The initial construction period would usually create considerable demand for manual and clerical labor and contractors' services, thus to a large extent fulfilling early expectations. But after basic construction was completed, dismissals would follow. This was especially true of the pipeline operation, which, after the pipes were laid, could be conducted very economically so far as manpower was concerned. The dismissed workers and clerks would automatically swell the ranks of the chronically unemployed or underemployed multitude so characteristic of practically every Middle Eastern country and thus form a new element of discontent and frustration.

Although benefits from oil industry operations percolated to the general public in varying degrees depending on the country, in no case were these benefits as spectacular, tangible, and widespread as the public expected them to be. Consequently even when the original public response to the projected concession was favorable (as, for example, in Lebanon), disappointment and disillusionment were likely to mark the next stage. Oil did not prove to be a cure-all; hence there must be something wrong with it. The public thus frequently passed from near-enthusiasm to almost chronic criticism. Its list of grievances fell into three categories: economic, industrial, and political.

ECONOMIC GRIEVANCES

The most frequent and typical complaint in this category has been what was believed to be the unfair division of profits between the company and the host country. The public outcry against AIOC in Iran is perhaps the most spectacular, but certainly not an isolated, example of this attitude. In every country other than the absolute monarchies the period preceding a revision of an old concession has abounded in vociferous charges of exploitation. The revisions of concessions in the producing countries providing for a fifty-fifty profit formula generally silenced these criticisms, although they did not eliminate them altogether. The prolonged delay in revising the transit agreements, stretching over at least five years, inevitably produced much criticism.

Against the background of these general complaints are a number of particular grievances voiced at one time or another in various countries. Linked to the over-all problem of exploitation was a frequent complaint, heard particularly in Iraq, about the gold conversion rate applied by the concessionaire company when calculating royalties due to the state.[8] Taxation of the company's profits by the home government before the payment of royalties has provided another grievance, which caused great public resentment in Iran and which would surely have been publicly aired in Saudi Arabia had the latter had a free press and a different political system.

Criticism that the company was artificially restricting production of crude or procrastinating in the development of new oilfields and facilities were particularly rampant in Iraq before 1952,[9] and they are likely to be revived at any time when production does not correspond to the expectations of the host country. This has been especially true in Iran since the conclusion of the Consortium agreement in 1954, when a major preoccupation of the Iranians has been to assure a speedy return to the pre-Mossadegh level of production and eventually to surpass it. In Saudi Arabia, too, complaints can be heard about the inadequacy, from the Saudi point of view, of Aramco's exploration program, whose few wildcatting operations are being rather unjustly contrasted with the multitude of similar developments in Venezuela, but this grievance has not gone far beyond well-informed government circles.

The transit states have had paralleled complaints that the companies were not employing so many people as they were expected to do when the concessions were initially granted. Typical was the

[8] Thus *Sada al-Ahali* (Baghdad), Feb. 6, 1951.
[9] For a typical critical article, see *Az-Zaman* (Baghdad), Aug. 25, 1950.

outburst of the influential Lebanese Maronite leader, Pierre Jemayyel, who in a press interview stated that his support of the Tapline project had been a grave mistake. "I was greatly deceived," he said. "When I was told that the company would employ over 30,000 persons and spend in the country about $83,000,000 within five years, I thought I was acting in the interest of Lebanon." Unfortunately, he continued, the company enrolled only 1,000 laborers "in a very miserable manner by mediation and recommendation [nepotism, political influence, and the like]" and started importing foreign goods such as meat, onions, and potatoes, as if Lebanon did not have similar commodities.[10]

Complaints about what the public believed to be unnecessary imports, not only of foodstuffs which could be procured locally, but also of such goods as prefabricated houses for company personnel recurred in the Lebanese and Syrian press in the early period of Tapline's construction and operation.[11] Linked to them was opposition to the importation of machinery for work which could be accomplished by local labor.[12] All complaints in this category had this in common that, regardless of what the state would receive in royalties and fees, the public expected the company to provide opportunities for employment and in general to stimulate the economic life of the host country. Having been exaggerated at the outset, these expectations were bound to receive a shock sooner or later.

Failure of the companies to provide Middle Eastern countries with more refining facilities has been another standard economic grievance in both the producing and the transit countries. Criticisms on this score vary between regrets that the companies have been slow in going ahead with such construction projects and outright accusations of deliberate malevolence allegedly calculated to keep the host countries economically retarded.[13]

The fiscal privileges and exemptions enjoyed by the companies

---

[10] *Beirut* (Beirut), April 21, 1950.

[11] For a Syrian criticism, see *Alif Ba* (Damascus), April 17, 1950. For Lebanese complaints, see *Ash-Sharq*, Jan. 3, 1950; *An-Nahar*, Sept. 3, 1949; *An-Nida*, Sept. 6, 1949 (all of Beirut).

[12] *An-Nida*, Sept. 28, 1949; *Al-Manar* (Damascus), Jan. 4, 1950.

[13] For typical criticisms, see *Ad-Diyar* (Beirut), Dec. 30, 1950, and *Sada al-Ahali* (Baghdad), Sept. 4, 1950.

likewise supply ammunition for nationalist critics. The latter often argue that by avoiding taxation the companies are making enormous profits and that such royalties as have been agreed upon in the concession agreements never give the state treasury as much income as would subjection of the companies to existing tax legislation.[14] The companies' privilege of importing necessary materials and supplies free of duty is also frequently challenged. Critics claim that not only is this privilege unjustified but that by a too-liberal interpretation the companies frequently import dutiable goods free of duty, thus violating the law and defrauding the government. The press in certain countries, especially in Syria, often seizes upon some minor difference of opinion between the company and the customs authorities and blows it up out of all proportion, thus casting further doubts on the company's honesty and fairness.

Another major problem is that of pricing the oil products sold locally by the companies. Critics are seldom satisfied with the level of prices, claiming that the companies could and should sell their products at a cheaper rate. Accusations of greed are here mixed with charges of monopoly. Before nationalization the Anglo-Iranian Oil Company was under especially heavy fire on this count,[15] but IPC and Tapline have also been criticized in this connection in Iraq and the Levant.[16]

INDUSTRIAL GRIEVANCES

Industrial relations provide a seemingly inexhaustible source of criticism. We shall examine the facts as to industrial and human relations in a later chapter; here we are concerned solely with public reactions to real or imaginary wrongs committed by the companies toward their employees. In general the companies do not enjoy charitable treatment by the press and the public. Unfounded accusations and proneness to accept unchecked statements as fact

[14] Himadeh, *op. cit.* See also *Le Commerce du Levant* (Beirut), Dec. 6, 1952.

[15] *A Report on the History of the Southern Oil of Iran*, presented by the National Oil Company of Iran to the Honourable Averell Harriman, special envoy of the President of the United States of America (Teheran, Aug. 1, 1951), p. 7 (hereafter referred to as *Iranian Report*).

[16] *Sada al-Ahali* (Baghdad), Sept. 4, 1950.

TYPICAL ATTITUDES

and to generalize on the basis of one or a few incidents are characteristic of public attitudes in this respect. The problem of human relations within the industry is complicated by four factors: (*a*) the difficulty of communications; (*b*) the novelty of steady habits of work for a majority of the laborers; (*c*) the clannish and family oriented outlook of the employees and those who claim the right to protect them; and (*d*) the marked differences in skills and competence between various national and religious groups represented among the personnel.

Disciplinary action or dismissals are a standard cause for public complaints. The Middle Eastern press, which seldom devotes time and space to the common man, often picks up a single incident of dismissal by the company and exaggerates it. In any disciplinary action the company is almost invariably presented as the evil-doer and the disciplined worker as an innocent victim of oppression. The matter is further complicated by the fact that the immediate supervisors of native workers, on whom the disciplinary burden rests, are frequently of British or American nationality. Any picturesque expression or gesture used to assure order and efficiency is often resented by the disciplined laborer as an "insult to national dignity." A hearsay story about such an alleged insult will be seized upon by a sensation-hunting paper, which will dwell upon the humiliations suffered by the national workers at the hands of the foreigners. As a rule no attempt will be made by the newspaper to verify the facts and to hear the other side of the story. Not infrequently the reporter or the editorial staff member writing on such a subject is a relative of the dismissed or disciplined worker. Family relations have much more influence on a writer than abstract truth. Complaints about individual or group discrimination often accompany such stories of grievances.[17] In Iraq, where IPC employs a

---

[17] For a typical comment, see "They Eat Dates and Throw Us the Kernels," *Al-'Amal*, Aug. 13, 1949. *Al-'Amal*, organ of the Lebanese Phalanges (a predominantly Maronite organization crusading for recognition of a Lebanese, as distinct from the general Arab, nationality), carried in the summer of 1949 a series of articles highly critical of IPC. Among the many grievances was a complaint that the company was employing Palestinian refugees to the detriment of Lebanese nationals. See also *Al-Fayha* (Damascus), June 6, 1950, for an article entitled "Oppression of Syrian Laborers by Tapline."

good many Christians (especially Assyrians) in skilled or semiskilled positions, accusations of favoritism are often voiced by the press.[18]

More general in nature are the complaints about the technical training programs of the companies. The standard grievance in Iran before nationalization and in Iraq before the 1952 revision was that the companies did not provide adequate educational opportunities for young nationals wishing to acquire technical skills and qualify for more responsible jobs. The companies were charged with deliberately hindering the advancement of nationals to higher managerial and technical positions in order to reserve them for Western personnel.[19]

The recruitment of personnel is a standard source of complaints, and the companies are caught between two fires. On the one hand, powerful leaders, often of a feudal type, in the territory of the company's operations lay claim to virtually exclusive patronage so far as employment of native personnel is concerned. A typical case was the influential Asad family in southern Lebanon, without whose recommendation it was difficult for Tapline to employ anyone in the Sidon terminal area. Similarly the influence of the leading Karame family in Tripoli in northern Lebanon could not be overlooked during local recruitment by IPC terminal and refinery installations. The tendency was for these local potentates to admit to employment only their own protegés, who would be recruited from their own districts rather than from other parts of the country. On the other hand, companies would be criticized for allegedly favoring certain "protected" elements and disregarding the merit system in the process of recruitment.[20]

In the early stages of the oil industry in the Middle East frequent complaints were heard about the inadequacy of the social services provided by the companies to their laborers or native white-collar workers. These complaints were especially strong during the prewar AIOC operations in Iran. The files of the disputes between Iran and the company, which have come to the attention of various international bodies and the public at large, are replete with Iranian

---

[18] *An-Naba* (Baghdad), Feb. 11, 1951.

[19] *Nida al-Watan* (Beirut), July 23, 1949. See also *Iranian Report*, p. 10, and "Iran Presents Its Case," pp. 79 ff.

[20] *Al-'Amal* (Beirut), Feb. 6, 1950; *Al-Akhbar* (Damascus), July 26, 1950.

TYPICAL ATTITUDES

grievances on this score.[21] One of the most persistent complaints was that the housing facilities were grossly inadequate both quantitatively and qualitatively. Poor housing for local employees would often be contrasted with fine homes for Western personnel. And complaints were made about the inadequacy of wages and salaries for the nationals of the host country in contrast to the generous treatment of the expatriate personnel. Both of these complaints were instances of a broader grievance, that of discrimination allegedly practiced between native and Western employees. In these complaints local critics were seldom willing to consider any mitigating circumstances. The demand of "equal pay for equal job" would be made without regard to the fact that a foreign employee had to be given fairly generous treatment in order to entice him to serve in a distant land under torrid conditions. Little or no attention would be paid to such problems as his need for home leaves or the necessity of educating his children in distant schools, naturally involving considerable expense.

The existence of a sizable Western colony composed of executives, professional specialists, and skilled foremen side by side with a large agglomeration of local workers generally employed in lower or intermediate ranks was bound to produce contrasts that would be seized upon by those anxious to criticize. Such criticism was, perhaps, sometimes justified, but often it stemmed from a basic political hostility toward the oil industry and the West as a whole, a hostility which certain elements assiduously cultivated.

POLITICAL GRIEVANCES

The most common political accusation against the oil industry has been that it is an instrument and a motive force of imperialism in the Middle East. Extreme nationalists, radical leftists, and Communists vie with each other in loud denunciations of the oil companies. An article in a leading Iraqi Socialist organ reflects the typical attitudes:

Oil has been the chief factor in consolidating British imperialism in Iraq and has been the basic factor attracting the new American imperialism to

[21] *Iranian Report,* p. 7; see also "Iran Presents Its Case."

the Middle East. . . . There is no area in the whole world richer in oil than the Arab countries, yet they are steeped in poverty and misery. They are, as an Arab poet said:
> Like camels which in deserts die of thirst
> With loads of water carried on their backs.[22]

Such views are not limited to Iraq. In practically every Middle Eastern country with some degree of freedom of the press, with the notable recent exception of Turkey, similar voices can be heard. A Lebanese nationalist paper called one of the major companies "an imperialist octopus," stating: "Foreign companies with their capital, employees, means, and objects are the devices of modern imperialism which save the imperialists the expenses of propaganda and the armies of occupation." [23]

The charge of imperialism serves as a general framework within which a number of subsidiary charges are formulated. Thus it is asserted that the interests of the oil industry determine the foreign policies of the great powers.[24] The companies themselves are frequently accused of interference in the domestic affairs of the host countries. AIOC in Iran, IPC in Iraq and in the transit countries (i.e., British-controlled corporations) were singled out as speecial targets of criticism, but American-owned Tapline has not been wholly immune to attack. Documents made public by the Iranian government, voluminously quoted and commented upon by the Iranian press, made a major issue of the role AIOC was playing as a political force in Iran as a whole and in the province of Khuzistan in particular.[25] The Iraqi press charged the oil companies with interfer-

---

[22] *Sada al-Ahali* (Baghdad), March 6, 1951.

[23] *Al-Ahad* (Beirut), Feb. 27, 1951. The Communist daily *Ash-Sharq* (Beirut), June 1, 1951, also used the word "octopus" in an attack on the oil corporations.

[24] "Why Do the Americans, the British and the Russians Quarrel? Oil Is the Source of Anxiety; Oil Which Comes Out of Arab Soil Should Remain Arab," *Al-Balagh* (Beirut), June 29, 1948.

[25] Iranian Embassy, *Some Documents on Nationalization*, p. 25; also *Iranian Report*, p. 20; *Text of the Report Submitted by Dr. Mohammed Mossadegh, the Prime Minister, to Majless Regarding the British Technical Staff and His Speech to the People Who Had Assembled in the Baharestan Square on the 27th September, 1951* (Teheran, 1951); and *Speech Delivered by the Prime Minister, Dr. Mohammed Mossadegh, at a Gathering of the Representatives of the American Press (Tuesday, March 18, 1952)* (Teheran, 1952).

ence in electoral processes, alliances with retrograde elements, and corruption of political leaders.

The companies' alliances with feudal elements are also occasionally condemned. An Iraq Socialist organ complained about IPC's cooperation with the tribal sheikhs along the pipeline linking Kirkuk with Syria to the detriment of local laborers;[26] an organ of the Lebanese Phalanges accused Tapline of "being entwined in the mesh of feudalists";[27] and a Syrian daily charged the same company with a patronage system designed to strengthen its "friendship with the leading influential personalities and feudal leaders among the Syrian population. . . ."[28]

Along with these charges go criticisms that the companies are courting certain political parties and that they are giving preference in employment to those parties' members.[29] Occasionally individual politicians are singled out as targets of vituperation for their alleged subservience to foreign oil interests.[30]

Sometimes the companies are charged with actual fraud and activities punishable under the laws of the countries. A recurring theme is an accusation that one or another company operating in Syria and Lebanon is smuggling oil from its Mediterranean terminals to Israel.[31] One paper went so far as to assert that, acting under Zionist influence, the companies were purposely refusing or delaying the construction of refineries in Syria and Lebanon so that the immobilized Haifa refinery in Israel might be reactivated.[32]

[26] *Sada al-Ahali* (Baghdad), May 1, 1952.

[27] *Al-'Amal* (Beirut), Feb. 6, 1950, in an article "A Whisper to Tapline."

[28] *Al-Akhbar* (Damascus), July 26, 1950, in an article entitled "Tapline Reveals Its Reality: Those Responsible in the Company Become Active in Spreading American Influence." See also the Lebanese Communist organ *Ash-Sharq*, Sept. 12, 1948, for its attack on the collusion of feudalists with the oil interests.

[29] *Al-'Amal* (Beirut), June 16 and 22, 1949.

[30] *Ad-Dhaffur* (Beirut), Dec. 11, 1950, in an article "Serious Secrets! About Britain's Friend No. 2, Nazim al-Qudsi" compares the latter to Iraqi leader Nuri as-Said and links him to oil.

[31] See *Az-Zaman* (Beirut), Dec. 11, 1950, for an article entitled "Saudi Oil Transported to Israel by Tapline," as well as pro-Communist *Ash-Sharq* (Beirut), Oct. 30, 1951, and *Al-Hadaf* (Beirut), Dec. 12, 1950, for attacks on Tapline. For similar charges against IPC, see *Al-Khabaz* (Latakia), Sept. 14, 1950.

[32] "Jewish Pressure Impeded Establishment of Refineries in Egypt and Lebanon," *Ad-Diyar* (Beirut), Dec. 30, 1950.

Somewhat similar to these accusations but differing in emphasis are occasional charges that the companies are indulging in espionage.[33] Some papers find fault with the companies' communications systems. The standard complaint in this respect is that through their wireless systems the companies are conveying coded messages which may be harmful to the interests of the host country.[34] Attacks on the companies' communications systems can actually be placed within a broader context of criticisms that the major oil companies operating in the Middle East are functioning like a "state within a state." Not only are the older, British-controlled companies in Iran, Iraq, and the Levant subject to this accusation, but the newer American ones are as well. Except for Iran, whose controversy with AIOC over this and related matters has been given considerable publicity around the world, Syria is the country most sensitive to the privileges which permit foreign companies to operate independently of the host country's transportation and communications systems. In line with this, in April 1949 Husni az-Zaim, then Syrian dictator, informed IPC's Damascus representative of three restrictions placed on the company. These were that official clearance must be obtained for IPC aircraft flights, that travel in border zones without permission was banned, and that company guards along the pipelines would be replaced by public security forces. The papers greeted such regulations with satisfaction and pointed out that company control of all such facilities was an infringement of national sovereignty.

In this array of charges and accusations it is not always easy for an untrained eye to distinguish between fact and fiction. By freely resorting to falsehood and distortion, critics often weaken legitimate demands, such, for example, as for better housing for

---

[33] *Barada* (Damascus), Nov. 4, 1948. In a similar vein a few years later Dr. Mossadegh referred to the personnel of the AIOC as "trained spies in the garb of oil technicians." "And here lies the main reason," he added, "for the staunch opposition of my Government not to admit any of those so-called technicians into the country who are in reality nothing but spies no matter what pressures are brought to bear on us" (*Speech Delivered by the Prime Minister,* Tuesday, March 18, 1952 [Teheran, 1952]).

[34] For this kind of accusation, see *An-Nasr* (Damascus), Nov. 18, 1948, and *Alif Ba* (Damascus), Dec. 12, 1948.

married laborers of the host countries. This is unfortunate, inasmuch as the blurring of contours and indiscriminate lumping of the fair and the unfair are bound to produce confusion and contribute to the creation of a negative and dangerous state of public mind. This in turn can and did lead to unjustified political excesses toward the companies. Such excesses provoke minor or major crises, which in the end hurt not only the companies but also the host countries. If some third party were wishing a disruption of harmonious relations between the oil industry and the host countries, it could have found much satisfaction in these emotion-ridden anti-oil campaigns. In the postwar period the third party is personified by Soviet Russia and its Communist allies. To say this is not to assert that all or most attacks on the oil companies have been due to Soviet influence. Most of them, perhaps a majority, were undoubtedly the product of assertive nationalism and would have occurred regardless of the Communists. But the role of the latter must not be ignored. Communist propaganda lays emphasis on points likely to appeal to nationalist emotions, thus increasing the irrational element already prevalent in the psychological makeup of certain ethnic groups in the Middle East. Because Communist activity is premeditated, whereas volatile native expressions of ill-temper generally are not, communism provides continuity and persistence to the otherwise spontaneous outbursts.

ARAB AND IRANIAN LITERATURE ON OIL

Rather unfortunately journalistic proneness to distortion and sensationalism is not checked by the balancing factor of a more serious literature. Whether in Iran or in Arab countries, very few books on oil can stand comparison with the studies authored in the West in terms of thoroughness, objectivity, and accuracy. Certain doctoral dissertations, particularly those produced by Iranians, can be considered an exception to this rule, especially if they were based on native language sources not easily accessible to Western researchers.[35]

---

[35] For example, A. Zangueneh, *Le Pétrole en Perse* (Paris, 1933); M. Nakhai, *Le Pétrole en Iran* (Brussels, 1938); E. Yeganegi, *The Recent Financial and Monetary History of Persia* (New York, 1934).

Studies produced by Iranians sometimes have the merit of unearthing piquant political details which throw better light on the complex relations between Iran on the one hand and British and Soviet oil enterprises and their respective governments on the other. For a historian of the political aspects of the oil industry in the Middle East Iranian works are of more value than the generalized studies by Arabs. The stormy history of Iranian oil supplied a lively subject to writers in the Mossadegh era and the periods immediately preceding and following. The authors ranged from young and relatively obscure individuals to well-known public figures, such as the "Hero of Abadan," Deputy Hussein Makki, one of the chief forces in the Iranian nationalization drive; [36] Senator A. Lessani, an outspoken critic of AIOC and later also of the Consortium; [37] Deputy Hussein Pirnia, formerly head of the petroleum department in the ministry of finance,[38] and Nasrollah Saifpour Fatemi, at one time governor of Isfahan and brother of Hussein Fatemi, Minister of Foreign Affairs in Mossadegh's cabinet.[39] Most of these works combined a historical account with a strongly nationalist plea for justice and recognition of Iran's rights. But though arguing Iran's case against the AIOC and British imperialism, they did not condemn the oil industry as a whole. In fact, some of them made a point of contrasting American oil companies and the American approach in general with British operations and political practices.[40]

Apart from these book-length studies, individuals, groups, and parties in Iran occasionally circulate brief pamphlets, usually of a polemical nature. One of the most noteworthy among them is a pamphlet entitled *Oilless Economy,* which was issued by the Iran Party, an organization largely composed of intellectual and profes-

---

[36] His book is *Kitab-i-Siah* (Black Book; Teheran, 1329 [1949]). This is largely a collection of documents purporting to prove British interference in the domestic affairs of Iran.

[37] His book is *Talai Siah ya Blay-i-Iran* (Black Gold or Disaster for Iran; Teheran, 1329 [1949]).

[38] His book is *Dah Sal I Kushesh Dar Rah I Hefz va Bast Huquq Iran Dar Naft* (Ten Years of Struggle for the Protection and Expansion of Iran's Oil Rights; Teheran, 1331, [1951]).

[39] His book is *Oil Diplomacy, Powderkeg in Iran* (New York, 1954).

[40] Thus Fatemi refers to the Saudi-Aramco 50–50 profit-sharing formula as "generous terms" (*ibid.*, p. 369).

sional people allied to Dr. Mossadegh during his term of office. In this pamphlet, published in 1954, the party argued that, in view of its numerous other resources, Iran should be capable of organizing its economy independently of oil if the latter's exploitation by the state proved impossible as a result of foreign boycott and opposition. Rather than submit to the dictation of foreign companies and powers and thus jeopardize its independence, Iran should renounce oil and concentrate on other sources of revenue. The same party was credited, in 1957, with authoring another pamphlet, entitled *The Way of Mossadegh*, which was circulated clandestinely and contained strong diatribes against both the Consortium and the post-Mossadegh regime of Iran.

Arab literature on oil has differed from the Iranian principally because, instead of concentrating on a single country, it tends to view oil operations as a regional phenomenon. Moreover, Arab experience with oil has been of shorter duration than has Iranian, and it has been free of the Russian factor. Consequently Arab works usually lack the detail characteristic of some Iranian studies. Furthermore, the fact that no Arab country has thus far resorted to cancellation of the concession (an event which has occurred twice in Iran, in 1932 and 1951) has deprived Arab literature of a specific issue, which has been an obsession with the Iranians for the last few decades. Despite these differences Arab books on oil have been characterized by an equally strong nationalist coloration. Apart from some descriptive works of an impartial nature (generally brief and inadequate as a source of serious knowledge),[41] the books have not differed markedly from the journalistic articles examined earlier in this chapter, sharing with them the same complexes, prejudices, and grievances. Their titles are suggestive of their authors' bias. *Our Stolen Oil*,[42] *Oil-Enslaver of Peoples*,[43] or *America Plunders Arab Oil*[44] are fairly representative samples. Themes repeatedly emphasized by the nationalist press reappear in these books. For instance, Hilu in *Our Stolen Oil* says:

[41] For a typical descriptive account, see Yusuf Mustafa al-Haruni, *Qadayat al-Petrol* (The Question of Oil; Cairo, 1950).
[42] Yusuf Khattar Hilu, *Naftuna as-Salib* (Beirut, 1954).
[43] Yusuf Ibrahim Yazbek, *An-Naft Musta'abat ash-Shu'ub* (Beirut, 1934).
[44] Adil Hussain, *Amirika Tanhab Petrol al-Arab* (Cairo, 1957).

The right national solution demands the confiscation and nationalization of the oil companies in the interest of the Arab peoples, as well as the elimination of all imperialist powers. There is no doubt that the unification of the Arab peoples will lead them to achieve their aspirations, namely, exploiting for themselves and by themselves their oil resources.[45]

Equally strong in his condemnation of Western oil enterprise is Dr. Rashid al-Barawi, an author of pronounced Marxist leanings serving as president of the National Bank of Egypt under the revolutionary regime. His book, *Oil War in the Middle East,* opens with the following dedication: "To the hard-working masses which moan from poverty, ignorance, and disease as a result of imperialist monopolies." [46] Like other writers and journalists he harps on the theme of a "state within a state," which, in fact, figures as the subtitle of one of his chapters. While thus castigating the Western oil companies, Barawi gives an ingenuous interpretation of the Soviet-Iranian oil agreement of 1946, which he tries to explain and justify. Russia, he asserts, has always opposed the penetration of any other power into northern Iran. In pursuance of this policy it has traditionally controlled the economy of northern Iran and, moreover, safeguarded its own interests through the treaty of 1921. When American influence began to be felt in Saudi Arabia and when Britain opened negotiations for a bilateral agreement with Iran, Russia believed a conspiracy was directed against itself. "For all these reasons," says Barawi, "Russia decided to cross the plans of others by securing participation in the profits from oil resources of the northern part of the country, by keeping this right exclusively for herself, and by thus depriving Britain and the United States of the opportunity [to move in]." [47] Furthermore, asserts the author, after World War II Russia was afraid that the balance of power might be upset in favor of Britain and the United States. By controlling the oil resources of northern Iran, she would be able to compete with Britain and America in world markets and would at the same time preserve the balance of power.

In conclusion Barawi emphasizes the psychological factor. It is essential, in his opinion, for "enslaved" peoples to be convinced of the need to be freed from the power of "imperialist oil companies"

[45] *Op. cit.,* p. 48. [46] *Harb al-Petrol fi ash-Sharq al-Awsat* (Cairo, n.d.).
[47] *Ibid.,* p. 146.

TYPICAL ATTITUDES

and of their ability to gain that freedom. Public opinion should be aroused against the companies, and all means of putting pressure on them should be used. He offers a nine-point program which calls for: (*a*) no new concession grants and encouragement of national, state-supported companies; (*b*) nonrenewal of existing concessions; (*c*) revision of the concessions to assure larger profits for Middle Eastern countries and the establishment of tighter control over the activities of concessionary companies; (*d*) national representation on the boards of the companies; (*e*) greater employment opportunities for national employees, to be assured by reports on the numbers and salaries of foreign and national employees; (*f*) equality of salaries and treatment of the national and foreign employees; (*g*) adequate housing for national employees; (*h*) keeping of accounts in the language of the host country; and (*i*) taxation of all pipeline companies operating in Arab lands.

This program is something of an anticlimax after the rather fiery oratory of the earlier parts of the book. It is as if the author were speaking with two voices: one of a crusading nationalist and the other of a somewhat more restrained economist who hesitates to recommend nationalization as a cure-all for the ills of Arab economies.

Another book that may be worth mentioning as illustrative of current nationalist thinking has been written by three men, of whom one, Major Amin Shakir, was formerly a prominent member of the Egyptian Free Officers Movement and secretary to Colonel Nasser. The book, *Oil and Arab Policy,* reveals the desire of the revolutionary regime of Egypt to assume a leading role in the formulation of general Arab oil policies.[48] The most interesting part is, perhaps, the foreword, written by Nasser himself. In it he delivers a highly nationalistic diatribe in which he speaks of God's revelation to mankind through Islam, abjures ignorance, and appeals for strength and unity of the Arabs from the Persian Gulf to the Atlantic. Referring to the Arabs' glorious past, he asks:

Why don't we rise upon the world once again with the message of peace and mercy and the principles of brotherhood and equality, and erase the darkness of falsehood that has enveloped the hearts and minds of those

---

[48] Amin Shakir, Said al-Arian, Tawfiq Maqqar, *Al-Petrol wa as-Siyasat al-Arabiyah* (Cairo, n.d.). A portrait of Colonel Nasser is the frontispiece.

223

who believe only in matter? Indeed, we have our Kaaba to which we make our pilgrimage and to which we turn in our prayers and which the others do not have. But the Kaaba in its spiritual reality is only a disciplinary symbol through which God wants to teach us, the Arabs, that humanity will not reach its perfection until hearts join together for one aim and goal and communities unite for a common purpose. And why don't we aim to create a Kaaba for lost humanity?

In this exhortation oil is mentioned only in the last paragraph, in which the author claims that it provides a "new proof" of the Arabs' material and moral capacities. These capacities, he asserts, will be realized only if the Arabs develop self-confidence, the latter to be based on belief that they are "a nation, one nation." To a rational Westerner this reasoning may appear unconvincing. But to convince him is obviously not Nasser's main objective. That seems to be to stir the Arab reader's emotions by appealing to his national and religious consciousness and by contrasting it with the presumably greedy materialism of the rest of mankind. As in the works previously cited, the reader is ushered into a world where the light of virtue and moral superiority is opposed by the darkness of evil and wickedness. It is in this climate of moral preconceptions mixed with emotions that he is expected to approach the study presented by the principal authors.

It should now be clear that Iranian and Arab literature on oil cannot be relied upon to provide its readers with thorough and dispassionate information and guidance. These books and pamphlets may have some patriotic merit in alerting the reading public to matters of great national importance, but in their superficial treatment of facts and their refusal to present an impartial discussion they fail in their basic mission of enlightenment. While it is debatable to what extent these books are responsible for the shaping of public attitudes in societies which are not highly literate, there is no doubt that, added to the journalistic output, they reinforce already existing biases and preferences.

CONCLUDING REMARKS

Although literature and the press are the most common vehicles of expression, public attitudes toward oil have been manifested through many other channels as well. The latter have ranged from politically motivated strikes and workers' demonstrations to perfectly peaceful parliamentary speeches and interpellations. Criticism of the oil industry has tended to emanate either from the strata having the most direct relationship with it or from politically motivated groups. Various political parties, especially those of a nationalist or Socialist character, have been most conspicuous. Such Socialist organizations as the Arab Socialist Renaissance Party in Syria (with a branch in Jordan), the National-Democratic Party of Iraq, the Socialist Progressive Party of Lebanon, and the Iranian Workers Party have been in the forefront of the companies' critics. Equally unfriendly have been the extreme nationalist parties that tended to interpret nationalism as virtually equivalent to anti-Westernism. The Iran Party in Iran and the Istiqlal Party in Iraq are representative examples.

In addition to the parties, a number of other social groups, whether formally constituted or not, have frequently taken a stand on oil affairs, generally of a critical nature. Such, for example, was the Lebanese Congress, a political study group headed by Abdullah Khadra which asked for radical revision of the pipeline agreements. Students, a highly inflammable and not unimportant group in Arab and Iranian politics, have also been prone to strike or demonstrate whenever confronted with some controversial issue between the companies and the host governments. Occasionally a special interest group arises, such as the Zahrani Landlords Defense Committee in southern Lebanon, which agitated in favor of higher rents and expropriation rates for the lands to be used by Tapline. From time to time a frustrated contractor who has failed to make good in his relations with the concessionaire company leads a campaign of revenge, resorting to subsidized press attacks and using whatever political influence he can command. Occasionally a secret terrorist group appears on the scene, acting usually as the instrument of self-seeking elements and attempting to intimidate a company and extort from it some special favor or concession. In 1953 private interests resorted to intimidating tactics vis-à-vis Tapline in the Sidon area. An increasing number of Pan-Arab and Afro-Asian conferences (of lawyers, university graduates, and so on) held in such nationalist centers as Damascus, Cairo, or Jerusalem either directly or indirectly (through support of the nationalization thesis) contribute to the creation of negative attitudes toward the Western investor in general and the oil industry in particular.

Despite the inclination of many groups to find fault with the concessionary companies, it would be an error to assume that a positive approach is totally absent. In the first place, some governments have received obvious advantages from the oil operations. Although they are not always constituted in accordance with democratic procedures, these governments are nevertheless an emanation of sociopolitical forces which cannot be disregarded in an analysis of public attitudes. It is the political and economic realism of such groups that prompts them to accept, and to insist on the maintenance of, the concession agreements. Moreover, it is wrong to assert, as Communists and certain other critics have done, that these positively inclined elements are nothing but a group of feudal reactionaries completely divorced from the people at large. The eagerness with which both manual laborers and white-collar workers seek employment in the oil companies and their adverse reaction whenever such opportunities decrease or terminate refutes the thesis that no one except the feudalists gains by the presence of the oil industries in the Middle East.

The argument does not end here. Many other elements, including those that are prone to attack the companies, are also anxious to see them continue their operations. This was eloquently revealed in Syria in 1957, when nationalist circles were very unhappy at the suggestion that the pipeline network might be expanded outside Syrian territory. Despite sabotage of the pipelines in the wake of the Suez crisis, there was a general desire in and out of the government for IPC to resume operations as soon as the proper political conditions were manifested. The same was true of certain political groups which ordinarily are unfriendly to the oil companies. Such, for example, was the case of the National-Democratic (Socialist) Party of Iraq, which despite its frequent and rather basic criticisms of the company preferred to advocate radical revision of the concession rather than outright nationalization in 1951.

Due account should also be taken of the effects of major revision of concession terms on public attitudes. Although it has not uniformly disarmed critics (especially if the complaints were due to some more basic motivation), any radical rise in a host country's revenue from oil has been likely to reduce the sharpness and frequency of attacks. In fact, in cases where the rise of revenue was truly spectacular, host governments have sometimes replaced the companies as the principal targets of criticism. This is true because for the time being the issue has ceased to be that of exploitation by a Westerner and has become instead that of proper utilization by the host government of the substantial revenue it was receiving. "What do they do with all that money?" has become the battle cry of the frustrated and the reformers, somewhat overshadowing the old slogan of imperialist greed and domination.

Consequently recognition, however grudging, of the benefits bestowed by the oil industry has been a political fact that should not be overlooked in an examination of the totality of public attitudes. This recognition has never been as articulate and as loud as has been the criticism. Individuals and groups favoring a stable and continuous relationship between their countries and the oil companies have either been reluctant to take a strong public stand on the matter or lacked the necessary skills to counteract the outspoken negativism of their opponents. As a result the burden of coping with public attitudes toward the oil industry has had to be shouldered by the companies themselves.

CHAPTER XII

Oil Industry's Reaction

THE basic question facing the companies in their relations with the public has been how they should react to the barrage of criticisms to which they were subjected. One way was to examine the grievances and see whether it was possible, by a change in behavior or policies, to remove some of the causes of complaints, even if the latter were exaggerated or based on false reasoning. Another was to undertake to defend their interests and policies by communicating with the public at large. The first method implied a degree of intelligent understanding of the sociopolitical environment in which the companies operated, of correct evaluation of the forces at work, and of adaptability to the ever-changing conditions in the host countries. The second called for willingness and ability to establish communications with the public and for a mastery of methods likely to assure the best possible response to whatever ideas the companies might want to convey. In practice, both approaches have been used to a varying degree, depending partly on the environment and partly on the attitude of the company.

REFUTATION BY DEEDS

With regard to the behavior of the companies, three specific criticisms hurt the companies' position especially. These are the charges of identification with feudal potentates, of acting like a state within a state, and of neglecting the economic well-being of the host country. Often unfairly exaggerated, these charges have nevertheless had some real basis, albeit the situation has differed with different companies. The first charge is perhaps the most difficult to deal with, assuming that a company wants to do something to remove this particular complication. The difficulty often stems from the basic sociopolitical structure of the country in which the company has to operate. If this structure gives a predominant position to feudal elements and if the latter actually hold the reins of power, the company can do little about it unless it does not mind getting involved in the internal politics of the host country. Nor can the company be held responsible for the fact that by dealing with a legitimately constituted government of retrograde tendencies it is supporting the nonprogressive elements in the country. There is, however, a possibility that a company may overlook vital social changes occurring in the host country and may conduct itself in a manner appropriate to an earlier, more conservative period. In a situation like this, it is not only the adaptation to new circumstances which counts, but also, and perhaps primarily, the pace of this adaptation.

Furthermore, even without improper interference in local politics, the companies by virtue of their economic power have often been able to advance the careers of individuals with political aspirations. By granting lucrative contracts or a local distributorship, the companies can contribute to the making or unmaking of public figures in societies where financial power accounts for a great deal in the political process. The habit of associating and doing business with easily identifiable strata or individuals, a belief that these and not other circles are the "safest" to deal with, are apt to hurt a company seriously should any radical shift occur in the political constellation of the host country.

The second principal charge, that of the existence of a state within a state, cannot be easily dismissed inasmuch as the companies were driven by the fact that they were operating in technologically retarded societies to achieve a maximum of self-sufficiency. It is unfortunate that what in practical life was necessary or inevitable should, to the jaundiced eye of a local nationalist, appear as premeditated wrongdoing. Certain aspects of the state-in-state charge are obviously devoid of a sound basis. This is true of the complaint that the companies do not pay any taxes or are exempt from customs duties. These features were an integral part of the concession agreement and were designed to safeguard the concessionaire against discriminatory taxation and similar abuses. As for the technical aspects of self-sufficiency, such as the right to possess aviation and communications systems, they were guaranteed by the concession and their purpose was to assure maximum efficiency in the company's operations. Here, of course, greater flexibility is conceivable, depending on the degree and pace of the host country's technological development. But by and large this particular complaint is unfair. The more retarded the society, the more need there is for independent technical facilities. Should the society achieve substantial technological progress, the existence of the companies' own facilities would automatically become less conspicuous and the psychological need for their replacement by national services less acute.

It should be admitted, however, that there are certain areas in the companies' operations where a less rigid separation from the surrounding environment could be practiced without sacrificing any substantial interests or efficiency. Such, for example, are the purchase policies, the construction and servicing contracts, and those employee policies relating to the surrounding communities and countryside.

In an attempt to close the gap between themselves and their environment certain companies—notably Aramco and IPC—adopted in the 1950's a policy of "integration." The underlying idea of this policy is to reduce company self-sufficiency to the minimum compatible with essential interests, while integrating company activities as much as possible with the social and economic life of the host country. More concretely the integration policy is expressed by the following measures:

1. Purchasing policies. Buying as they do large amounts of various materials and supplies for their operations, the companies are obliged to import most of their requirements. In doing this, they can—and do—avoid intermediaries by purchasing at the source. For the convenience of their Western personnel they have established commissaries where most of the articles needed for daily use,

OIL INDUSTRY'S REACTION

including foodstuffs, may be obtained. So long as the companies operate in primitive areas devoid of essential economic services, such practices are understandable and unavoidable. But with the gradual growth of neighboring communities, such as Al-Khobar or Dammam near Aramco operations, many needs on both company and employee levels can be satisfied by merchants of the host country. With regard to company purchases the obvious factor to be considered is that of prices, which are likely to increase with the introduction of a middleman. Another factor is that of efficiency and reliability in meeting specifications and deadlines. Despite inevitable drawbacks and inconveniences, at least in the initial stages, companies have increasingly begun to avail themselves of local purchase opportunities. The following table illustrates this upward trend in the value of local purchases in the case of Aramco: [1]

Year	Local purchases	Total purchases
1954	$511,141	$45,300,000
1955		54,804,000
1956	2,349,000	76,475,000

Although a complete conversion from direct overseas buying is still a distant, and perhaps never attainable, goal, there is evidence of growing trade between the company and local suppliers. In pursuance of the integration policy Aramco in 1956 abolished its commissary services, compelling its personnel to buy at local stores. The effect of this measure was widely felt in Al-Khobar and Dammam, whose growing prosperity received a powerful stimulant.

2. *Contractors' services.* What has been said of the purchasing policies applies to contractors' services also. The companies have a choice of relying on their own organization or of entrusting certain less complicated tasks to local contractors. Here again the problem of compatibility—of making what was socially desirable jibe with what was technically necessary—has to be solved. As time passes, local ability to perform useful tasks for the companies is increasing. Independent contractors are used in construction, maintenance, transportation, and operations. They are gradually

[1] Based on the annual *Report of Operations,* 1954–1956, to the Saudi Arabian Government by the Arabian American Oil Company. No figures for local purchases in 1955 are given.

being entrusted with more and more complex tasks involving greater responsibility and skill. Aramco in particular makes considerable use of contractors' services in transportation and in construction of school and health buildings. The company's increasing reliance on contractors' services is illustrated by Table 4.

Table 4. Independent contractors servicing Aramco

Year	Total paid by co. (in Saudi rials)	No. of contractors	Monthly avg. of workers
1951	21,000,000	117	6,340
1952	47,495,000	149	11,170
1953	41,115,000	160	7,500
1954	32,273,343	135	6,300
1955		126	8,900
1956	44,585,000 *	135	7,700

Source: Aramco, *Report of Operations*, 1951–1956.
* Equivalent to $11,889,000.

3. Housing. Although housing is primarily a company-employee problem and as such is treated in Chapter XV, it is considered here as an element of the integration policy. If the companies felt responsible for providing housing for their employees, as most of them did, a question was apt to arise as to the character and location of the housing developments. Should the houses be company-owned or employee-owned, and should they form separate company towns or merge with existing communities? In early periods the trend was to build camps or company towns near the work sites. Such camps could be managed rather easily with regard to sanitation, discipline, and related matters. They did not require the organization of a transportation system, and in general they tied the laborer closely to his work. Sometimes, especially when the distance between the nearest population center and the work site was substantial, this was the only practicable way. When long distances were not involved, two other solutions were possible: either to build a camp adjoining the nearest town or to help employees establish themselves within the existing community. About 1952–1953 both IPC and Aramco made an effort to subordinate their housing policies to the larger ideal of integration. This did not mean that they definitely abandoned the construction of separate camps. Such camps continued to be erected if there were compelling reasons to do so. For example, with the development of the Ghawar field in the vicinity of the oasis of Al-Hasa from 1953 onward, Aramco decided to establish a camp for its laborers in Udhailiyah near the work site, some fifty kilometers from Hofuf. The decision was reached only after the Arabian Research Division of the company had made exhaustive studies of the problem, with full awareness of the issues at stake.[2]

Similar wisdom has been displayed by IPC in its housing policies, especially since the revision of its concession agreement in 1952. The company had earlier erected a complex of dwellings for its employees, known as the Arrapha Estate, near the operating installations in the neighborhood of Kirkuk. Despite the many fine qualities of the Estate's houses the average Moslem employee preferred to live in Kirkuk, where he felt more at home. The Arrapha Estate gradually began to attract members of the Christian minorities employed by the company and it eventually was so identified with them as to gain the popular nickname of "Assyrian Village." This was obviously not what the company had intended. "It is . . . recognized," said a company brochure in 1955, "that anything that tends to segregate the employee from the natural life of the country should as far as possible be avoided. By building a house in the township to which he is affiliated, the employee is enabled to make a greater contribution to the development and prosperity of his own community."[3]

The principal way chosen by both IPC and Aramco to implement their housing integration policies was to launch schemes designed to help their employees acquire houses of their own. Known as the "home ownership scheme" in IPC and as the "housing loan plan" in Aramco, these schemes provide either for financing by the com-

[2] An excellent by-product of this research was a socioanthropological study written by a member of the Arabian Research Division, F. S. Vidal, *The Oasis of Al-Hasa* (priv. pr. 1955).

[3] *The Employee of the Iraq, Basrah and Mosul Petroleum Companies in Iraq*, a booklet put out by IPC in London in 1955, p. 14.

panies of the houses constructed for employees by local builders or for the construction and subsequent sale of the houses by the companies to their employees.[4]

The third main criticism addressed to the companies is that they do not contribute to the economic development of the host countries. This criticism is based on the assumption that the companies' responsibility does not end with the payment of royalties to the host government. For instance, the Anglo-Iranian Oil Company was often charged with neglecting to develop the province of Khuzistan, where most of its operations centered. Although such criticisms had no legal basis so far as concession agreements were concerned, socially they were relevant. In trying to respond positively to this challenge, the companies can follow one of two paths: either they can help the host governments in their development activities or they can encourage private civic and economic endeavors aimed at increasing the prosperity and well-being of the country and people. If they are imaginative enough, they can do both simultaneously.

Aramco is pre-eminent in such endeavors. It extends substantial technical assistance to the Saudi Arabian government and local municipalities,[5] and it has adopted the policy of aiding private initiative to develop industries and business enterprise in general, especially in the Eastern Province. In pursuance of this policy a special company unit, the Arabian Industrial Development Division, was established. Tendering advice and technical assistance to private Saudi citizens or corporations, it has been instrumental in helping the establishment of an impressive number of enterprises.

The latter include three electric power plants, in Al-Khobar, Dhahran, and Dammam; ice plants in Al-Khobar, Dammam, Qatif, Jubail, and Hofuf; machine and carpenter's shops, brick, block, and pipe plants, manufacturing and bottling works for soft drinks, and a liquefied petroleum gas plant in Dammam; oxygen, acetylene, carbon dioxide, and dry-ice plants in Al-Thaqbah; automatic laundries, a foundry for manufacturing cast-iron pipes and brass castings, a private hospital, stores, office buildings, and apartment

[4] For further details, see pp. 302–305 below.   [5] See p. 120 above.

OIL INDUSTRY'S REACTION

houses. In 1954 alone the company gave technical assistance to nearly a hundred privately financed industrial and commercial projects. "The amount of aid ranged from complete, detailed design of an entire project to advice on operational procedures." [6]

Other companies have thus far not institutionalized their technical assistance programs, and none can point to a record of stimulating local industry comparable to that of Aramco. IPC has been working on a more informal basis, giving technical advice when requested to do so by interested individuals or providing assistance to local municipalities, such as constructing a water supply system in Kirkuk.[7] During its operation in Iran, AIOC also performed services for the municipalities of Abadan and Khorramshahr. These were generally related to the maintenance of minimum sanitation standards, and in view of the proximity of the company's own settlements could not be construed as entirely disinterested gestures.

To summarize: The actual behavior of the companies, based on well-conceived and well-implemented policies, refutes much of the criticism directed at them. Intelligent awareness of the points exacerbating public opinion and a genuine desire to adjust their conduct to the needs of the time are two indispensable prerequisites of any policy—be it integration or technical assistance—that is to succeed.

THE OIL INDUSTRY'S PUBLIC RELATIONS

While the deeds are eloquent, words are equally important. *Les absents ont tort*—this old adage can be applied to any human group, and oil companies are no exception. In the preceding pages we have seen that many charges leveled at the industry are based on ignorance, emotion, or willful distortion. Even the most exemplary behavior of the companies in various fields of activity cannot dispel these multifarious exaggerations and falsehoods, unless it is aided by an effort to refute them and to present the truth to the

---

[6] Aramco, *Report of Operations*, 1954, p. 49.

[7] It may be added here that Iraq has launched a comprehensive development plan of its own (see Chapter III). Consequently there is no compelling need for the Company to duplicate the government-sponsored activity.

public. Recognizing this need, practically every company has set up a public relations unit.

This development did not come easily: obstacles, both internal and external, had to be surmounted. Internally it took some time to convince technology-oriented executives that the companies would benefit by having their practices and policies clarified and publicized. In this respect some companies were not unlike military establishments in trying to make a classified matter out of subjects which could be divulged without harm to their interests and which were already known to the public, however imperfectly, anyway. While such opposition stemmed from the basic reluctance to admit the usefulness of public relations, another no less important internal obstacle was often due to underestimation of the political consciousness of the society in which the company operated, or of the pace of the changes the society was undergoing. The external obstacles were generally due to reluctance on the part of host governments to tolerate the companies' public relations activities. Certain Middle Eastern countries were suspicious of any foreign-sponsored information programs, treating them as propaganda likely to confuse the public and injure vital national interests. Moreover, some governments consciously adopted a policy of preventing contacts between their nationals and foreigners. Iran is an example of both of these attitudes. At times in the interwar period Reza Shah strictly enforced the policy of separation between his subjects and aliens. Similarly during his brief but stormy career as premier Dr. Mossadegh once decreed the closing of all foreign information agencies and branded AIOC's Bureau of Press and Information as

a centre of espionage [which] has been the rallying centre for all those unfortunate individuals who, through the Company's temptations and through its baneful financial and political influences over a period of 50 years, have deviated from the path of honour, and have offered their services to the Company against the interests of their own country.[8]

Apart from these unfriendly official attitudes there is always a possibility that the host government may take exception to one

[8] Iranian Embassy, *Some Documents on Nationalization*, p. 25.

OIL INDUSTRY'S REACTION

or another medium or method of communications. For example, Syria passed a decree in 1952 prohibiting advertising by foreign corporations in the local press. (The ban was later removed.) A country may be concerned lest its traditional way of life be disrupted by the exposure of its citizens to images and ideas emanating from a center of foreign culture. This is clearly the case in Saudi Arabia, where there are no public movie theaters because of religious injunctions. The company cannot establish a theater for its workmen, limiting such facilities to intermediate and senior staff categories. A potential problem arose in connection with a projected installation of a television station by the company. As an effective means of spreading ideas, television is likely to increase the area of contact between local and foreign ways of life. Rigid censorship is a difficulty in some countries.

These external obstacles are often counterbalanced by favorable factors. For example, some of the host governments are anxious to avail themselves of the companies' public relations organization for their own benefit. This is especially true of Saudi Arabia and Iraq, whose governments either commission or simply allow the companies to carry out publicity tasks in their behalf. Notable in this respect was Aramco's role in setting up the Saudi Arabian exhibit on behalf of the government at the international Damascus Fairs in 1956 and 1957. In the same category has been the production of a number of documentary films for the Saudi government, such as "Modern Saudi Arabia" and "The Pilgrimage." These films were made at the express desire of the host government. In addition to such commissioned works, Aramco's Public Relations Department from time to time produces a film on its own initiative, showing the general advance of the country since the discovery of oil. One film, "Leap of the Isle," was shown with considerable success at the Damascus Fair and to numerous other audiences in Saudi Arabia and abroad. A similar approach is being followed by IPC in producing documentary films of life and progress in Iraq. Noteworthy among them is a series entitled "Biladuna" (Our Country) portraying Iraq's culture and achievements under the Development Plan. Although sometimes entirely divorced from the subject of oil, these films are valuable from the companies' point of view because they are a bond of common interest and good will between the concessionaire and the host country.

The primary aim of the message that the industry has been anxious to convey to the public has been to provide authoritative factual information which can put an end to rumors and unfounded speculations and supply material for anyone wishing to write or speak on the subject of oil. In addition, the message may contain a certain amount of persuasion or argumentation of a positive or a negative character. Positive persuasion stresses the benefits that the host government and the country as a whole or some specific groups derive from the existence of the oil industry in their midst. Such "propaganda" can be direct or indirect. By emphasizing the government-sponsored development schemes in the host countries, the companies indirectly suggest a link between their own activities and general progress. Work openings in areas chronically suffering from underemployment, the stimulation of private business enterprise, benefits and career opportunities for their own employees—all of these are positive themes which are featured in a variety of media. The object of negative persuasion is to refute criticisms and generally to defend the companies against all kinds of attacks. This task is infinitely more difficult to perform. Sometimes an outright denial of a misstatement is necessary, but in many cases it is unprofitable to refute fantastic allegations and the companies' interests are better served by ignoring instead of emphasizing them. The companies often gain more by concentrating on the positive aspects of their story and thus restoring the much-needed balance in the minds of the public.

In their choice of themes for public consumption the companies always have to keep in mind the groups to which their message is to be addressed. These groups can be roughly divided into four categories: (*a*) the government and officialdom, (*b*) the companies' own employees, (*c*) the politically conscious educated groups, such as students, journalists, and members of the professions, and (*d*) the public at large. Effective reaching of the target groups depends on the selection of right methods and media and on the general quality of organization of public relations units.

OIL INDUSTRY'S REACTION

By the mid-1950's all companies operating in major producing or pipeline transit countries had developed large and diversified public relations organizations. Aramco in Saudi Arabia has built up a Public Relations Department directed by a manager working under the over-all supervision of a vice-president in charge of public relations. The department is divided into four major divisions: Press and Publications; Television; Photo, Radio, and Advertising; and Public Activities. Each division is subdivided into a number of units and sections. Apart from its central headquarters in Dhahran the department has representatives in Ras Tanura, Abqaiq, Dammam, and Jeddah. In 1957 it employed over ninety persons.

Similarly developed though more modest in numbers is the organization of IPC's Public Relations Department in Baghdad. It is headed by a public relations adviser (to IPC's chief representative in Iraq), assisted by a public relations officer, a film producer, a display and photographic officer, a press and information officer, an employee magazine editor, a general assistant, and an office superintendent. Including its local representatives in Kirkuk, Mosul, and Basra, the department employs thirty-two persons.

The Consortium tends to place less emphasis on public relations in view of the mixed character of the Iranian operation, in which responsibilities are divided between the Consortium and the National Iranian Oil Company. It is felt that, inasmuch as NIOC undertakes to carry out the nonbasic operations, mostly of a social service type, it should assume the responsibility of facing the public on any questions that may arise in this connection. Moreover, because the Consortium operates on a different legal basis than do the other corporations in the Middle East, performing functions for the nationalized Iranian oil industry, it is believed that its presence and role should not be overly emphasized before the public. However, it is debatable how long a major industrial enterprise, notwithstanding its special status, can avoid having broader contacts with the public and answering questions that may be addressed to it. The question arose soon after the conclusion of the Consortium agreement when in 1955 public interest was aroused over the disposal of the 12.5 per cent share of crude oil production guaranteed to Iran. If not properly handled, the matter might have become a subject of unnecessary public controversy at a time when it was imperative that nothing mar the harmonious government-company relationship so recently established. Furthermore, the taking over of the nonbasic functions by NIOC turned out to be a longer process than was originally intended. By mid-1957 only the medical services had been transferred to NIOC's care, all the rest being administered by the Consortium. Consequently it is still up to the Consortium to give information based on first-hand knowledge in case of inquiries or criticism.

Diversification of Media

These public relations offices are practically unanimous in employing as wide a variety of media as modern technology and the absorptive capacity of their host countries permit. They issue publications, make radio and press releases, advertise in the local newspapers, print posters, organize exhibits, arrange conducted tours for groups and individuals, produce motion pictures, and, in one case, established a television station. Although the framework is alike in all major companies, they differ in several respects. The differences are partly due to the environment in which the companies operate and partly to the temperament, concepts, and habits of the two major executive groups, British and American, active in the Middle Eastern oil industry. There is, in the first place, a difference in the general approach to any operation: British executives tend to favor informal over formal arrangements, their plans and projects are more modest, and they are more economy-minded. A good reflection of this attitude can be found in the annual reports on the companies' operations. Aramco makes a point of publishing every year a bilingual (English and Arabic) *Report of Operations to the Saudi Arab Government* of impressive size and format, richly illustrated and abounding in maps, charts, and statistics. A typical annual issue of the report contains chapters on petroleum operations, the people of Aramco, industrial training, health services and industrial, community, and agricultural developments. These in turn are subdivided into sections, each easily traceable in a table of contents. Such a report makes an excellent impression on readers and apart from offering pleasant and informative reading matter conveys the idea

OIL INDUSTRY'S REACTION

that the company's operations are open to public scrutiny. Noteworthy is the amount of space—slightly over 50 per cent—devoted to social topics.

No other company operating in the Middle East has published such reports. The Anglo-Iranian Oil Company used to publish in English, for the benefit of its shareholders, the *Annual Report and Accounts,* the bulk of which was devoted to balance sheets, profit and loss accounts, and reports of the auditors. In addition, these reports contained statements by the chairman of the board in which he usually emphasized the financial and technical aspects of AIOC's operations, with an occasional brief reference to nonbasic activities. Although the reports gradually grew in size (from a thin unbound brochure of 12 small pages in 1945 to a paper-bound pamphlet of 33 large pages with black-and-white illustrations in 1951), it was evident that they differed from Aramco's reports both in their concept and execution and could hardly be considered as an instrument of public relations vis-à-vis the host country.

IPC stands somewhere between AIOC and Aramco. Each year the company publishes a pamphlet entitled *Iraq Oil in 1951* [*1952, 1953,* etc.] in English and Arabic versions. About 30-odd pages, it contains colored illustrations, maps, and graphs not unlike those in Aramco's reports. IPC's pamphlets definitely emphasize technical aspects, such as exploration, production, and pipelines. Social problems, however, are not ignored, and each issue, especially since the adoption of an integration policy in 1952, contains references to the home ownership scheme and similar employee benefits. The latter are interspersed somewhat at random among technical items and are seldom, if ever, given subtitles. The pamphlets are not provided with tables of contents and in their briefness and informality contrast with the elaborate publications of Aramco. There is no doubt, however, that they are a product of a conscious public relations policy, and the existence of an Arabic version testifies to the company's desire to make them available to Iraqi readers. The company also publishes from time to time special pamphlets supplying information about the conditions of its personnel and related social matters. Of these, *A Hundred and One Facts about the IPC* devotes as much as 50 per cent of its brief but well-organized text to topics not directly concerned with oil technology,[9] while the brochure entitled *The Employee of the Iraq, Basrah and Mosul Petroleum Companies in Iraq* deals entirely with employee benefits. Such publications do a great deal to fill the gaps in the annual pamphlets.

Press Relations

Another area where one can observe a divergence between British and American methods is in press relations. Through their press officers the British-managed companies maintain an informal personal relationship with local editors in the hope that the latter will refrain from printing material hostile to the companies and instead publish either favorable editorials and reports of their own or such materials as the companies supply. Although not rejecting the method of personal contact, the American-managed companies —Aramco and Tapline—tend to rely heavily on institutional advertisements. The latter are placed in the papers of all those countries whose political attitudes directly affect Aramco's and Tapline's operations, i.e., Saudi Arabia, Lebanon, Syria (with the exception of a brief period when foreign advertising was banned), Jordan, and Egypt. Their general aim is to correct whatever distorted image of the oil industry has taken hold of the public mind. Consequently they emphasize the multifarious benefits that the industry is bestowing upon the host countries, while stressing the complexity of the operations and the special skills that are required. They dwell particularly on the following major themes:

1. The revenues flowing to Middle Eastern countries from oil operations and the employment opportunities for their nationals.

Government income from oil operations [said one of them] has jumped from $100,000,000 in 1948 to over $900,000,000 in 1955—and total income is expected to pass the $1,000,000,000 mark in 1956. Besides, the oil industry in the Middle East employs nearly 200,000 Arab nationals and allied industries provide a livelihood for many thousands more.[10]

2. The stimulation of economic development by the oil industry and the latter's role as an integrated part of the local environment.

[9] *Miat Zuruf wa Zarf an Sharikat Naft al-Iraq* (Baghdad, 1950).
[10] Tapline advertisement in *Ar-Rawad* and other Lebanese papers, Oct. 21–27, 1956.

OIL INDUSTRY'S REACTION

The following excerpt is a typical illustration of this type of advertisement:

A STONE CAST INTO WATER

A stone cast into a pool creates a series of ripples across the whole surface of the water. Similarly, the introduction of a new industry in any area results in benefits that spread beyond the immediate goal of the enterprise. During the eighteen years of commercial production of crude oil by the Arabian American Oil Company, many new businesses have been established in the Kingdom of Saudi Arabia. Each has helped to stimulate the growth of the country's whole economy by filling the demand for its services, increasing employment and providing a need for further services.[11]

3. *Benefits to the employees.* Aramco's advertisement of a newly constructed dining hall in Dhahran carried the title "Good Food at Low Cost" and extolled the advantages of a modern industrial cafeteria. "Powerful machines and giant steel workshops are necessary to the oil business," read the text, "but it's the men who run them that count." [12]

4. *The virtues of free enterprise.* These, rather paradoxically, are carried not only in the regular "bourgeois" newspapers, but also in organs known for their pronounced leftist proclivities. The general pattern of these advertisements is to point out the better life enjoyed by the present generation in comparison with its forebears and to give credit for it to the free enterprise system.[13] Deeper values are also emphasized on these occasions:

A wise man has defined human happiness as having things a little better each year, both materially and spiritually, than they were the year before. The free enterprise system is the best method yet devised for making things a little better each year because it is the only system that makes full use of teamwork. . . . Tapline is a good example of teamwork.[14]

[11] Aramco advertisement in *Al-Ish'a'a* (Al-Khobar, Saudi Arabia), Nov., 1956, and in *Al-Yaum* (Beirut) and other Lebanese papers, Oct. 28–Nov. 3, 1956.

[12] Aramco advertisement in *Al-Majalis al-Musawara* (Beirut) and other Lebanese papers, Dec. 16–31, 1956.

[13] A Tapline advertisement in *Raqib al-Ahwal* (Beirut) and other Lebanese papers, Feb. 10–16, 1957.

[14] Advertisement in pro-Communist *Ash-Sharq* (Beirut) and other Lebanese papers, Feb. 17–23, 1957.

5. The complexity of oil operations in both technical and economic phases. To possess oil in one's subsoil is not enough, is the argument employed. Risk capital, technological know-how, and well-developed marketing organizations are other necessary ingredients of a successful oil enterprise. The world is economically interdependent and the "oil reserves of the Middle East countries constitute but half a necessary combination." "Without a market, the oil has only a theoretical value." [15] The interdependence is stressed by such titles as "Links in a Chain" [16] or "The Refinery." "Crude oil," said the latter advertisement, ". . . is valueless until transformed by the modern magic of refining into fuels, lubricants, and the materials for hundreds of diverse petroleum products." [17]

6. The role of the pipelines as one of several means of transportation and their essential character as carriers rather than as oil companies. The need to maintain their competitive ability vis-à-vis tanker transportation is repeatedly stressed.[18]

By and large these themes reflect noncontroversial matters or at least are couched in noncontroversial terms. An exception to the rule occurred in 1956, when Lebanon passed a law taxing concessionary companies and Tapline decided to bring its case before the public.[19]

Although institutional advertising has been practiced almost exclusively by the American-managed companies, occasionally—perhaps more as an experiment than as a general policy—a British-managed company has resorted to a similar device. For example, IPC's Public Relations Department published a series of graphs which presented in the form of a puzzle the growth of the company's home ownership scheme for its employees.[20]

[15] Tapline advertisement in *At-Tarbia* (Damascus) and other Syrian papers, Feb. 3–9, 1957.

[16] Tapline advertisement in *Ash-Sharq* (Beirut) and other Lebanese papers, Nov. 25–Dec. 1, 1956.

[17] Aramco advertisement in *Sada Lubnan* (Beirut) and other Lebanese papers, Nov. 25–Dec. 1, 1956.

[18] See such Tapline advertisements as "Tapline and the Transport of Oil," *An-Nidal* (Beirut) and other Lebanese papers, March 24–30, 1957; and "Transportation of Oil," *Al-Ayyam* (Damascus) and other Syrian papers, March 24–30, 1957.

[19] *Sada Lubnan* (Beirut) and other Lebanese papers, July 29–Aug. 4, 1956.

[20] In six issues of *Ash-Sharq* (Baghdad) between July 30 and Aug. 3, 1957.

OIL INDUSTRY'S REACTION

The reluctance of the British-managed companies to employ institutional advertising did not prevent them from freely using press releases. The latter varied in content from current production statistics to general articles about the oil industry. Employee benefits and all the advantages to the economy of the host country flowing from the policy of integration were a recurrent theme of IPC's handouts,[21] which in 1956 accounted for 29 per cent of all items published on the subject of oil in the press of Iraq.

While these differences between British and American approaches should be recorded as indicative of different concepts and operational habits—not to speak of more profound divergences stemming from the totality of cultural and psychological backgrounds of the two executive groups—there were also many similarities due to the general "internationalization" of public relations methods. Two such similarities deserve special mention, namely, the growing use of motion pictures as a communication medium and the publication of employee magazines in the native tongues of the Middle East.

Employee Magazines

Employee magazines are a medium to the development of which both IPC and Aramco devoted major efforts in the 1950's. IPC's Arabic-language magazine, *Ahl an-Naft* (The People of Petroleum), began in 1951, the year which in many respects opened a new era in the company's relations with Iraq. At the time of the conclusion of the revised concession agreement the company embarked on a number of new policies generally aiming at maximization of the advantages inherent in the new profit-sharing formula. As a result the company became much more alert than before to the social and political aspects of its operations and launched serious efforts to establish a closer rapport both with its own employees and the

---

[21] The following are typical examples of IPC's releases: "Oil and the Local Market" (a feature article stressing the fact that in 1956 forty-two different contractors undertook building work for the Basrah Petroleum Company for almost half a million dinars); "Basrah Housing Scheme Forges Ahead" (about the 200th house built under the home ownership scheme); "The new Arrapha Estate Canteen in Kirkuk"; "I.P.C. Awards Prizes for Essays on Oil Industry by Kirkuk Students"; "U.K. Training for Iraqi Staff at Mosul Petroleum Company"; and "New Building Techniques in Iraq" (refers to the home ownership scheme in Kirkuk).

public at large. The policy of integration represented one facet of this new approach, and the stepped-up public relations activity added another indispensable element to the over-all policy aiming at the establishment of closer and friendlier relations with the host country. A well-conceived and properly executed employee magazine in Arabic was a logical need if the new invigorated atmosphere was to be sustained. Such a magazine could aim at three major objectives: (*a*) providing the employees with superior yet easily digestible reading matter which would both fill the gap in existing Iraqi periodical literature and successfully compete with politically tinged magazines produced in outside xenophobic centers; (*b*) stressing the advantages the host country was obtaining from the operation of the oil industry; and (*c*) creating general good will toward the company by publicizing its employee benefits and its progressive spirit.

The editors of *Ahl an-Naft* seem to have these objectives in mind. The fine format distinguishes it from practically everything else published in the Arab East. Its editors are careful not to make it like the typical employee magazine produced by a big corporation in the West. A regular issue of *Ahl an-Naft* contains very little about oil or the company. The subject matter is varied, including topics usually found in ordinary illustrated periodicals. There are items of general cultural interest, belles-lettres including poems (this is a popular literary form in the Middle East), current news and features, articles popularizing medicine and child care, adventure stories, and geographical descriptions. Unobtrusively, yet frequently, the magazine carries stories about the development of Iraq, stressing the magnitude of the dams and the irrigation and electrification projects executed by the government. This implies, of course, funds for such projects, yet direct references to the oil industry as the chief provider of these revenues is generally avoided. In the spring of 1957 when the government of Iraq organized country-wide celebrations under the name of "Development Week," *Ahl an-Naft* devoted a special issue to this event, covering the government's achievements probably better than any other private or official publication in Iraq.

Aramco's employee magazine, *Qafilat az-Zait* (Oil Caravan), was first published in 1953. Its relatively late appearance was not due so much to lack of company interest in this medium as to Saudi Arabian environmental conditions, which necessitated both a careful preliminary study and official clearance before the magazine could be launched. Whereas the publication of a magazine was considered by Iraqi authorities as a matter of course, in Saudi Arabia, where only three newspapers existed in the whole kingdom in 1954 [22] and where the regime adhered to strict Wahhabi orthodoxy, publication of a magazine by an alien institution was enough of an event to justify careful examination on the part of the authorities before clearance was granted. Even after it was given, the magazine was subjected to censorship and any deviation, however unintentional, from established routines is apt to result in adverse consequences, such as temporary suspension. The editors have to be wary of two stumbling blocks: opposition by conservative religious circles to anything not issued under their auspices and the inclination of the employees to treat as "propaganda" anything published by the company. Aramco's Public Relations Department found understanding and co-operation in the Saudi Directorate of Press, Publications, and Broadcasting, whose chief and officials perceived the advantages to the government, partly in the field of labor relations and partly in the general sphere of publicity, that such an employee magazine could offer.

In content and execution *Qafilat az-Zait* is similar to *Ahl an-Naft*. Richly illustrated, it carries articles on a variety of topics not usually connected with the oil industry. To the extent to which it is compatible with the general concept of the magazine as a medium of information and entertainment, emphasis is laid on the development of the country. But whereas IPC's magazine can devote considerable attention to government-sponsored projects, Aramco's organ is inclined (and obliged) to lay more stress on private initiative resulting from the general stimulation of Saudi economy by oil revenues. In its choice of topics, *Qafilat az-Zait* has to take into account the peculiarities of Saudi society, which is much more conservative than the Iraqi public. Thus while IPC's organ can publish

---

[22] These were the official paper *Umm al-Qura* and two privately owned papers, *Al-Bilad as-Saudiyah* and *Al-Madinat al-Munawwarah*.

pictures of art works (including fairly exposed human forms), fashion shows, and females, as well as short romantic stories, Aramco cannot. It does, however, have articles on religious subjects such as "Islam and the Fast" and does a good deal to popularize Arab history and cultural attainments among its readers. Articles of historical, cultural, and religious interest are frequently solicited from Saudi or other Arab authors. An interesting and thoroughly American feature of *Qafilat az-Zait* is the attention paid to Aramco's Arab worker as an individual. Every issue of the magazine carries columns entitled "Personality of the Month" or "Sportsman of the Month" and pictures of the "Men of Aramco." Individuals for these columns are selected on the basis of merit. Consequently many a worker clad in overalls and the traditional Arab headgear has found himself photographed, described, and praised in these articles. Devices like this cannot remain without effect on employee morale.

It would be difficult to underestimate the importance of these magazines in the companies' public relations. In content and form they can compete with the best privately published periodicals of the Middle East. Their circulation is considerable. It can be taken for granted that they are reaching not only the employees but their relatives and friends as well. By offering their columns to native literary talent they are establishing friendly relations with the Arab class most open to extremist ideas and most likely otherwise to direct its energies toward negative objectives. Most important of all, they serve as a channel of information and comment which is free of the bias and emotionalism characteristic of ultranationalistic papers and literature.

SUMMARY

It is clear that an oil company operating in the Middle East cannot restrict its contacts with the host country to a formal relationship with the latter's government without incurring grave risks to its status and the continuity of its operations. Because of its size and uniqueness as one of the main providers of revenue, such a company can easily become a target for the distressed and frustrated elements in the host country as well as for those groups which have made it part of their political program to discredit Western institutions and enterprises in the eyes of native publics. Critics of the oil industry repeat standard grievances over and over, but the intensity of these attacks fluctuates in accordance with two distinct sets of circumstances. One set, largely outside the companies' control, consists of the actions of the companies' home governments in the field of international politics and of the repercussions that such actions evoke abroad. The Palestinian and the Suez crises are good examples of such developments. The other set depends largely on the companies themselves. Their basic decisions and policies concerning the division of profits with the host governments, their willingness to contribute above and beyond royalties to the economic development of the host countries, and their readiness to integrate with the local environment, may lay the foundations of a new relationship between themselves and the host countries based on trust and mutual advantage.

In thus adjusting their conduct to changing times the companies must take into account the existence of a new factor, public opinion. It may not express itself the way it does in the West and may run a wide gamut ranging from sophisticated articles in the local press to mob demonstrations in times of acute political crisis. It is still immature and lends itself to twisting and exploitation by demagogues. But it exists, and to deny it is courting disaster, as events in Iran during Mossadegh's era have so eloquently proved. This being the case, it is not illogical for companies to rely more and more on public relations as an additional safeguard of their interests. In employing various media of communications with the public, the companies acknowledge the fact that their security and well-being depend on the image of themselves that they succeed in creating in people's minds. An indispensable psychological weapon, public relations techniques can be abused as well as used. Abuse occurs when the torrent of words and ideas poured forth does not conform to the contours of truth. Ideas must be expressed, because the truth, despite pious assertions, will not always speak for itself. But there is also need for vigilance lest the words stray from reality.

Thus we come back to the problem we discussed at the beginning of our chapter—the companies' reaction to unfriendly criticisms. We ventured the opinion that some of these charges could be destroyed or mollified by the behavior of the companies, and we have singled out two basic policies likely to affect the companies' status and prestige vis-à-vis the host communities, the policy of stimulating economic growth and the policy of integration. Inasmuch as our frame of reference has been the relationship of the companies with the host public at large, we have purposely omitted matters that are of direct interest to the companies and their native employees. These must not be neglected if one aims at a rounded picture of the oil industry's position in the Middle East. It is therefore toward this sector of relations that we will turn in the following chapters.

PART FOUR

COMPANIES AND EMPLOYEES IN THE HOST COUNTRIES

CHAPTER XIII

Oil Workers: A New Force

LAUNCHING an extractive industry has always contained a considerable element of risk. In the case of the Middle Eastern oil enterprise the risk has been compounded by the emergence of labor problems of hitherto unknown proportions. The employment of thousands of workers by the oil industry has had multifarious effects on the socioeconomic processes of the Middle East. It has meant a massive drift of the population from the villages to the oil camps or cities. The uprooting of old loyalties and the shaking of traditional values have been an inevitable consequence of this movement. Once harnessed to an industrial machine, the former villager has become acquainted with new material standards. He has, moreover, become exposed to contact with Western men, their technology and their customs. He has tasted the benefits of high wages and steady income, but he has had to pay for these advantages by diminution of his freedom and by acceptance of the heavy demands of discipline, orderliness, and punctuality. His individuality has been submerged in a mass effort and he has lost a great deal of whatever personal prestige he enjoyed in his old village. On the other hand, he has found himself living and working in the largest single agglomeration of industrial labor in his native country (see Table 5), and if organized he has discovered the tempting possibilities of concerted political action.

Table 5. Employment in oil industries in relation to general industrial employment (in round figures)

	Petroleum Industry (1)	Total Industry (2)	Percentage of (1) to (2)
Iran	49,000	155,000	35%
Iraq	12,000	57,000	21%
Saudi Arabia	16,000	40,000	40%

Sources: UN, *Economic Developments in M.E.*, p. 30; Iraq, Ministry of Economics, *Report on the Industrial Census of Iraq* (Baghdad, 1956); also George L. Harris, *Iraq* (New Haven, 1958), p. 324; and Aramco, *Report of Operations, 1957*; Saudi Arabia, Ministry of Commerce, *Al-Mamlakat al-Arabiyah as-Saudiyah, Tasjil wa Ta'arif (Kingdom of Saudi Arabia, Records and Data;* Damascus International Fair, 1376 A.H. [1956]).

LABOR LEGISLATION

Both the governments and the companies have become aware of the magnitude and complexity of the labor problem and, especially in the period following World War II, have expressed their interest in various ways. The governments have naturally had recourse to regulatory legislation. While labor laws have usually purported to affect industrial labor as a whole, they have frequently been drafted with an eye to the petroleum workers, in view of the latter's numerical and qualitative importance.

Of the three major producing countries—Iran, Iraq, and Saudi Arabia—Iraq was the first to recognize the need for labor legislation by enacting in 1936 Labor Law no. 72.[1] Amended by Law no. 36 of 1942 and supplemented by a number of ministerial regulations and notifications, the law dealt with conditions and hours of work, compensation for death, injury, or disease, the formation of trade unions, conciliation and arbitration procedures, government inspection, and a variety of miscellaneous matters. It was accompanied by a schedule of disabilities and another schedule of dangerous and unhealthy industries, and it had all the characteristics of modern labor legislation. Of particular interest to the oil industry were the following major provisions, contained either in the amended law

[1] The text is in *Al-Waqai' al-'Iraqiyah*, no. 1511, April 30, 1936.

OIL WORKERS

or in the subsequent regulations, many of which were enacted in the 1950's:

1. The principle of the 8-hour day and 6-day week was established.

2. The right to form trade unions was recognized. This right was, however, subject to licensing by the Ministry of the Interior, which in the Middle East invariably exercises political surveillance. The Ministry of the Interior was given authority to refuse a license, in which case appeal could be made to the Council of Ministers for final decision. Article 28 of the law further emphasized the government's political authority in this respect by stating:

> In the event of its being proved that an association is conducting its affairs in a manner which would lead to a breach of public security, or of the safety of the State, or which would prejudice the interests of the work and the proper performance thereof, the Council of Ministers may, at the request of the Minister of the Interior, cancel the permit.[2]

3. The law authorized the government to issue regulations concerning the construction of houses for workers. In 1947 Law no. 29 proclaimed that the Minister of Social Affairs "may obligate owners of industrial undertakings employing more than one hundred workers, to build special houses for their workers in certain localities, provided that public places such as hospitals, etc., are constructed there." [3]

4. The law also authorized the government to fix minimum daily wages for workers according to their occupational classifications. Pursuant to this provision, the government issued a regulation in 1953, amended in 1954, fixing the minimum daily wage for an unskilled worker aged sixteen years and over at 250 fils, and for the young person (*murahiq*) aged twelve to fifteen years at 180 fils.[4] This is still the minimum wage in Iraq.

5. Pursuant to provisions of the law, the government in 1954 enacted Regulation no. 63 concerning conciliation and arbitration procedures for the settlement of disputes between employers and

---

[2] *Ibid.*   [3] The text is in *Al-Waqai' al-'Iraqiyah* no. 2498, July 24, 1947.
[4] In U.S. currency 250 fils is approximately 70 cents. For the text of the regulation see *ibid.*, no. 3333, Dec. 17, 1953, and no. 3367, March 8, 1954.

workmen. "A strike or lockout will not be considered legal unless the Minister [of Social Affairs] has been notified at least fourteen days before its occurrence" and unless reasons for such action have been given. If a dispute is not settled by conciliation, the Minister of Social Affairs is authorized, subject to the consent of both parties, to refer the dispute to arbitration by either (*a*) appointing a single arbitrator or (*b*) setting up a board of arbitration "consisting of one or more persons nominated by the employer and an equal number of persons nominated by the workmen party to the dispute and an independent experienced person as chairman." Decisions rendered by the arbitrators are to be final and enforceable.[5]

The year 1936 marked the beginning of official interest in labor problems in Iran also. Regulations issued on August 10, 1936, by the Iranian Council of Ministers dealt with industrial hygiene and safety, the obligations of workers and employers, wages, and the inspection of working conditions but made no provision for the formation of trade unions.[6] The war and its aftermath produced a good deal of social unrest in Iran, which directly affected industrial workers. Spurred by these developments, the Council of Ministers on May 18, 1946, adopted a labor law which with later amendments and numerous regulations dealt comprehensively with various aspects of labor conditions in the country. The law was not formally enacted by Parliament until March 1949. With its official promulgation on June 7 of that year (Khordad 17, 1328), it became the basic labor legislation of the postwar period.[7]

Three sections of the new law deserve special attention in view of their probable impact upon the oil industry, namely, those dealing with minimum wages, trade unions, and the settlement of labor disputes. In contrast to the unitary wage system set up in Iraq, the Iranian law permitted differences in the minimum wage depending on the industry or the area of the country. The law itself did not decree any rigid minimum. Instead it provided that minimum wages should be determined once a year and "as occasion arises" by a committee composed of "the local Farmandar, the President of the Mu-

---

[5] For the text, see *ibid.*, no. 3477, Oct. 5, 1954.
[6] The text is in International Labour Office, *Legislative Series, 1936—Iran 1*.
[7] *Ibid., 1949—Iran 1*.

nicipal Council, the governor of the Bank Melli, or their representatives, a representative of the Ministry of Labor, two employer representatives, and two worker representatives." The committee's decisions acquired validity after approval by the High Labor Council.

One innovation was that for the first time the law officially recognized the right to form syndicates and trade unions. Although the law spoke only of the representation of unions, a subsequent regulation, issued by the High Labor Council in March 1947, established a full-fledged system of licensing. Applications to form a union (accompanied by the statutes proposed) were to be submitted to the Ministry of Labor and Propaganda and if found contrary to the law could be rejected by the Ministry. The Ministry's decisions in this respect were subject to appeal to the High Labor Council. Federations of two or more unions were permitted, but the union shop was forbidden. The law did not ban political activity by unions. It provided, however, for suspension or total dissolution of any union established in contravention of the law, or exceeding its specified rights and limits, or disturbing public peace. In all such contingencies authority to act was vested in the local courts, which were to issue a decision at the request of the Ministry of Labor acting through the Prosecutor General. In case of suspension the court's verdict was to be final; in case of dissolution it was subject to appeal.

With regard to the settlement of disputes the law stipulated that conciliation should be sought by a conciliation committee composed of a worker representative and an employer representative. Should this method fail, the dispute was to be referred to arbitration by an adjustment board consisting of the worker and employer representatives and a delegate from the Ministry of Labor. If unanimous, the adjustment board's judgments were to be final and binding. If not unanimous, they could be appealed to the Board for Settlement of Disputes. This board was to be composed of the *Farmandar* or his representative, the head of the Justice Department in the *Shahrestan* (district) or a judge representing him, a representative of the Labor Ministry, two employer representatives, and two worker representatives. The three groups, i.e., the government's representatives, the employer's representatives, and the workers' representatives, were to have only one vote each. The board's awards were to be final if based on a unanimous or majority vote. Strikes were lawful only if proclaimed after the expiration of specified periods of time. The law provided also for the establishment of a High Labor Council composed of representatives of various government agencies, trade unions, and employers. The Council was to act as the highest regulatory agency handling labor conditions. At a later date unsuccessful attempts were made to adopt a new law providing for collective bargaining. The Labor Law was supplemented by workers' social insurance legislation in 1943, 1955, and 1957.[8]

The last of the major producing countries to pass labor legislation was Saudi Arabia. On October 10, 1947, a high decree sanctioned the first "Labor and Workman Regulations" in the history of the kingdom.[9] Composed of sixty articles, the act covers various aspects of labor conditions. It provides for an 8-hour day and a 6-day week, decreeing, however, that daily working hours must be "interrupted by one or more intervals for prayers." Among other obligations employers must "furnish suitable living quarters for workmen employed in localities remote from centers of population" and set up "a plan for thrift and savings for the workmen." Disputes between an employee and an employer are to be submitted to arbitration, which is to be handled by a committee consisting of two members, one to be appointed by the employer and the other by the government. Should these two members disagree, the Finance Minister is to appoint a third arbitrator to act as an umpire. This paternalistic approach to labor relations is further emphasized by a clause providing that "no workman shall be allowed to work in an industrial project unless he is in possession of a work permit . . . issued by the Government." No mention is made of trade unions.

The law has an interesting nationalistic aspect in that it specifically refers to disagreements between foreign and national workmen and to "personal offenses committed by a foreign individual against a subject of the Kingdom of Saudi Arabia either by hand or by tongue," including such manifestations as scoffing, "scorn, or contempt." In such cases the accused individual is to be brought to

---

[8] A new labor law is now (April 1959) being drafted in Iran.

[9] The text is in Saudi Arabia, Ministry of Finance, *Nizam al-'Amal wa al-'Ummal* (Labor and Workman Regulations; Mecca, 1367 [1947-48]).

trial (in the Sharia, i.e., the religious courts, the only ones in the kingdom). If the accused is found guilty, "due punishment must be imposed, including the assurance of personal liability, in addition to deportation from the Kingdom, should this be necessary." [10]

This basic law was supplemented on June 11, 1956 (2 Dhu al-Qa'adah 1375 A.H.), by a royal decree concerning strikes and disturbances.[11] Both are expressly forbidden to employees and workers of concessionaire companies and private firms "performing work of a nature benefiting the public or undertaking a public project for the Government" under penalty of imprisonment. Incitement to such acts "by word, deed, sign, writing, drawing, or any other device" is liable to even heavier penalties than participation itself. Moreover, the employer is given the right to discharge any worker guilty of such offenses, and the amir (governor) of a province where a strike or disturbance has occurred is empowered "to compel the employer to discharge any employee or worker who has been penalized for committing any of the violations referred to" above.

This then is a brief summary of the labor legislation adopted in the three leading producing countries. The Iranian laws seem to be the most advanced, setting up an elaborate arbitration procedure, providing for flexibility in minimum wage procedures, and establishing a High Labor Council, in which trade unions as well as government and employer elements are expressly represented. At the other end of the spectrum stands Saudi Arabia, whose legislation, though signifying official awareness of labor problems, is still largely of a paternalistic character. This is especially evident in those provisions which exclude workmen from the arbitration bodies and forbid strikes and disturbances. This gradation in legislation from the more to the less advanced corresponds in general to the advancement of the countries themselves and the complexity of their labor problems.

This observation extends to the transit countries as well. Syria and Lebanon, with their rather advanced labor movements, have adopted fairly progressive labor legislation, recognizing trade union-

[10] *Ibid.*, art. 39.

[11] Royal Decree, no. 17/2/23/26 39. The text is *Umm al-Qura* (Mecca), 13 Dhu al-Qa'dah 1375 (June 22, 1956).

ism and providing for elaborate arbitration machinery designed to safeguard the workman's interests.[12] In Syria the basic labor law has been supplemented by a number of *ex parte* laws affecting the Iraq Petroleum Company specifically. These laws were enacted in the mid-1950's in connection with labor disputes that arose between the company and its workmen as a result of the destruction of the pumping stations in the wake of the Suez crisis. This *ad hoc* legislation is reviewed in Chapter XVI within the general framework of the political situation then prevailing in Syria.

The observation that labor legislation in these countries has, on the whole, kept in step with the degree of their advancement and social consciousness is of course general in nature. Laws may play one of three roles: they may anticipate trends and pioneer in setting up new principles; they may reflect the *status quo;* or they may check spontaneous social developments by imposing restrictions. It is impossible, a priori, to say which of the three roles is morally superior and socially most useful. In any case laws by their very nature must contain a certain degree of restrictiveness. This does not necessarily signify that they are reactionary in character. Recognition of trade unionism may be cited as an example. Trade unions represent a form of social organization which fills a definite social need in societies possessing a certain level of development. But are they a positive phenomenon in any society, regardless of its social and educational level? Are they the right form of organization for illiterate and superstitious people in a semitribal society, a people easily swayed by outside leaders whose motives may have little to do with the welfare of the working class? If trade unions were to become an instrument leading to the establishment of a fascist or Communist totalitarianism, could their existence be justified? It would appear, therefore, that questions of this sort should not be resolved by positing an abstract principle without regard to sociopolitical reality. Consequently it may be appropriate at this juncture

[12] See the Lebanese Labor Code of Sept. 23, 1946; the text is in suppl. no. 40 to *Al-Jaridat ar-Rasmiyah* (Official Gazette), Oct. 2, 1946; also see the Syrian Labor Code of June 11, 1946; the text is in *Al-Jaridat ar-Rasmiyah* (Official Gazette), no. 25, June 13, 1946.

to have a brief look at labor conditions in the countries reviewed and see how they square with the legislation previously cited.

IRANIAN INDUSTRIAL RELATIONS

Among the producing countries Iran has had the longest and the most turbulent record of industrial relations. This is partly due to the fact that its oil industry is the oldest in the Middle East, having passed its first half-century of operations. Even more important are such factors as the presence of the largest single aggregation of industrial labor, both in relative and in absolute figures, in any country of the Middle East (as much as 62,000 during the peak employment period in 1950); the location of the industry in a torrid zone far removed from any natural urban center; the attitudes of management; and the politically inspired unionization of the oil workers.

The last two points were, of course, crucial. The attitudes of management found expression in concrete company policies affecting the material welfare of the workers. They will be examined more systematically in a later chapter. It suffices here to say that industrial relations in the old Anglo-Iranian Oil Company had been fairly harmonious until the end of World War II. Workers found many causes of complaint prior to that date, but until then AIOC's labor-management relations were generally devoid of dramatic upheavals and violence. The situation changed drastically under the impact of the war. This was due to a combination of objective and subjective factors. Objectively wartime shortages and inflation rendered the life of workers more difficult. Subjectively unionization influenced the relations of workers and company.

Ever since the creation of the Communist Tudeh Party in 1942 its leaders had made major efforts to organize trade unions in the country as a whole. The Tudeh-sponsored Federated Trade Unions of Iranian Workers, headed by a Central Council, was established on May 1, 1943. Its secretary-general, Reza Rusta, had been one of the original founders of the Tudeh Party and in 1944 became a member of the party's eight-man Control Commission. In 1945 the

Central Council of the Federated Trade Unions became affiliated with the Soviet-dominated World Federation of Trade Unions. Later it protested vigorously when its representative was given a subordinate position in the Iranian workers' delegation designated by the government to attend the first postwar meeting of the International Labor Organization. The council's secretary-general, Reza Rusta, toward the end of the war became most active in the movement to unionize the oil workers in Khuzistan. When the war ended, the oil workers' union was already in existence as a separate organization but connected through its leaders both with the Tudeh Party and the Central Council.

The formation of the union coincided with a general deterioration in the political conditions of the country, characterized by an upsurge in Communist activities, considerable military and political penetration by Russia, and a corresponding intensification of Soviet-Western rivalry. When Russia made a bold bid to subject Iran to her domination by instigating a Communist rebellion in Azerbaijan in 1945–1946, the Tudeh-infiltrated oil workers' union resorted to strikes and violence in Khuzistan in what appeared to be well-synchronized strategy. First, a strike occurred in May 1946 in the oilfields. The government answered by proclaiming martial law in the affected areas. On July 16 another strike broke out, in Abadan. Company and contractors' workers as well as other laborers—estimated at 100,000—were involved in this strike, which abounded in violence and resulted in 17 killed and 150 wounded among AIOC's European and native personnel. About 300,000 tons of oil were lost, and a number of acts of sabotage were committed. The political character of these strikes and their destructiveness led the government to adopt stern measures to prevent a recurrence. The union was broken up, and punitive steps were taken against Reza Rusta and other organizers.

After the evacuation of the Red Army from Iran the government dealt a decisive blow to the rebel regime in Azerbaijan, which collapsed in December 1946. Thus strengthened internationally, the government was in a position to restore law and order inside the country. In 1947 the Tudeh-dominated Central Council of Federated

Trade Unions was disbanded, and the leadership of the labor movement was taken over by the two rival organizations: the ESKI (Federation of Trade Unions of Iranian Workers) and the EMKA (Central Council of Unions and Workers and Peasants), the first of which enjoyed active government support. A fresh start was made in Abadan and the fields, where two new unions, independent of the above-mentioned federations and recognized by the government, were created.[13] Both were free of Tudeh influence, at least insofar as their executive bodies were concerned.

During the next four years relative peace reigned in labor-management relations. In September 1949 a small and orderly strike of restaurant workers occurred in Abadan. Purely economic, it was settled without much delay. It was followed in October by a strike in favor of an unsuccessful candidate in the parliamentary elections. Despite the political motivation no Tudeh influence was discernible in this strike, which was also of short duration. During the next seventeen months there were no labor disturbances.

The situation changed in the spring of 1951 when the Nationalization Law was passed by the Iranian Majlis. Urban centers were seized by a frenzy of nationalist fervor, a state promptly exploited by the Communists. Late in March the Tudeh (outlawed since 1949) called for a general strike in Khuzistan, ostensibly in a demonstration of solidarity with the workers in one of the fields, who had struck because of a reduction in some "outstation" allowances by the company. The strike not only spread to other fields and Abadan, but on April 12 degenerated into riots which took the lives of three Britons and nine Iranians. The strikers resorted to intimidation and picketing. The huge refinery was compelled to shut down, the airport was closed, and communications with the outside world were disrupted. Having proclaimed martial law on March 26, the government made arrests among the Communist strike leaders, but it was not until the army seized the strikers' headquarters in Abadan in mid-April that the refinery could resume production. Considerable restlessness persisted among the workers, and in midsummer the operations ceased

[13] The Trade Union of Workers of Khuzistan and the Trade Union of Petroleum Workers.

altogether as a result of the deadlock in the negotiations over nationalization.[14]

It would appear from the record of the six postwar years that intensification of union activity was directly related to the political situation in Iran and, more particularly, to the upsurge in Communist agitation. It is no wonder therefore that, when the turbulent Mossadegh era came to an end as the result of the August 1953 coup and the new government of General Zahedi dedicated itself to the restoration of normalcy, union activity should resume a more orderly pattern and in some cases cease altogether. This was reflected in the labor situation in the oil industry after the conclusion of the Consortium agreement in 1954. Whereas trade unionism in the country as a whole was under the aegis of the government-sponsored Congress of Workers' Syndicates (an heir to ESKI and like its predecessor affiliated with the International Confederation of Free Trade Unions), oil workers remained unorganized.

To fill the gap created by the lack of organization, the Consortium —together with the National Iranian Oil Company—promoted the formation of conciliation committees within the operating companies. A number of such committees were established in the fields and at Abadan. A conciliation committee was composed of management and labor representatives, the latter appointed by the Ministry of Labor. No regular schedule of meetings was provided for. Although appointments were initially made only in connection with particular cases, eventually—by mid-1958—the Ministry appointed workers' representatives for all areas of operation. In the same year the Consortium tried another experiment in industrial relations, dictated by the fact that many of its staff positions were held by Iranians. Because favorable attitudes among these employees were as important as among laborers, an institution known as "staff consultative committees" was launched in Abadan. This provided for a

---

[14] For various aspects of the Communist movement and trade union activity, see George Lenczowski, *Russia and the West in Iran, 1918–1948* (Ithaca, N.Y., 1949), with a *Supplement*, 1954. For the development of trade unions, consult also International Labour Office, *Labour Conditions in the Oil Industry in Iran* (Geneva, 1950) and M. A. Djamalzadeh, "An Outline of the Social and Economic Structure of Iran," *International Labour Review*, vol. LXIII, Jan. and Feb., 1951.

monthly meeting of Iranian staff representatives with the top management of the operating companies. Staff representatives consisted of five members chosen by the management from a list of twenty-five submitted by the Iranian staff. It was an innovation in the industry's practices in the Middle East, revealing as it did a greater awareness of the problems of the white-collar nationals employed by a foreign-managed company.

Mention should also be made of the informal monthly luncheons for groups of about fifteen Iranian engineers from Abadan and the fields. These functions, which were usually presided over by the general managing director, gave the best-educated (and most sensitive) group of Iranians in the companies' employment direct contact with the Consortium's top officials and enabled them to identify themselves more closely with its policies and interests.

The conciliation committees conformed to a pattern established earlier by the AIOC when so-called "joint departmental committees" had been in existence. The principal difference between these and the Consortium's conciliation committees lay in the fact that workers' representatives on AIOC's committees were elected by the workers themselves instead of being appointed by the Ministry of Labor. On the other hand, their frame of reference was somewhat narrower: they were "to consider questions affecting the persons employed in the department except those which are dealt with at a higher level or which come under the Labour Law." [15] This meant that questions such as wages or the length of working hours were excluded from consideration.

Thus far industrial relations under the Consortium have been devoid of the unrest and turbulence characteristic of the years 1945–1951. A strike of brief duration affecting less than 6 per cent of the employees occurred in workshops in Agha Jari and Masjid-i-Suleiman at the end of June 1957. The chief complaints of the strikers were the low level of the minimum wage and the high level of prices in native stores. The strike was free of disturbances and production was not slowed down. The men resumed work after hearing explanations by management and certain government officials. Soon afterward a committee was established under the chairmanship of

[15] ILO, *op. cit.*, p. 55.

the governor-general of Khuzistan. Composed of representatives of NIOC, the Ministry of Labor, and management in Abadan, the fields, and Teheran, it considered ways and means of improving the conditions of oil workers and other laborers in Khuzistan. Its deliberations led to minimum wage negotiations and eventually resulted in an agreement to increase the minimum level from 82 to 99 rials a day, effective as of July 23, 1957. Since all other classification rates were based on the minimum rate, this brought about an over-all increase of approximately 20 per cent to the whole labor force.

IRAQI INDUSTRIAL RELATIONS

Although permitted by the Iraqi labor law, trade unions were never organized in IPC and its affiliated companies. In this respect the company did not differ from the rest of the country, where trade unionism was discouraged by the government. IPC's labor force was smaller and more manageable than Iran's. It seldom exceeded 12,000, a mere one-fifth of that employed in Iran. IPC's main operations centered on Kirkuk, an old, established urban community, thus lessening the tension inherent in AIOC's isolation from its natural social environment. Extremist elements tried from time to time to stir up the workers, but they lacked the organizational skills and facilities afforded by the Tudeh Party and its ubiquitous agents in Iran. Whereas the Iranian government was often subjected to direct Soviet pressure and intimidation and sometimes, in the intricate game of politics, allowed Communist influence to spread here and there, the Iraqi government felt no need to disguise its anti-Soviet feelings. It did not hesitate to repress rather vigorously anything even remotely resembling a purposeful Communist activity. Moreover, between the quelling of Rashid Ali's nationalist uprising in 1941 and the revolution of July 14, 1958, the successive cabinets were dominated by a pro-British statesman, Nuri as-Said, who was equally opposed to any manifestations of extreme nationalism. During seventeen years of government-enforced stability Iraq followed conservative and moderate policies which left little scope for extremist tendencies and put a fairly effective check on labor turbulence.

OIL WORKERS

Yet even with this heavy lid on the organized labor movement the government did not entirely avoid an occasional flare-up. Kirkuk experienced a strike of oil workers in 1946. Rather unfortunately from the company's point of view government police showed excessive zeal in breaking up a strikers' meeting in the town, with attendant loss of life. The easily excitable Baghdad newspapers thereupon launched attacks on the company for its alleged brutality. However, the government and the company were unanimous in attributing the strike to Communist inspiration.[16]

The subsequent period of peace was broken by major disturbances in the Basra oilfields early in December 1953. On December 5 workers of the Basrah Petroleum Company went on strike demanding higher pay and better working conditions. Their manifesto contained seventeen demands covering thirty detailed points. Chief among them were the right to form trade unions; increases in wages; improvement of transportation facilities for workers and their families; improvement of canteen food; extension of medical services; stricter application of the Iraqi labor law; and no punishment of strikers by reduction of pay. The strike was not confined to oil workers alone. In a gesture of solidarity 700 members of the Cigarette Workers Union also went on strike, soon to be joined by Basra students. The latter were eager, as usual, to demonstrate against the government and foreign imperialism. Their ranks were soon swelled by city mobs, which included outside elements. Before long the strike grew into a series of riots severely impairing public security in the Basra area. Convinced that Communist agents crossing from Iran had had a hand in instigating the trouble, the government placed a number of strike leaders under arrest and ordered the police to break up any demonstrations. In the ensuing clashes one striker was killed and several were severely injured. Police action was personally supervised by the Minister of the Interior, Said Qazzaz, whose arrival on the scene testified to official concern with the political aspects of the situation. On December 16 martial law was proclaimed in the Basra area. At the same time negotiations were carried on with the strikers' committee by the Minister of Social Affairs and his delegates. This led to a settlement which involved acceptance by the company of a good many demands voiced by the

[16] Longrigg, *op. cit.*, p. 177.

workers. The company refused, however, to increase wages on the ground that they were already above the Basra wage level. The government agreed to the formation of an oil workers' union—a promise which has not yet materialized. Work was resumed in company installations on December 17.

The Basra strike had a twofold effect. It led the government to impose tighter security controls through such devices as the proclamation of martial law, the suppression of political parties, and the suspension of newspapers. Liberal tendencies in evidence before the strike tended to be stamped out by a policy dedicated to preservation of the political *status quo*. Instead of political liberalism the government offered—and began implementing—a bold plan of economic development.[17]

The Company, on the other hand, began applying with greater energy certain new policies—such as integration and home ownership schemes—which had been inaugurated shortly before the strike. Ample use was made of joint consultation committees to discuss grievances and assure better communications between management and workers. Meeting once a month, these committees covered a wide range of welfare, safety, and technical matters. But like their counterparts in AIOC, they could not discuss wages.

The combination of these two factors—governmental and industrial—led to a general pacification on the labor front so that no significant labor dispute has marred company-employee relations in the years following the Basra strike.

SAUDI ARABIAN INDUSTRIAL RELATIONS

The record of Aramco's industrial relations has been, of course, shorter in time than that of the companies in Iran and Iraq. The general setting has also been somewhat, though not entirely, different. Aramco started in Arabia with a clean slate so far as its position vis-à-vis the government was concerned. There were no political connotations in what was obviously intended to be a purely economic venture. Saudi Arabia was a sovereign state, not only not subject to American mandate or influence, but actually largely ig-

[17] See p. 49 above.

nored by official Washington. Even as late as 1943–1944 Washington was reluctant to consider any closer links with the Saudi kingdom and when a loan to its government was urgent preferred to let Britain lend what had ultimately been American funds.[18] The social environment in which the company began its operations in Al-Hasa was so primitive and unsophisticated that organized labor activity could hardly be expected to appear for some time. If we add to this that Americans came upon the scene as bearers of a humane and egalitarian tradition and as believers in a progressive approach to labor questions, we may venture the opinion that industrial relations could not have started under a better augury.

Despite these auspicious beginnings there were hard facts of physical location, structure of society, and form of government from which there was no escape. These weighed so heavily in the overall balance that the company's own policy was frequently overshadowed. Yet these environmental factors in certain important aspects resembled those prevailing in some older producing countries, thus inevitably facing Aramco with the problems that had earlier beset such companies as IPC and AIOC.

For example, there was no escape from the fact that the company had to begin its operations in a desert area far removed from any major urban center and utterly devoid of industry. Virtually everything, including nails, had to be imported. In the early days when efforts were concentrated on discovery and commercial production of oil, only essential services could be managed. This meant less than perfect housing facilities for local labor, and thus the first major problem in industrial relations arose. If not encamped near the company's installations, laborers had to commute to work from the villages and towns in which they lived. This posed the second inevitable problem—maximum efficiency with a minimum of friction in transportation. However different its American labor philosophy might be from the older European approach, the company found itself sharing the problems already experienced by the British-

[18] For details of how President Roosevelt instructed Jesse Jones, administrator of the Federal Loan Agency, to induce Britain to grant a loan to Saudi Arabia, see U.S. Senate, *Hearings before the Special Committee Investigating the National Defense Program, Part 41, Petroleum Arrangements with Saudi Arabia* (80th Cong., 1st Sess.; Washington, 1948), p. 25415.

managed corporations in the neighboring countries. Another inescapable fact was inflation, which the steady influx of money paid as salaries or contractors' compensation was bound to produce in the eastern provinces and eventually in the country as a whole. Consequently an oil worker soon learned that the purchasing power of the rials of his initially attractive pay tended to decrease with time, sometimes with discomforting rapidity.

Moreover, it does not take long for a human being to discover, and develop an appetite for, the advantages that his neighbor enjoys. Company officials (like their counterparts in Iraq and Iran) could contrast the steady work and high wages of an oil worker with the misery of a villager or a nomad, but a man is not so much inclined to compare his present status with his earlier poverty as to compare it to the position enjoyed by more fortunate contemporaries in his immediate environment (i.e., in company employment), whether his countrymen or foreigners. And here the contrast between an expatriate American and a native worker was both inevitable and impressive. Furthermore, certain efficiency requirements were also conducive to the ultimate rise of social ferment. Such was the necessity of employing the better-educated nationals of other Arab countries, such as Palestinians and Lebanese. These men performed vital functions, bridging the linguistic gap between the Americans and the Saudis and providing much-needed intermediate skills. But they also brought with them social sophistication, ambition, and political attitudes which could not fail to affect the Saudi rank and file.

Nor was the management unaware of these problems. More than any other company in the Middle East, Aramco has developed departments and research units whose task is to carry out systematic studies relating to the social aspects of the operations. Experts in all sorts of tests and social measurements are employed, and the company has accumulated an impressive store of information pertaining to the nontechnical side of its activities. But as social scientists well know, human behavior depends on intangible factors that are beyond the grasp of any measurement technique. Accurate identification and correct evaluation of these factors call for special skills, a knowledge of sociology, history, and politics, and a breadth of judgment born of experience. An observer entrusted with such a task of analysis should be placed at a vantage point, have easy access to all relevant information, possess good contacts in all strata of the society, and be as free as possible of routine hierarchical considerations. But even if the company was imaginative enough to avail itself of the services of such men, the best that could be hoped for is that errors of judgment would be reduced to the safest possible minimum rather than completely avoided.

With all the attention that the company paid to social problems, it undoubtedly experienced a shock upon learning how quickly the erstwhile tribesman has learned the advantages of organized action. Early in the summer of 1953 a workers' committee came into being in Dhahran. Claiming to represent 6,500 company laborers, it submitted a petition demanding higher cost-of-living allowances, better housing, additional transportation facilities, and schools for children. In the absence of a free press in Saudi Arabia it availed itself of Bahrein newspapers to attack the company for ignoring its demands, attempting to sabotage its efforts, and exhibiting "its abominable hostility to the Saudi people." [19] On September 21 the committee addressed an open telegram to Crown Prince Saud and followed it up with a petition on October 16 in which it asked that it be granted official recognition by the company and that the workers be permitted to form a union. The government reacted by appointing a committee of investigation, and upon the latter's recommendation promptly incarcerated the workers' leading spokesmen.

This precipitated a strike, which broke out on October 17 and affected about 13,000 Saudi Arabian employees in Dhahran, Abqaiq, and Ras Tanura. Seriously alarmed by the dimensions of the strike and by what they believed to be its advance planning, the Saudi authorities decided to resort to energetic measures to put an end to it. On October 18 armed troops were dispatched to Dhahran, and martial law was proclaimed. Two days later Crown Prince Saud, in his capacity as Prime Minister, issued a decree ordering the strikers to return to work under the penalty of dismissal. For about a week the workers defied the order. Further arrests were made, until the number of jailed strikers was about one hundred. These

[19] *Middle East Journal,* Winter, 1954, p. 71.

stern measures brought about a change in the character of the strike: from a purely industrial dispute it was transformed into a challenge to the established political order. The patriarchal Saudi government was not prepared to tolerate such a defiance, and it became a matter of self-esteem (if not actually self-preservation) for it to reassert its authority. In October the strikers resumed work.

Although no formal concessions were made to the workers' demands, the strike did affect company policies. Alerted to the urgency of certain social problems and acting partly in response to the requests made by the Saudi Royal Commission of Investigation, the company carried out a number of reforms, chief among which were the speeding up of the housing program, especially in outlying areas; an increase in the minimum daily wage from 5 to 6 Saudi rials and a corresponding raise in the pay rates of all Saudi employees by 12 to 20 per cent; improvement of promotion procedures; restoration of food and clothing subsidies (eliminated in 1952 as a result of a wage raise); construction of the first schools for the workers' children; shortening of the work week; and introduction of "communications committees," through which laborers and intermediate employees could channel their suggestions and questions to management and receive from the latter information on company matters.

The Aramco strike and the previously described Basra strike, which occurred some seven weeks later, marked the end of a period of upheavals in the Middle Eastern oil industry, a period which began with the Iranian nationalization bill of March 1951. These two years of actual or potential strife produced new programs or acceleration of those already in existence with a view to improving labor conditions and generally enhancing the companies' positions in the host countries.

While the degree, if any, of subversive political influence on these disturbances remains a matter of conjecture, there is no doubt that some workers' grievances had their root in real deficiencies in their social and economic condition. The companies tacitly acknowledged this by inaugurating new policies and improvements. The Dhahran and Basra strikes may therefore be categorized as economically motivated disturbances.

Since 1955, however, the Arab world has been increasingly agitated by a number of political issues and incidents which tend to widen the rift between itself and the West resulting from the partition of Palestine. These include the signing of the Baghdad Pact, the Soviet-Egyptian arms deal, the frustrated plans for the Aswan Dam, and the resulting nationalization of the Suez Canal. Arab nationalism has received a powerful boost from Nasser's regime in Cairo and has found new expression in the ideas of Pan-Arabism. A degree of unity could be achieved by co-operation among existing Arab governments. But this process is slow and has its obvious limitations. Inflamed nationalist thinking points to full unity as a desirable objective. This can be obtained only if the principal obstacle—the existence of old-fashioned sovereign governments—is removed. To the restless Pan-Arab mind such regimes have the double stigma of being socially too conservative and too accommodating to the West. Thus, somewhat inevitably, the slogans of unity and revolution have become closely interwoven.

This evolution in nationalist attitudes is reflected in the position of the oil companies in the producing countries. It did not take long for the new attitude to make itself felt in the triangular relationship between the Saudi government, Aramco, and labor. By mid-1956 the company's workers had evolved a long way from the atomized mass of tribesmen and villagers initially employed. They had become conscious not only of matters affecting their welfare as laborers, but also of a number of political notions. On June 9, 1956, during a visit by the king to Dhahran, a strike and a demonstration took place. This time the demonstrators did not limit their demands to adjustment of labor conditions. In fact these were played down. Their placards displayed nationalist and anti-imperialist slogans. No overt criticism of the government was voiced, and some signs hailed the king in an ostensible gesture of loyalty. Yet the suddenness and the massive character of the demonstration represented such a definite challenge to public order and security that it could not be viewed with equanimity by the Saudi authorities. Their reaction was swift and stern. A number of strike leaders were thrown into jail, and an ultimatum was issued by the provincial governor, Saud ibn Jiluwi, demanding immediate return to work. The workers obeyed. Soon the royal decree mentioned earlier was published, banning all strikes and demonstrations in industrial establishments.[20] The speed and severity with which the government had acted were not without their effect on industrial relations: political restlessness soon subsided and normal habits of work were resumed. The events of June 9 had doubtless strengthened the government's conviction that trade unionism was unsuited to a country like Saudi Arabia. In fact, the continuation of the company-sponsored communication committees became questionable from the government's point of view.

This account of industrial relations in post-Mossadegh Iran, in Iraq, and in Saudi Arabia shows that there were long periods of tranquillity. If the Dhahran and Basra strikes of 1953 were an indication of deficiencies, these were rather promptly remedied by the managements. Disruption of work in IPC and Aramco installations on account of labor turbulence was rarer and amounted to considerably fewer man-hours than in some major industries in the United States. Although labor conditions and other social aspects of oil operations require continuous attention, solutions are not beyond the ability of an enlightened management.

There are a number of common denominators in the labor situation of the three producing countries reviewed. One is the current absence of trade unions in the oil industry, usually reflecting the general condition of the country in this respect. This absence has largely been the result of the conscious policies of the governments concerned, regardless of what the laws say. While the companies may have concurred, the policies themselves invariably originated with the governments. In fact, the companies have preferred to remain noncommittal and, if queried, like to point to their harmonious record of co-operation with trade unionism in the transit countries (of which more will be said below) as evidence of adaptability to any sociopolitical system under which they may have to function. Another common denominator is the proved ability of even a primitive group of workers to organize and act once they are massed

---

[20] See p. 259 above. The decree was published on June 22, but its official date was given as June 11. It is not clear whether its enactment was spurred by the events of June 9 or whether it had already been planned.

together in a community. The third major feature is the increasing influence of political factors upon the shaping of industrial relations. Political considerations are constantly interwoven with purely economic or social factors, and it requires considerable analytical skill to determine the primary motivation in a labor disturbance.

The political motivation constitutes one of the most baffling problems for the companies. Although capable of coping with industrial issues, they are not in a position—and are not supposed—to solve the political grievances of the workers. These range from purely domestic matters (such as the ban on certain forms of entertainment, and a general political rigidity) to international and inter-Arab questions. Thus the companies are sometimes penalized by a disturbance the true target of which is the government. Moreover, since 1955–1956 Pan-Arab consciousness has been playing a steadily increasing role in the Arab countries. This creates the possibility of labor strife unconnected with the local situation but linked to larger issues affecting a sister country or the Arab world as a whole. It also makes the workers more amenable to the guidance of inter-Arab labor organizations based on centers outside the producing countries.

INDUSTRIAL RELATIONS IN THE TRANSIT COUNTRIES

The labor situation in the transit countries has been quite different from that in the producing states both in its general aspects and on a number of specific issues. In the first place, the numbers of workers employed by the oil companies have been much smaller, both in absolute figures and in relation to other industries. Consequently labor's behavior has never been of such crucial importance to the transit countries as it is in Saudi Arabia or Iraq. Secondly, the transit states' oil laborers, with the minor exception of those employed at the pumping stations, are generally distributed in or near the natural urban centers, thus avoiding the artificiality of the desert oil towns. The social composition and educational level of the transit workers have also been different. Instead of being nomadic, they represent an old, settled element, partly rural and partly urban, often literate and not lacking a certain degree of political sophistication. Their appreciation of steady employment is as a rule much greater than is that of the Arabian bedouin, and this is eloquently expressed by the considerably lower rate of labor turnover in the transit states than, for example, in Saudi Arabia. They are also more jealous of their regional and local right to employment, resenting the intrusion of outside elements from another province or town. Furthermore, as nationals of more advanced states (notably Syria and Lebanon) and in keeping with their higher degree of education, the transit workers tend to be unionized, though exceptions can be noted in this general tendency.

As to their responsiveness to political appeals, it is not uniform. It is more pronounced in Syria than in Lebanon, especially in connection with Pan-Arab issues. All in all, the typical labor problems in the transit states are different in several respects from those in the producing countries. Housing, for example, has never been a great issue either in Syria or in Lebanon, but regional recruitment and employment priorities have been major problems.

A brief summary follows of salient facts and issues with regard to trade unionism and the general course of industrial peace and strife in the companies operating in Lebanon and Syria.

In *Lebanon* oil workers fall into one of three categories: (*a*) those employed by Tapline; (*b*) those employed by IPC; and (*c*) those employed by a variety of other companies, principally distributors but including also the refinery-owning Medreco. By far the largest number is employed by IPC, which in addition to its pipeline organization operates a terminal and a refinery in Tripoli. The following table shows the distribution of workers in various oil enterprises (in approximate figures):

IPC	2,300
Tapline	420
Medreco	280
Distributing companies (Socony Mobil, Shell, Esso, etc.) and U.S. Point Four organization	1,000
	4,000

Tapline's workers have not as yet been unionized. A number of upper-middle-bracket Lebanese employees of the company considered organizing a union in 1956, but the plan did not materialize. Instead in the summer of that year Tapline's personnel formed two employee associations, one for Beirut and another for Sidon. As associations they were not subject to the Lebanese labor law and were not eligible for affiliation with regular trade unions. Their leadership is composed of high-level employees, such as engineers and physicians, and their principal objective is to assure better communications with the management. In Sidon up to 38 per cent of the membership is made up of the sailors, whose problems are somewhat different from those of the rest and who display somewhat more radical tendencies than the "ground" personnel. They play an important role in the company's operations inasmuch as ships calling on Sidon cannot, as a rule, be berthed without tug service.

Tapline's record of industrial relations is generally satisfactory. In 1953 the company was somewhat embarrassed by feuds between certain leading families in southern Lebanon, which were reflected in local vendettas affecting the company's employees and in the sporadic emergence of so-called "Black Hand committees" issuing threatening handbills and attempting to provoke unrest. Their efforts were largely fruitless, and with the subsiding of the feuds these committees vanished from the scene. With a small but fairly select labor force, which is in many ways Lebanon's labor aristocracy, Tapline can look forward to continuous harmony in its industrial relations.

IPC's workers were unionized in 1949. Originally their union formed part of the League of Syndicates of Workers and Employees of Lebanon (*Jamia*),[21] but in 1954 it withdrew and it has since remained independent. Outside Lebanon the union became affiliated with the International Federation of Petroleum Workers. Both the monthly (i.e., white-collar and artisan) and the daily rate workers make up its membership, which at the peak of IPC's construction activity in 1953 consisted of over 5,000 members. IPC's industrial relations were fairly harmonious, with an occasional minor

---

[21] One of the six labor federations in Lebanon and affiliated to ICFTU. The other five are United Union of Employees and Workers, Federation of Unions of Workers and Employees of Lebanon (affiliated to WFTU), Allied Independent Unions (a splinter group from the League), Federation of Labor Unions of the North (Tripoli), and Artisans Federation.

dispute over wages or working conditions. A 4-day strike in mid-May 1950 was ended by mutual accommodation. Difficulties were experienced in labor-management relations in 1956–1957 as the result of the destruction of the pumping stations in Syria, which led the company to suspend some of its operations in Lebanon. In protest against anticipated discharges IPC workers in Tripoli went on strike in January 1956. A number of them were subsequently relieved of their work for one reason or another, to be reinstated at a later date when the flow of oil resumed. The principal grievance of the reinstated workers was the unwillingness of the company to treat them as they did the reinstated workers in Syria, who were paid for the period of idleness in conformity with special Syrian legislation. By the spring of 1958, however, these difficulties had been eliminated by the adoption of compromise solutions.

Among the other companies, Medreco's refinery workers in Zahrani are unorganized, while Shell's and Socony's laborers have formed unions. Esso's personnel has set up an employee association. Oil workers in Lebanon remain fairly cool to Pan-Arab appeals, showing little inclination to join inter-Arab labor organizations. In 1956 efforts were initiated by one of the pro-Western Lebanese labor federations, the United Unions of Employees and Workers, to form an all-Lebanese oil workers' union. Negotiations were conducted in 1957 and 1958 with a likelihood of success.

In *Syria* trade unionism dates back to the mid-1930's, but officially unions came into being in 1946, the year of the promulgation of the Syrian Labor Code. Consequently Syria has traditionally been with Lebanon in the forefront of the organized labor movement in the Arab East. Syrian trade unions do not restrict themselves to union activities proper, but frequently play a political role as well. This was recognized rather early in the postwar period when various political parties, groups, and dictators tried to influence the unions and to secure their support. As a result elections to the unions' executive bodies have often reflected political struggles. As in Lebanon, Syrian oil workers constitute only a modest fraction of the totality of industrial labor. The bulk of them—about 4,500 in 1956—are employed by IPC at its pumping stations, the headquarters in Homs, and the terminal in Banias. Tapline's labor force after the conclusion of the initial construction work has been negligible as there is no pumping station in Syrian territory. Other oil workers are employed by distributing companies.

IPC's workers organized late in the 1940's into three unions, one for each of three provinces, Homs, Deir ez-Zor, and Latakia. Contractors' workers were entitled to join them, but seldom did so. Nevertheless the unions have frequently acted on their behalf. The oil workers' unions became affiliated with the General Federation of Workers of Syria, a body established in 1949 and by 1958 comprising 135 unions in various parts of the country. Presided over for a number of years by a Damascus textile worker, Subhi Khatib, the General Federation published a press organ, *Saut al-'Ummal* (Voice of the Workers), and constituted a significant factor in Syrian domestic politics. The General Federation's executive committee usually had among its members a representative of the oil workers' union. The latter's chief spokesman for a long time has been Omar Sibai, president of the oil workers' union in Homs, reputedly of leftist tendencies. Until 1957 the General Federation's Executive Committee was dominated by elements with moderate political tendencies, but in that year it was captured by labor leaders either affiliated with the Arab Socialist Renaissance Party (*Ba'ath*) or showing pro-Communist tendencies. The change in executive personnel has been reflected in increased emphasis on the General Federation's political activities, following mostly a Pan-Arab course.

Because of its scrupulous observance of Syrian labor legislation, IPC's industrial relations record was until the mid-1950's, rather uneventful. An occasional minor dispute would revolve around demands for better wages or benefits. Syrian workers were anxious to be treated on an equal footing with their counterparts in Lebanon. A strike of brief duration occurred on this account in the late 1940's, to be promptly settled by the company's agreeing to apply equal treatment to workers on both sides of the border. Since 1955 the company has experienced somewhat greater difficulties on its labor front because of the greater than usual involvement of Syrian trade unions in domestic and Pan-Arab politics and the tendency of Syrian cabinets to take the side of the workers rather than to act impartially. Moreover, the Syrian government has developed the legally questionable habit of passing *ex-parte* laws, either implicitly or explicitly aimed at IPC. This happened in the spring of 1956, when in response to an illegal strike [22] of oil workers in Homs the Parliament voted to compel concessionary companies with a capital above a specified amount (obviously applicable to IPC) to increase their workers' salaries by 16 per cent.

The destruction of the pumping stations in November 1956 not only disrupted IPC's activities in Syria, but also ushered in a period of major labor disputes. These centered on the complex problem of mass dismissals and partial reinstatement later of the company's labor force. These events took place against a background of considerable political restlessness, which only deepened existing difficulties. Because of the close interrelation of industrial and political factors, which makes separation of the two impracticable, a fuller and integrated account of these developments is presented in a special case study at the end of the book.

[22] Articles 166–177 of the Syrian Labor Code of June 11, 1946, specifically determine the conditions under which a strike can take place. The strike in question did not conform to these provisions.

CHAPTER XIV

Pan-Arab Labor Movement

IT should be evident from the preceding chapter that some of the labor disturbances in the Middle Eastern oil industry were politically motivated. There seems to be no doubt that, in a number of instances such as sabotage of the pipelines in Syria, considerations of Pan-Arab solidarity and a desire to square accounts with certain Western powers for their real or alleged imperialism were at the bottom of workers' actions. Consequently it may be appropriate at this juncture to have a look at an organization which has made major efforts in recent years to organize workmen in Arab lands and subject them to centralized political direction.

The Federation of Arab Labor Unions (FALU) was established on March 24, 1956, in Damascus by labor representatives from Egypt, Syria, Lebanon, Jordan, and Libya. The founding members represented single or dominant labor federations in their respective countries [1] except in the case of Lebanon, where no single federation exists. Two out of six Lebanese federations sent delegates to the meeting. FALU was set up largely on the initiative of the Socialist-dominated Jordanian Federation of Labor Unions. It had the full blessing of the revolutionary regime in Cairo, and its found-

---

[1] These were the Permanent Congress of Labor Unions of the Republic of Egypt, the General Federation of Labor Unions in the Syrian Republic, the Federation of Labor Unions of Northern Lebanon, the Federation of Labor Unions in the Hashimite Kingdom of Jordan, and the General Federation of Labor Unions in Libya.

ing members largely represented the brand of militant Arab nationalism promoted by President Nasser of Egypt. Because of the nonexistence of trade unionism in the major Arab oil producing countries, Saudi Arabia, Iraq, and the Persian Gulf principalities were not represented at the founders' meeting.

FALU'S CONSTITUTION AND OBJECTIVES

FALU's constitution, adopted at its formative meeting, contained a lengthy enumeration of its objectives. Listed under thirty-two points, they were a mixture of industrial and political goals, the latter forming a substantial part of the whole. The general tone was set by the preamble, which read as follows:

The Workers of the Arab world, being convinced that they cannot attain dignity and their natural rights, nor assume a position commensurate with their Arab heritage, except by being associated in a democratic federation of Arab labor unions built upon a firm basis of co-operation for the purpose of co-ordinating their efforts, consolidating their ranks, developing specialization among themselves, improving the position of the Arab worker, and guaranteeing for him freedom of opinion and expression, have resolved to form a Federation of Arab Labor Unions, which is to provide the means of their attaining glory, freedom, and co-operation among themselves.

The Federation hereby declares its support of the principles of self-determination and of the right of nations to govern themselves. The Federation also declares its support for every nationalist movement aiming at the eradication of imperialism, and for every effort made in the interest of peace.[2]

One of the first aims was "to strengthen the bonds of brotherhood, friendship, and co-operation among the workers of the Arab world in particular and among Asian and African workers in general, as well as to establish permanent liaison between all workers of the world." Another stated aim was "to affirm the right of nations to self-determination," to eradicate imperialism, and "to support na-

[2] From the text of the draft constitution published by *Al-Waqt* (Beirut), March 3, 1956.

tional movements seeking such objectives in the Arab states and other parts of the world." Parallel to it was the resolve "to strive for the removal of the barriers placed by imperialism between the various Arab states" and "to pool the natural resources of the Arab states for the service of the Arab community as a whole." These objectives were to be sought aggressively. The Federation was "to urge military training for the Arab people in order that they may be ready to expel aggression against any part of the Arab homeland" and was "to strive to enable Arab workers to participate effectively in national and regional political movements." One of the final aims has a familiar ring: the Federation was to "support all efforts aimed at establishing world peace, suppressing all movements designed to bring about war among nations, suppressing the use of atomic energy for war, and urging its peaceful uses." [3]

Membership in FALU was to be open to all national labor union federations and organizations in the Arab states, with the explicit mention of the founding members' countries, as well as those not yet affiliated, namely, Sudan, Tunisia, Morocco, Algeria, Aden, and Palestine. The constitution admitted the possibility of membership for labor organizations not yet legalized in their own countries. The formula was that such organizations could become members upon approval of the Executive Council of the Federation even if they were to "become operative at some future date."

Basic policy decisions were to be made at regular sessions of the Federation at stated intervals, but much power was vested in FALU's Executive Council, which was entrusted with the day-to-day operational direction. The Damascus meeting ended in the election of FALU's officers. Subhi Khatib of Syria became president and Fathi Kamel of Egypt secretary-general. This combination, together with the decision to establish the headquarters in Cairo, gave the Egyptians a commanding voice in the Federation's affairs. In this respect the Federation did not differ substantially from the Arab League, whose over-all conformity to Egypt's leadership is a matter of historical record. A Beirut newspaper, *Al-Waqt,* became the Federation's press organ.

[3] *Ibid.*

MEMBERSHIP AND UNIONIZATION DRIVES

FALU'S activities branched out in three general directions. One was expressed in a persistent effort to increase its membership by drawing into it the yet-unaffiliated federations in various Arab countries. Another was to unionize workers in those countries where no labor organizations were in existence. And the third consisted of multifarious political actions having little or nothing to do with industrial problems.

FALU succeeded in adding Sudan to its ranks in May 1957. But so far its activities aiming at the inclusion of the Tunisian, Algerian, and Moroccan labor federations have not been successful. These organizations prefer to continue within their own regional group, the North African Federation of Labor Unions.

It is clear that FALU considers the unionization of workers in trades and states where unionism is either outlawed or nonexistent as one of its major objectives. At the first regular meeting of the Executive Council in Amman, June 17–19, 1956, it was decided that FALU's delegation would visit Iraq and Kuwait to assist in the formation of unions in those two countries. This ostensibly technical undertaking had political undertones. Iraq was ruled at that time by a government opposed to President Nasser's leadership of the Arab nationalist movement, and Kuwait was a virtual British protectorate. Trade unionism was discouraged in both areas. A FALU mission with its stated objective could be interpreted as a political move aimed at undermining legitimate authority in the two countries in question. Why was Saudi Arabia left out? The reason is that the Federation is sponsored by Egyptian authorities. At the time when the Amman meeting was held, outward cordiality prevailed between President Nasser and the House of Saud. It would have been impolitic for the Egyptian leader to promote unwanted labor activity in the domain of his ally and friend.

Nevertheless, FALU, as a body dedicated to the active support of trade unionism in the Arab world, could not remain indifferent, without losing face, to the situation in Saudi Arabia. This delicate dilemma was solved by steering a middle course between the extremes of conformity with the Saudi union ban and the open advocacy of unionization. The method adopted was that of dealing with a man named Nasser Said, a former Aramco employee living in Damascus, who claimed to represent the Saudi oil workers. Thus Said was reportedly included in the FALU delegation which visited the Soviet Union in 1957 (of this visit more will be said later), and his name appeared from time to time in connection with the Federation's activities. When, however, the Executive Council held its session in Cairo in May 1957, Said was refused participation as an observer. In addition to these facts, other circumstantial evidence gives ground to suspect the existence of some clandestine organization among Aramco's workers which maintains liaison with the outside world. In August 1956 a Jordanian paper, *Ar-Rai,* carried a message from a group styling itself "The Central Committee of the Workers of Dhahran." Pro-Communist papers in Damascus and Beirut, such as *Nur, Talia,* and *Talligraf,* occasionally print news items about the labor situation in Saudi Arabia, which, despite the usual distortions, have some factual foundation. Considering the speed with which these items of news are reported, there is strong reason to believe that they must be conveyed by special channels.

FALU moved to appear officially on the Saudi scene by deciding in April 1958 to send a delegation on a pilgrimage to Mecca. As a tactical step it was not devoid of ingenuity inasmuch as it would penetrate a hitherto forbidden territory under the guise of a pious enterprise. In fact, it bore strong resemblance to the Russian stratagem of sending to Mecca every year during the month of the *Hajj* a group of properly indoctrinated Soviet Moslems. Although the group usually came via Cairo, the Egyptian government could always dissociate itself from all such activities. Its interest in these developments was confirmed when it was announced in March 1957 that the first Egyptian labor attaché was about to be appointed to Jeddah.

FALU's special role in areas where unionism was outlawed was further emphasized in connection with the momentous events in Jordan of April 1957. In a move to frustrate an antiroyalist coup King Hussein abruptly dismissed the cabinet of Premier Suleiman Nabulsi, a nationalist known for his advocacy of rapprochement with Egypt and the Soviet bloc. Martial law was proclaimed, followed by the jailing of various elements of the opposition and a suspension of trade union activities. If the letter of the law had been observed, the Jordanian members of FALU should have withdrawn. They did not withdraw. Instead the secretary-general of the Jordanian Labor Union Federation, Zaidan Yunes, went into exile to Damascus and from there continued to represent Jordanian workers at the meetings of FALU's Executive Council. On April 25, 1957, the council met in Damascus in an extraordinary session, principally to consider the situation in Jordan. It issued a call to King Hussein to restore parliamentary democracy, release imprisoned leaders, and lift martial law. A few days later FALU's secretary-general, Fathi Kamel, cabled the heads of state of Saudi Arabia, Syria, Lebanon, Egypt, and Libya, urging them to put an end to the crisis in Jordan and to stop imperialist conspiracies. This was followed on May 4 by his personal appeal to King Hussein to rescind the decree banning trade-union activities.

Several attempts were made, either directly by FALU or by its affiliated Egyptian organizations, to establish an all-Arab oil workers' federation. Notable in this respect were the efforts made by a representative of the Egyptian Federation of Petroleum Labor Union Councils, who during a visit to Tripoli, Lebanon, in 1956 contacted local political figures and IPC union leaders, trying to induce them to join the projected oil federation. Thus far the results of his mission have been negative. Knowledge of FALU's intentions spurred the leaders of the United Unions of Employees and Workers (a right-of-center organization) to start a movement toward the creation of an oil workers' federation in Lebanon, partly with a view to forestalling the Egyptian initiative. Neither the government nor the companies expressed any objection to this plan.

Should the Egyptian Petroleum Federation succeed in its schemes, its political ideas would of course influence the newly created body. These ideas are far from reassuring to the oil industry. This became clear when the Federation, at a conference held in Cairo in August 1957, declared that the "Egyptianization" of the oil industry in Egypt was the ultimate policy of the state and that it would exert all its efforts to help the state achieve it.

FALU'S POLITICAL ACTIVITIES

It should be evident from the foregoing account that even in such matters as membership and unionization FALU's activities are strongly influenced by political considerations. As a matter of fact political issues, in FALU's thinking, seem to enjoy primacy over industrial questions. FALU follows a policy of "Arab liberation" and "positive neutralism," both concepts propagated by the revolutionary regime in Cairo and adopted at one time or another by nationalist and socialist groups in other Arab countries. The Federation's "positive neutralism" is a virtual replica of the policy followed by Egypt. Ostensibly it means official nonalignment as between the West and the East, but in reality it consists of close collaboration with the Soviet bloc. The numerous statements, messages, and appeals emanating from FALU have closely followed the pattern of official pronouncements of the Cairo government. These generally have denounced the West for its imperialism, its alliance with Zionism, and its hostility to Arab national aspirations.

FALU's political activities are punctuated by the regular and extraordinary meetings of its Executive Council. Although the agendas of these meetings are replete with a number of points, each meeting tends to deal with one or two major issues, which distinguishes it from others and gives it a special political flavor. It will be recalled that the first regular meeting of the Executive Council took place at Amman on June 17–19, 1956, at which time the decision to unionize workers in Iraq and Kuwait was taken. Another major decision was to open the Federation's branch office in Beirut, largely with a view to counteracting the activity of the Beirut office of the International Confederation of Free Trade Unions (ICFTU). Inasmuch as FALU's Lebanese membership consisted of two small federations only, the Executive Council aimed at inducing larger and more representative Lebanese federations (such as the *Jamia* and the United Unions) to join FALU's ranks. Despite its policy of nonalignment FALU did little to disguise its hostility to the Western-influenced ICFTU. When in January 1957 the latter was about to hold a meeting at Accra in the Gold Coast to emphasize its interest in African labor problems, FALU's initial reaction was to boycott it despite an invitation to attend. Eventually FALU's Executive Council decided to send an observer but made it clear that the only reason for doing so was to prevent ICFTU from counteracting FALU's plans to set up an Arab-sponsored federation of Asian and African labor unions. On its part, ICFTU was critical of FALU's emphasis on political questions to the detriment of industrial problems. This was especially evident at ICFTU's meeting in Tunis in July 1957, when serious concern over and criticism of FALU's activities were voiced by several delegates. FALU's record was defended by the Moroccan delegation to the meeting.

The Executive Council's second session, held in Cairo on August 10–18, 1956, was truly historical. Its agenda included such matters as the mobilization and recruitment of workers for trade unions, advocacy of uniform labor legislation in Arab countries, and a decision to send a special commission to Saudi Arabia to intercede with the authorities in favor of workers who had been jailed as a result of the disturbances in the oilfields in midsummer of that year.[4] Furthermore, the council discussed the problem of affiliation with the World Federation of Trade Unions (WFTU, a Soviet-influenced body) and decided in the name of "positive neutrality" to abstain from joining either this organization or the ICFTU. By far the most important decisions were the secret ones (since become public) concerning oil. They provided for the destruction of Western-operated oil installations in the event of foreign aggression against Egypt. Such aggression was anticipated as a retaliatory measure for the nationalization of the Suez Canal which had taken place a few weeks earlier. Sabotage of foreign military bases in Arab lands and refusal to load, unload, and service ships and airplanes belonging to the aggressor countries were the two other measures decided upon. The plans called for co-ordinated action by all of FALU's affiliated organizations.[5] The same meetings resolved to create within the Federation a special office for oil workers, and according to some reports plans to create an all-Arab oil workers' union were also discussed.[6]

These secret plans, prepared in anticipation of hostile actions

[4] See p. 273 above.
[5] Information based partly on an article in *Al-Qabas* (Damascus), Aug. 22, 1956.
[6] *Akhbar al-Yaum* (Cairo), Aug. 18, 1956.

against Egypt, seemed to provide further proof of the synchronization of FALU's activities with those of the Egyptian government. In fact, when the Anglo-French invasion occurred early in November 1956, Egyptian broadcasting facilities, such as Radio Cairo and the Voice of the Arabs, were made available to the Federation. It promptly issued a call for sabotage. However, as will be seen in Part Five, there are strong reasons to believe that it was not the workers but the military who eventually carried out the demolitions at IPC's pumping stations in Syria. Aware of the fact that many people both inside and outside the Middle East shared this belief, the Federation took great pains to assume responsibility for it. Its motivation was dual: on the one hand, it wanted to uphold the Syrian government (at that time strongly influenced by a pro-Egyptian group of army officers) in its stubborn disclaimer of any responsibility in the matter; on the other, it was afraid of losing face with the Arab masses by exposure of the fact that instead of spontaneously performing this patriotic feat itself it had to relinquish its task to the regular army.[7]

Also politically oriented was the third regular session of the Executive Council, held in Cairo on May 18–21, 1957. Coming closely after the extraordinary session in Damascus, to which reference was made earlier in connection with the Jordanian situation, it ended in the adoption of twelve resolutions, seven of which pertained to

[7] In an article entitled "Who Blew Up the Pipelines?" in *Rose al-Yussef* (Cairo), Dec. 31, 1956, Su'ad Zuhair forcefully claimed that the Arab labor movement was responsible. In the conclusion Salah Zanzul, president of the Federation of Labor Unions of Hama, Syria, was quoted as saying: "In an attempt to discredit the unity of the Arab working class, imperialists and imperialist stooges say that Syrian military authorities blew up the installations. They make this claim for two reasons: (a) to make the Syrian government bear the brunt of responsibility for the sabotage, and (b) to discredit the Arab working class. The Syrian government, however, cannot be held responsible for actions committed by the Syrian people as represented by the Syrian working class. Arab resentment against the imperialists led the workers to sabotage the oil installations, and it is those who created this resentment by their aggression on the Arab nation as personified by Egypt, who must bear the responsibility for what has happened. When the Arab people and the Arab workers enter into a real fight with the imperialists, the latter will not find one single inch of pipeline intact in Arab territory. The Federation of Arab Labor Unions is not merely a symbol of unity but represents in microcosm the unity of the Arab nation."

political and five to union matters. The political ones covered such items as Israel, Suez and Aqaba, Algeria, military pacts and foreign aid, Cyprus, the atom bomb, and neutralism. The union items included a proposal to convoke an Afro-Asian Labor Conference, a renewed plea to unify Arab labor laws, a decision to admit as a member the Sudanese Labor Federation, a call to form unions in all Arab countries, and a resolution to organize training in unionism and arrange for mutual visits among Arab labor leaders. In a statement to the press FALU's secretary-general, Fathi Kamel, termed this third meeting of the council the most important since the historic session of August 10, 1956.[8]

Ever since its creation FALU had tried to bring about the formation of an Afro-Asian labor federation. This policy corresponded to two guiding ideas. The first was Gamal Abdul Nasser's concept of three concentric circles around Egypt: the Arab, the African, and the Islamic. In each of them Egypt was to play a leading and unifying role.[9] The second idea was the concept of solidarity and neutralism as expressed by the Bandung Conference of Asian and African nations in April 1955. What Bandung attempted to do at the governmental level, the Afro-Asian labor federation was to emulate on the working-class level. In pursuance of this policy a FALU delegation visited the People's Republic of China in the fall of 1956. This visit was returned by a Chinese Communist mission in the spring of 1957. The Chinese spent about two months in the Arab countries, visiting Egypt, Jordan, the Sudan, Syria, and Morocco. On March 24 in Amman (where Premier Nabulsi was still in power) they attended celebrations of FALU's first anniversary. At the end of their tour they issued a joint statement with FALU pledging co-operation in setting up an Afro-Asian labor federation, as well as support for the Bandung resolutions, the policy of positive neutrality, and the Egyptian struggle against imperialism. Moreover, it was generally believed in Cairo that, as a matter of strategy in organizing the Afro-Asian labor group, the Chinese were to secure adherence of the Far Eastern and South Asian labor organizations, and the Sudanese were

[8] *Al-Ahram* (Cairo), May 17, 1957.
[9] Gamal Abdul Nasser, *Egypt's Liberation: The Philosophy of the Revolution* (Washington, D.C., 1955), pp. 85 ff.

to work in the African sector. Cairo was to serve as a co-ordinating center.

Although China was the principal Soviet bloc country with which FALU maintained close contact, Russia and its European satellites were not neglected either. During 1957 several visits were arranged, especially in connection with May Day celebrations, for FALU's leaders in Moscow, Prague, East Berlin, and other Communist capitals. FALU acted as host to a number of Soviet bloc labor delegations, including one from Russia early in 1958. There is no doubt that these visits benefited Russia more than the Arab countries. Although it is highly unlikely that any Arab during these visits ever persuaded a Soviet or a satellite Communist of the justice of his national cause, no such assurance can be given regarding the penetration of Communist ideas into the less tutored minds of Arab labor leaders. The Soviet undoubtedly regards the liaison with FALU as another stepping stone in the general political offensive which the Soviet bloc has been waging from 1955 onward in the vast continents of Asia and Africa. The liaison is in line with such manifestations as the Afro-Asian Peoples' Solidarity Conference, which was held in Cairo in December 1957 as an unofficial heir to Bandung. FALU's protestations that as a purely national force it does not lend itself to Communist penetration and that the accusation of Communist proclivities is the work of imperialists bent upon discrediting it are rather unconvincing. To hobnob continuously with Communist leaders from China, Czechoslovakia, East Germany, and Russia, to sign manifestoes couched in terms taken from the Soviet vocabulary, and to pretend that this means "positive" neutrality—all this is an obvious exercise in hypocrisy and dissimulation. The record of deeds is more eloquent than the argumentative rhetoric.

True enough, not all of FALU's affiliated groups give positive neutrality the same pro-Soviet connotation. Occasionally a voice will be heard urging a strictly neutral line and warning of the dangers of identification with either of the two major camps in world politics. For example, Salah Zanzul, vice-president of the Syrian Federation of Labor Unions, in an article published in the organ of the Ba'ath Socialist Party, opposed a contemplated move by the Syrian Federation of Builders' Unions to join the Soviet-sponsored

WFTU. Arab workers, he argued, should stay clear of both the Communist and the imperialist alignments and remain faithful to their own Arab liberation policy.[10] But such words of moderation are either too weak or too rare to affect the course of FALU's policies.

FALU's policy is to draw toward itself as many Arab countries as possible. We have already mentioned FALU's attempts to expand into North Africa. The Federation is also paying considerable attention to the Arabian Peninsula, where the stake—the unionization and domination of oil workers—is perhaps greater than in any other part of the Arab world. Although FALU's tactics toward Saudi Arabia are largely determined by the political status of Saudi-Egyptian relations, it is doubtful whether the Federation will wish to renounce further action in this field. It can be taken for granted that no moral or political scruple will deter the Federation from attempting a thorough penetration of oil labor in the British-protected Persian Gulf principalities.[11]

How shall we evaluate the Federation's role and importance? There is no doubt that the Federation has given the Arab nationalist movement a new weapon with considerable potentialities. Hitherto Arab nationalism has been expressed either through individual governments or through the Arab League. The League is rent by inner dissension and often proves ineffective in action. FALU's organizers and sponsors apparently act on the assumption that it is easier to achieve unity among the Arab working masses than among the ruling classes. Moreover, the workers, as an illiterate group easily swayed by emotion and having little to lose, may be a better instrument of action than vacillating politicians. That FALU is regarded by its leaders as primarily a political instrument has been amply proved by the activities it carries out and the tenor of its many statements and resolutions. Labor questions are obviously of minor importance as compared with political goals. FALU's leaders seem to be working on the theory that Pan-Arab political appeals are a weightier factor in influencing labor's attitudes than industrial advantages offered by foreign or domestic employers.

[10] *Al-Ba'ath* (Damascus), Sept. 13, 1957.
[11] For the reaction of Kuwait oil workers to FALU's call for sabotage at the time of the Suez crisis, see p. 337 below.

Although this confidence may be justified, it does not follow that the Arab world in general or the workers in particular will gain by allowing a labor federation to be used chiefly as an instrument of a specific political policy. With its appearance on the Middle Eastern scene FALU undoubtedly brings a revolutionary specter to the foreign-owned oil industry. Its revolutionary edge could turn against the existing social order in the Arab world itself. Often unjust and antiquated, this order needs radical reform. Yet it is by no means certain that the Federation in its present form and with its present methods provides the best way for achieving reforms. If by negligence, ignorance, or willfulness it allows itself to serve as the main channel of Soviet influence in the Arab East, its assumed role as a national and liberating force may degenerate into that of a forerunner of a new totalitarian order which will put an end to Arab national aspirations with irrevocable finality.

◈◈ CHAPTER XV ◈◈

The Human Side of Operations

THE foregoing review of peace and strife in the oil industry in the Middle East shows that workers' attitudes have often been determined by political factors. These factors do not, of course, exist in a social vacuum, and they could not have played their important role without some measure of concurrence by economic factors. A person whose economic needs are well provided for and whose morale is high does not easily follow the path of political extremism. Consequently it may be in order now to review those practices and policies of the companies which affect the work, the material conditions, and the welfare of their employees.

WORK AND WAGES

The hours of work in the major producing companies are in conformity with the maximums prescribed by law, which is a 48-hour, 6-day work week in Iraq, Iran, and Saudi Arabia. Since the war the work week in Iran under the regime of AIOC has not exceeded 43.5 hours in summer and 44.5 hours in winter. Beginning with September 1958 Aramco voluntarily reduced its work week to 42 hours while maintaining the same level of wages. Moslem holidays with pay, usually amounting to seven days a year, are observed by the companies. Paid vacations, up to two weeks a year, are standard in all three countries.

With regard to wages it may be said that (a) the minimum company wages are generally above legal minimums;[1] (b) they are higher than prevailing wages in local enterprises; (c) they account for only a part of the current compensation of the workers, the rest being provided in the form of subsidized food and clothing and miscellaneous benefits; and (d) they show an ascending tendency with the passage of time. The tendency to rise is illustrated in Tables 6–8 showing the progress of wages in IPC, Aramco, and the companies in Iran.

Table 6. Level of annual earnings of national employees in Iraq Petroleum Company

Year	Average earnings per worker	
1951	ID 250–300	$700–840
1952	ID 300–350	840–980
1953	ID 350–400	980–1,120
1954	Above ID 400	1,120

Source: [IPC group], *Iraq Oil in 1954* ([London, 1955]), p. 9.

This ascending trend is a function of two variables: on the one hand, there are wage increases because of promotion to higher grades or for merit because of improved work performance; on the other, the inflationary tendency (largely generated by the presence of the oil company itself) demands periodic readjustment of wages to keep pace with steadily rising prices. Without accurate price indexes in

Table 7. Minimum daily wages in Arabian American Oil Company

Year	1945	1950	1951	1952	1953	1954	1955	1956	1957
S.R.	1.5	3	3.5	5	5	6	6	6.5	7.5
	(0.40)*	(0.81)	(0.94)	(1.35)	(1.35)	(1.62)	(1.62)	(1.75)	(2.02)

Source: Aramco, *Report of Operations,* 1950–1957.
* The amounts in parentheses are U.S. dollar equivalents.

the areas of company operations, it would be hazardous to venture an opinion as to the adequacy of the minimum wages paid by the oil industry. The procedure followed in Iran, namely, that of a commission's fixing minimum wages for each district from time to

[1] Except in Iran, where minimum legal wages are determined for each province. See p. 256 above.

time provides perhaps the best solution to the problem. Actually, relatively few workers have ever belonged in the minimum wage category. During the last year of AIOC's operations in Iran fewer than 1 per cent of its laborers were receiving the legal minimum wage. In a single year (1957) Saudi Arabian employees of Aramco received 10,165 promotional and merit increases out of a total of 12,729 Saudi workers. Moreover, any increase in the minimum wage resulted in a general upward revision of wages on all levels. Consequently, if these general increases were in step with the mounting price index, the simultaneous merit and promotional increases represented a real gain, which allowed them to move steadily upward and away from the mere subsistence level.

Table 8. Minimum daily wages in Iran under AIOC and the Consortium management

Year	1949–1951	1955–1956	1957
Rials (and U.S. equivalents)	40 ($0.48)	82 ($0.98)	99 ($1.19)

Source: ILO, *Labour Conditions in the Oil Industry in Iran*, p. 21; *The Anglo-Iranian Oil Company and Iran*, p. 13; *The Iranian Oil Operating Companies*, 1955, p. 21; and personal communication from the Consortium management in Teheran.

Labor turnover is another factor to be taken into account in this connection. The more stable the labor force, the more workers are likely to be affected permanently by the merit and grade wage increases. With a big turnover, however, a correspondingly high proportion of laborers is bound to be kept in the low wage brackets. Generally turnover is considerably higher than it is in Western industry. This is partly due to seasonal interest in work on the part of an Iranian tribesman or a Saudi bedouin and partly to the nature of the operations conducted by a company in a given year. A new recruit is frequently not anxious to work for more than a few months, when there is little or nothing to do in his village or tribe. On the other hand, there are periods when the companies go through a phase of major expansion. During such periods they are willing to hire quickly large numbers of workers for construction purposes and to forego the usual screening processes and aptitude tests. For a number of years after World War II turnover in the two older companies—AIOC and IPC—was about 21 per cent.[2] But when IPC undertook a major construction program in 1949, this rate rose to 28 per cent, later to recede again.

An even more vivid illustration of this phenomenon can be found in the history of Aramco's separation rate:

Year	Annual separation rate
1951	34.3%
1952	26.5
1953	26.0
1954	9.7
1955	12.4
1956	12.3
1957	10.6

These turnover percentages correspond to periods of expansion and stabilization in Aramco's history. Prior to 1954 the company was rapidly expanding its plant and facilities. A "boom camp" type of operation was flourishing, with a "labor gang" method of recruitment. Since 1954, however, with its most pressing construction needs satisfied, the company has developed a more stable kind of operation, with attendant changes in its recruiting system. This is reflected in the turnover figures.

IN QUEST OF A DECENT LIVING

Food, Clothing, and Workmen's Compensation

In addition to wages, workers' remuneration covers a variety of other benefits. These include subsidized meals in company cafeterias, assuring the workers one or two hot, high-caloried meals a day. Aramco offers the workers breakfast and lunch at one-fourth of a Saudi rial (7 cents) a meal. At IPC's canteens in Kirkuk, Ain Zalah, and Basra meals can be obtained at 30 fils (8.4 cents) apiece. Subsidies vary according to the locality. At Kirkuk they amount to 20 fils (5.6 cents) per meal. In addition to meals, some companies sell their employees food and clothing at subsidized prices. The Consortium established (or inherited from AIOC) several company

---

[2] Turnover is measured by the number of separations divided by the average strength during the year and multiplied by one hundred.

shops in Abadan and the fields. Stocking over one hundred items of food, clothing, kitchenware, school supplies, and the like, these shops are managed on a nonprofit basis. Because of the size of its labor force (around 45,000, the largest in the Middle East) and the remoteness of its installations from natural urban centers, this method is probably best suited to ensure an adequate supply of food and household articles at reasonable prices. Food sold in these shops is additional to the basic foodstuffs—such as flour, rice, ghee, sugar, and pulse—that the operating companies have always provided at subsidized prices.

Aramco's practice is somewhat different. Trying to do its utmost to encourage local free enterprise, the company is reluctant to compete with it by making these goods available in company stores. In 1952 Aramco's declared policy was to give its employees full cash value for their services, and to this end the company raised wages while eliminating food subsidies.[3] The employees' reaction was unfavorable. Complaints about scarcity and exorbitant prices began to pour in and soon became one of the major grievances leading to the strike in the fall of 1953. As a result, the company reverted to the policy of subsidies, making a total of six staple food items (coffee, tea, rice, sugar, cooking oil, and flour) available to general employees at a 20 per cent discount on local prices. In addition, each general and intermediate employee became eligible to purchase a limited number of shirts and trousers a year at 50 per cent discount. In this way the company hoped to strike a balance between the interests of the local shopkeepers and those of the workers. This new policy was carried out with the full concurrence of Saudi Arabian authorities.

All the companies as a matter of course conform to national laws with regard to severance pay and death and disability benefits. Certain companies, however, go beyond the provisions of the law. Thus Aramco in 1957 introduced a plan providing for a full year's wages to Saudi Arabian employees who became totally incapacitated whether by reason of sickness or off-the-job accidents.

Health Services

Modern health services, including preventive medicine, dispensaries, and hospitals, constitute another benefit provided by the com-

[3] Aramco, *Report of Operations*, 1952, p. 33.

panies. Comparisons between the companies are difficult to make in this respect, inasmuch as some—for instance, the companies of the IPC group in Iraq—operate near urban centers like Kirkuk, Mosul, and Basra, where municipal or private medical facilities are readily available. The task of these companies is merely to fill the gaps in the existing organizations. IPC's principal medical services consist of a large modern hospital in Kirkuk. Opened in 1937 with 44 beds, it was expanded in 1955 to accommodate 125 beds. In addition, the company has established eleven clinics. Fourteen doctors are employed by IPC's medical organization. BPC's health facilities consist of a 16-bed medical center at Makina and a 10-bed center at Fao. Otherwise the company is using the Basra municipal hospital. In the Mosul area two clinics and two dispensaries cater to the needs of MPC's labor force.

Inasmuch as the Consortium and Aramco are somewhat similar in terms of their remoteness from major urban centers, it is not perhaps unfair to compare the figures of their respective medical organizations (see Table 9). With respect to Iran it should be pointed out that, beginning January 1, 1956, the management of medical services was transferred, as a nonbasic function, from the operating companies of the Consortium to the National Iranian Oil Company.[4]

Table 9. Company-operated major medical facilities in Iran and Saudi Arabia (1957)

Company	No.	Hospitals Location	No. of beds	No. of employees
Consortium-NIOC	3	Abadan (1) Fields (2)	1,000	45,475
Aramco	3	Dhahran (1) Abqaiq (1) Ras Tanura (1)	327	18,325

In addition to the hospitals, both the Iranian and the Saudi Arabian companies operate a number of clinics and dispensaries. All of these services are available free of charge not only to company employees, but also in varying degrees to their dependents and to local tribes about the desert pumping stations in Saudi Arabia. Pre-

[4] However, capital expenditures for new medical facilities are made by the operating companies.

ventive medicine, especially the fight against malaria, is standard practice in all producing companies. Aramco's medical services are highlighted by research on trachoma, conducted jointly with the Harvard School of Public Health. Furthermore, the companies have developed their own medical-training programs, the aim of which is to prepare qualified hygienists, dental assistants, and nurses.

Housing

Next to food subsidies and health services, housing is undoubtedly the most important amenity the employees expect from the companies. From the companies' point of view this is perhaps the most costly and difficult benefit to assure. Practically every producing company has passed through a stage when its housing facilities were woefully inadequate for local employees. This was especially true of early periods or of periods of major expansion when the rapidly hired laborers had to be content with tents and similar primitive arrangements.

In an earlier chapter we have already discussed one aspect of the housing problem—the problem of building company towns versus the policy of integration. No matter which of these two policies was adopted, genuine desire and effort were necessary in order to make a major contribution to employees' needs. The writer is convinced that the right solution of this problem is the key to gaining the good will of employees and thus to achieving a labor-management relationship free of continuous friction. It should be realized—as the companies did in due time—that barrackslike quarters with adequate plumbing are not enough. What if the Arab villager or tribesman has to share his room with numerous other persons and beasts? This is his normal social environment in which he has always found security, recognition, and companionship. Lack of toilet facilities and pure water are of minor importance in the over-all scale of things. Neither good wages nor sanitation can compensate for the loss of home life, especially if the worker is a married man compelled to live in the company's bachelor quarters. And if even the barracks are lacking and the workers are obliged to live in a shanty town near company headquarters, the contrast between it and the houses provided for the company's Western personnel is bound to be too great to be socially safe and psychologically sound.

HUMAN SIDE OF OPERATIONS

The companies' positions with regard to housing have varied greatly. Companies operating in Iran—whether AIOC or the Consortium—faced the most formidable problem of all. The number of workers and the remoteness of the installations from civilized centers made any quick solution to the housing problem most difficult. This circumstance was usually invoked by AIOC when it was criticized by the Iranians. According to an International Labor Organization report, "at the end of 1949, about 90 per cent of all salaried staff have been given accommodation in Company houses. On the other hand, out of 31,875 wage earners only 5,298, or 16.6 per cent, were in Company houses."[5] Gradual progress was being made by AIOC as is shown in Table 10. Yet there was a big gap between the workers'

Table 10. Progress of housing under AIOC in Abadan

Year	Houses for married salaried staff	Rooms for bachelor salaried staff	Houses for married wage earners	Spaces for bachelor wage earners
Before 1934	476	774	28	33
1936–1940	875	54	1,995	709
1942–1944	80	1,229	1,484	136
1945–1949	883	187	2,271	78
Loss on conversion			−199	—
Total	2,314	2,244	5,579	956

Source: ILO, *op. cit.*, p. 33.

expectations and the pace of progress. This was emphasized in the ILO report, which said that "housing is the most serious problem in the Company's areas and the one which gives the most cause for concern." The report added: "The conclusion can hardly be avoided that a large and rapid increase in the construction of houses is both necessary and possible. The shortage of housing accommodation is one of the most serious causes of discontent in the Company's areas."[6]

Aware of the seriousness of the problem it had inherited, the Consortium launched a frontal attack on it at once as Table 11 indicates. In 1958 the Consortium's capital budget allocated 2,100,000,000 rials (£10,000,000) for 1,400 additional dwelling units and a wide

[5] ILO, *op. cit.*, p. 33. [6] *Ibid.*, p. 76.

301

range of clubs, hotels, guest houses, theaters, and sports grounds. Simultaneously the Consortium began studying the possibility of instituting a home ownership scheme.

Table 11. New housing constructed under the Consortium from the effective date (October 29, 1954)

	1954–1956		1957	
	Fields	Abadan	Fields	Abadan
Labor		684	354	330
Staff	330	107	168	34
Total	330	791	522	364
Pending completion			391	610
Grand Total	330	791	913	974

Source: *Iranian Oil Operating Companies*, 1956, 1957.

IPC's situation was of much more manageable proportions. IPC's labor force was the smallest of the three major companies under review. Its operations were closer to inhabited areas, reducing the need for company housing. The need existed however and, had it not been for the company's alertness, could have caused considerable difficulties. For the number of employees provided with houses and bachelor quarters by IPC and its associated companies at the beginning of 1955, see Table 12.

Table 12. Housing provided by the IPC group in Iraq, 1955

	Houses for married personnel	Bachelor quarters	Total accommodations
Staff	423	405	828
Monthly rate	478	321	799
Daily rate	509	975	1,484

Source: IPC, *The Employee in Iraq*, p. 5.

These figures indicate that approximately one-fifth of the Iraqi personnel in the IPC group was housed in company accommodations. This was not sufficient, despite the presence nearby of urban communities. Instead of greatly expanding the company-provided housing, the IPC group launched (beginning in 1952) a home owner-

HUMAN SIDE OF OPERATIONS

ship scheme, which was mentioned earlier in relation to the policy of integration. Progress under this scheme has been considerable, as is shown in Table 13.

Table 13. Houses constructed under IPC's home ownership scheme

Year	Kirkuk	Basra
1952	16	—
1953	82	—
1954	104	3
1955	137	15
1956	148	122
1957	131	127
Total	618	257

Source: *Iraq Oil in 1955, 1957.*

The home ownership scheme was initiated in the Mosul area in 1957. IPC's plans called for the construction of one thousand houses in Kirkuk alone by the end of 1961. Hundreds of applications were being received every year from employees anxious to own a house. There is no doubt that the scheme was very popular and that IPC's record in this respect was the best of the companies operating in the Middle East.

Aramco stood somewhere between the Iranian companies and IPC with regard to housing. Its physical environment and distance from civilized centers made its position like that of the Iranian companies. Its labor force was about twice the size of IPC's. Consequently the housing problem presented a major challenge to its over-all social policies. The company met this challenge by a program that was divided roughly into two phases. In the first phase stress was laid on providing as large a proportion of its labor force as possible with company-owned permanent quarters to replace the tents and other temporary arrangements. Progress during this phase is illustrated in Table 14. As the figures indicate, the percentage of Aramco's workers housed in permanent accommodations was steadily increasing. By 1953 all the workers employed in the company's three principal centers (Dhahran, Ras Tanura, and Abqaiq) were living in brick and concrete structures. Only a minor proportion assigned to temporary camps at Haradh and Uthmaniyah were living in tents, which were steadily being replaced by fireproof portable buildings.

Table 14. Company-owned quarters for general employees of Aramco

Year	Employees eligible for accommodations	Accommodations	Per cent of employees housed
1947	14,392	2,106	14.6
1948	16,070	3,780	23.5
1949	13,113	5,796	44
1950	14,519	7,798	53.7
1951	19,064	9,038	47
1952	17,022	13,132	77
1953	*	14,568	

Source: Aramco, *Report of Operations*, 1947–1953.
* No figures available.

This first phase was successfully concluded by 1953, but these were only bachelor quarters, provided to all laborers married or unmarried. Realizing that such a situation was untenable, the company in 1952 launched a housing loan plan, which in its essentials was very much like the home ownership scheme of IPC. Originally limited to qualified employees of grade 4 and above, the plan was gradually broadened to include married workers of grades 1, 2, and 3 of at least ten years' service. It provided for interest-free loans to be repaid by installments and subsidization by the company of 20 per cent of the building costs. Its progress is shown in the following table: [7]

Year	Dwellings built or acquired
1952	12
1953	67
1954	79
1955	34
1956	104
1957	367
Total	663

[7] From Aramco, *Report of Operations*, 1952–1957.

In view of the fact that some of the houses built by local contractors were defective and had to be demolished, the ultimate total of houses built or acquired by the end of 1957 was 612. Aramco's figure was almost identical with that of IPC, which built 618 in the same period. However, since Aramco has about twice as many workers as does IPC, there is no doubt that IPC is making better progress and that Aramco still has much to do.

Saving for the Dark Hour

Of the three major companies, IPC and Aramco have introduced saving schemes for their employees. These schemes play a truly pioneering role in initiating oil workers into regular thrift habits. When newly introduced in 1951, IPC's 5.5 per cent interest-bearing scheme met with great enthusiasm, expressed by the participation of as many as 75 to 80 per cent of the employees. Gradually, however, participants tapered off, and only about 30 per cent of the eligible employees persisted in regular saving. At the same time the savings rose both in total figures and in average amounts per member. The table below is illustrative of these trends: [8]

	Total members	Members as % of eligible employees	Avg. savings per member	Total saved
1954	4,583	42.6	£58	£267,056
1955	4,320	38.1	70	302,000
1956	3,603	30.8	71	256,800
1957	3,454	30.7	88	302,654

Aramco's employee thrift plan differed technically from IPC's saving scheme in that on account of religious injunctions against "usury" observed in Saudi Arabia it did not provide for an interest rate on deposits. Instead Aramco paid awards for thrift to the "savers" when they ended their service with the company. These awards depended on seniority. They amounted to 50 per cent of the savings of 5-year employees and ranged as high as 100 per cent for 15-year employees. Instead of releasing the figures of the total amounts saved, the company's practice has been to publish the sums saved by and rewards paid to the employees who left Aramco's

---

[8] Information by courtesy of IPC management.

employment in a given year. This is illustrated in the following table: [9]

	Total Saudis employed	No. of Saudi participants	Amts. saved by those leaving work	Amts. given as awards
1953	13,555	7,465		
1954	14,182	8,162		
1955	13,371	8,245		
1956	13,213	8,715	$75,000	$56,000
1957	12,729	9,510	95,000	78,000

Aramco's thrift plan is supplemented by a system of cash awards for continuous service with the company. Under this system a Saudi employee receives SR3,780 ($1,008) during his first fifteen years of uninterrupted service. In a single year (1956) 13,221 Saudi Arabian employees received payments totaling SR2,529,000 ($674,500).

CAREER OPPORTUNITIES

In preceding sections we have dealt with benefits for local employees, most of whom are laborers only. We shall now turn to those more ambitious employees who are interested in a career as well as in wages. Company policies toward this group are no less important than are those relating to simple laborers. Company relations with the career-minded strata involve both tangible and intangible elements. The tangible ones, i.e., those that can be expressed in definite policies and measured in figures, can be narrowed down to two: opportunities for training and the possibilities of promotion to more responsible positions.

Company Training Schemes

Each major producing company at one time or another developed a program to train its employees in mechanical, clerical, and professional skills. These programs were generally the result of two influences. On the one hand, the companies were anxious to increase the efficiency of their laborers, artisans, and white-collar employees. On the other, the governments of the host countries, echoing the demands of the more ambitious employees, insisted that the companies should provide engineering and professional education to

[9] Data compiled from Aramco, *Report of Operations*, 1953–1957.

eligible candidates. Formal stipulations to this effect were sometimes contained in the concession agreements, notably those of IPC and the Consortium. As a result company programs included training at three levels: lower, intermediate, and higher.

The earliest training program was introduced by the oldest company in the area, AIOC. Rather paradoxically it started with training on a high rather than on a low level. In 1939 the company opened the Abadan Technical Institute to provide engineering education of both the trade-school and the lower academic type. Although financed and administered by the company, the Institute was under the academic direction of the Iranian Ministry of Education. Prior to the nationalization of oil its enrollment was close to 800. The next step was to develop training at the lower levels, and with this in view the company in 1947–1948 inaugurated a training within industry scheme in Abadan and the fields. As for the higher academic level the company was sending to England for further engineering studies candidates selected from among the best students of the Abadan Technical Institute. The number of such students amounted to nearly eighty in the immediate pre-Mossadegh period.

When the Consortium took over at the end of 1954, it promptly set up a comprehensive training program at all levels. Training was given a high priority in the operating companies' social policies. This reflected the management's own convictions and also the awakening of the Iranians to the importance of training. It should be remembered that one of the major criticisms leveled at AIOC by Dr. Mossadegh was its alleged denial of adequate educational and career opportunities for Iranian nationals. As will be shown later, the Consortium began with a much smaller contingent of foreign employees than AIOC had in its final stages. Thus many positions had to be filled by Iranians, with the attendant need for proper training. Table 15 shows the number of employees receiving instruction in various kinds of skills in Abadan and the fields.

Recognizing the aspirations of its more ambitious employees, the Consortium gave special emphasis, beginning in 1957, to the establishment of a long-range program for Iranian career development. To quote from a company report, the program included: "(1) careful selection and appraisal (by committees that are nominated to assist the Training Departments in respect of Iranian Career Development) of all Iranian employees who show a potential capacity to hold responsible positions; (2) organization of specialized training tailored to fit each individual; (3) periodic reappraisal of the employee's progress by the same committees." [10] A major role in the career development plan was assigned to the reorganized institute, which after its inauguration in October 1956 as the Abadan Institute of Technology began offering a 2-year preparatory course, followed by a 4-year engineering curriculum.

Table 15. Progress of training in the Iranian operating companies

Skill	Enrollment	
	1956	1957
Supervisory training	920	803
Overseas training	35	30
Employee Relations Advisors training	14	—
Apprentice training	336	432
Training for transfer (mostly for construction)	1,400	—
On-the-job training	—	1,115
Technical Institute	123	—
Abadan Institute of Technology	78	100
Literacy classes (Persian)	2,620	5,000

Source: Iranian Oil Operating Companies, 1956, p. 21, and ibid., 1957, pp. 33 ff.

IPC's training program bore a broad resemblance to that conducted in Iran under both the old and the new regimes. It gained greater impetus with the opening in 1951 of the Industrial Training Centre at Kirkuk. The curriculum was composed of seven courses of instruction, as follows:

1. Apprentice training scheme of five years' duration. Sixty boys aged sixteen were enrolled every year to study Arabic, English, mathematics, science, engineering, and laboratory and workshop practice. They were paid an allowance ranging from ID10 ($28) monthly in the first year to ID20 ($56) in the fifth year.

2. Artisans' two-year course, less academic in its emphasis than the five-year training scheme. Open to twenty entrants every year,

[10] Iranian Oil Operating Companies, 1957, p. 33.

this course was designed to prepare young men as foremen and supervisors. Six months were spent on full-time instruction in the center and the remaining eighteen months in on-the-job training.

3. Nine-month commercial course for young men who have attended primary and intermediate schools. This course offered twenty vacancies a year.

4. Background evening courses for adult employees in various subjects, of forty weeks' duration. These were designed to service 400 to 500 employees.

5. Training within industry courses for foremen, supervisors, and executives. These courses, attended by twenty-four men, emphasized job instructions, job methods, and job relations.

6. Scheme for sending company employees to the United Kingdom for specialist trade courses of six to nine months' duration. A dozen trainees benefited from this scheme annually.

7. Arabic courses for foreign staff members.

In addition, the company undertook to pay for the education of fifty Iraqi students annually at institutions of higher learning in Britain. The maximum number of such students was not to exceed 250 at any time. When accepting the stipend, the students pledged to work for either the government of Iraq or the IPC companies for double the length of time that they spent at their studies abroad.

Thus IPC's program assured instruction at artisan, clerical, and academic levels. The main difference between IPC and Consortium schemes lay in the method of providing higher education. Whereas all the Consortium efforts centered on instruction in Iran itself, the IPC made use of British institutions for academic instruction, confining local training to the artisan level. IPC's program catered to a smaller work force and was geared to a stricter correspondence between the number of students and the needs of industry than did the Consortium's.

Aramco's training program reflected the fact that its operations underwent an extremely rapid expansion in the postwar period. The host country was able to supply only the raw recruits at the totally unskilled level. Practically all other positions, intermediate and senior, had to be filled by alien expatriates. Consequently, emphasis was on the training of the largest possible number of general laborers so as to allow them to pass from the unskilled to at least the partly skilled category. Training for advanced industrial and supervisory positions was necessarily restricted to the few who could profit by it, and provisions for professional or academic instruction affected only an insignificant number of employees in view of the dearth of qualified candidates. At this stage there was no need to emulate either Iran or Iraq in setting up technical institutes comparable to those in Abadan and Kirkuk. The heavy emphasis on training in practical (i.e., job-related) skills is shown in Table 16, though there were many "voluntary" or off-the-job trainees also.

Table 16. Progress of training under Aramco, 1951–1954
(figures represent monthly averages)

Type of training	Enrollment			
	1951	1952	1953	1954
Job-related	3,109	4,853	3,032	2,348
Voluntary	1,149	1,374	1,429	1,216
Supervisory	257	142	148	114
Advanced industrial	74	132	99	23
Professional	12	20	25	15

Source: Aramco, *Report of Operations,* 1951–1954.

In 1954–1955 Aramco's training system entered a new phase with the inauguration of two new programs. On the one hand, detailed procedures were set up for selecting and training Saudi *muqaddams* (or *mushrifs*) and *muraqibs,* i.e., first- and second-line supervisors. The company's target was to prepare Saudi employees for 600 positions available on the *mushrif* level. On the other, the company launched the Saudi Employee Development (SED) program. Similar to the Consortium's career development program (which it preceded by two years) the SED program was designed to help Saudis realize their leadership potentialities. Its inauguration corresponded to a general rise in the skill level of Saudi employees, expressing a new emphasis on more advanced types of training. In 1956 sixty Saudis were selected for personal development programs suited to individual needs, and in 1957 the number rose to 142. There was also an increase in the number of those sent abroad for academic instruction, from fifteen in 1954 to fifty in 1957.

The increased emphasis on more advanced types of training found institutional expression as well. Curricula of the industrial training centers in Dhahran, Abqaiq, and Ras Tanura were broadened by the inclusion of more courses at higher levels and the addition of senior staff instructors to their teaching personnel.

This brief review of the companies' training systems shows that, with the passage of time and due regard to the local conditions, each major producing company began paying increased attention to programs which would provide advancement and career opportunities. If this was true with training, i.e., with preparation for careers, it may not be illogical to inquire at this juncture about the careers themselves.

Advancement and Promotions for National Employees

Although the presence of a sizable team of foreign specialists and managers is taken for granted by host governments and public in the initial stages of a concession, this attitude is subject to change. Consequently, when the early concession agreements were either revised or replaced by new ones, the host governments have insisted on the inclusion of clauses assuring their nationals adequate representation on all levels of administration. Even where no such formal provisions exist, the companies have become aware of mounting pressure on the part of the host countries. The question is not a simple one. There is a lurking suspicion in the host countries, sometimes fanned by demagogues to exaggerated proportions, that the companies deliberately keep national employees at lower levels so as to prevent them from being able to run the industry when the concession comes to an end. This is coupled with accusations of selfishness and discrimination in favor of alien employees. A standard criticism is that in the usual classification of jobs into general (or daily rate), intermediate (or monthly rate), and senior (or staff) categories Western employees invariably belong to the senior one even though their positions, such as driller or stenographer, would not qualify them for the upper category in their home countries. Even when it can be proved that national employees have attained high management grades, criticism continues on the ground that this is generally restricted to administrative positions and that the companies are reluctant to entrust technical tasks to self-sufficient teams of local citizens.

When queried on these issues, the companies deny discrimination and deliberate restriction of local nationals from assuming responsible technical and other tasks. The presence of alien employees, they point out, is dictated by the needs of efficiency. And as for their classification in the senior category, no other way can be devised to attract them to employment in remote areas and a difficult climate and to compensate them for severing ties with their home environment. The career has to be made attractive for Western employees; otherwise only second-class people would volunteer for service, and their general morale would be low, which in turn would be reflected in costly turnover.

In this never-ending debate neither side is wholly wrong. The companies' pleas for efficiency and morale are genuine and reasonable. So too are the host countries' claims that the advancing education of their nationals entitles them to recognition and promotion. The matter can ultimately be reduced to three main and measurable questions: (*a*) Does the company use purely national teams to perform certain autonomous technical tasks? (*b*) Is the ratio of aliens in the total work force decreasing, increasing, or remaining stable? (*c*) What is the progress in advancing local nationals to the intermediate and senior positions?

With regard to the first question gradual progress could be noted in the 1950's, especially in exploration. All-Iranian and all-Saudi drilling crews made their appearance in the Consortium and Aramco, respectively.

As for the ratio of aliens in the companies' personnel, the general trend is toward a gradual decrease. This is true of IPC and Aramco, and prior to nationalization it was true of AIOC also. The percentage of foreigners in the total labor force of AIOC decreased from 10.7 per cent in 1934 to 7.3 per cent in 1950. In the latter year there were 4,503 foreigners, of whom 2,725 were British, as against 57,237 Iranians.[11] Although the services of the British personnel were of

[11] British Information Services, Reference Division, *Anglo-Iranian Oil Company: Some Background Notes* (I.D. 1059; New York, 1951), p. 13.

HUMAN SIDE OF OPERATIONS

a high order and could by no means be called superfluous, there is no doubt that many of these British employees could have been replaced by Iranians, perhaps with some slowing down of efficiency but without serious damage to the company's interests. This seemed to be proved when the Consortium took over the operations in Iran. Its alien personnel was about 500, with the rest of the positions formerly held by the British now filled by Iranians. This is illustrated in Table 17. Although the figures for the overseas staff show

Table 17. Iranian and foreign personnel of the operating companies in Iran

	1955	1956	1957
Iranian staff	5,920	5,080	5,272
Overseas staff	321	453	510
Labor	42,850	40,391	39,693
	49,091	45,924	45,475
Contract labor (approx.)		6,000	4,000
		51,924	49,475

Source: *Iranian Oil,* 1956, 1957. Data for 1955 by courtesy of the management.

an increasing trend, it should be remembered that in 1955 the companies were in their first year of operation, when the staff was not yet fully assembled. It did not reach its necessary strength until 1957, when its numbers corresponded to the initial estimates made when the Consortium was formed.

The decreasing trend in the percentage of foreign personnel was also discernible in IPC, as Table 18 indicates.

Table 18. IPC's Iraqi and foreign personnel on the staff level in Iraq

	Iraqis	Aliens	Aliens as % of total staff
1952	81	677	89
1953	104	634	86
1957	246	619	72

Source: *Iraq Oil in 1952, 1953.* Figures for 1957 by courtesy of the management.

A similar tendency could be observed in Aramco's personnel situation. From the beginning Aramco's work force was composed of three distinct groups: Saudis in the general employee (labor) category, Amercians on the senior staff, and a variety of other nationalities (mostly from the Middle East, India, Pakistan, and Italy) in the intermediate category. The company's objective, in terms of personnel, was to become as purely a Saudi-American partnership as conditions would permit. With the passage of time and the rise in Saudi skill levels, the alien element began to be supplemented by Saudis, as is shown in Table 19.

Table 19. Saudi and alien personnel in Aramco

	1950		1952		1954		1956		1957	
	No.	%	No.	%	No.	%	No.	%	No.	%
Saudi Arab	10,767	62.1	14,819	59.7	14,182	64.9	13,213	67.3	12,729	69.5
American	2,826	16.3	4,067	16.4	3,141	14.4	2,878	14.7	2,676	14.6
Palestinian	826	4.8	1,083	4.4	858	3.9	489	2.4	360	2.0
Italian	693	4.0	1,046	4.2	479	2.2	182	0.9	110	0.6
Adenese	682	3.9	663	2.6	345	1.6	175	0.9	140	0.7
Indian	667	3.8	1,110	4.5	1,036	4.7	1,052	5.4	969	5.3
Pakistani	455	2.6	1,320	5.3	1,415	6.5	1,334	6.8	1,185	6.5
Sudanese	343	2.0	508	2.0	222	1.0	83	0.4	75	0.4
Dutch			91	0.4	54	0.2	33	0.2	20	0.1
Others	86	0.5	131	0.5	126	0.6	193	1.0	61	0.3
Total	17,345	100	24,838	100	21,858	100	19,632	100	18,325	100

Source: Aramco, *Report of Operations*, 1950–1957.

The third major question—whether the national employees were being promoted to positions of greater responsibility—could be answered in the affirmative for all companies. Although there was an initial drop in the number of Iranians serving on the Consortium staff, this was merely a general reduction in personnel made necessary by overstaffing during the three idle years between the Nationalization Act and the beginning of operations by the Consortium. Once the surplus was eliminated, the number of Iranians in staff positions began to increase as Table 15 indicates. Moreover Iranians found themselves in several top managerial jobs. Thus the administrative co-ordinator—the third highest position in the refinery—was an Iranian, as was his deputy. Directly below this level were the department managers, of whom several were Iranian, notably those in charge of personnel, accounts, estates, and health.

HUMAN SIDE OF OPERATIONS

Table 18 likewise shows the progress of Iraqis in staff positions. The distribution of the 246 Iraqi staff employees among the various staff grades was as follows: grade 7 (next to the highest), 3; grade 6, 2; grade 5, 4; grade 4, 7; grade 3, 33; grade 2, 87; and grade 1, 190. The 246 Iraqi staff employees constituted 28 per cent of the staff. The high-ranking positions held by Iraqis included those of the legal adviser, his assistants, and the manager of general services.

The most notable feature of Aramco personnel policies was the spectacular rise in the number of Saudis advanced to and within the intermediate category. There was also a marked increase percentagewise in the senior staff, although, on account of educational factors, the numbers were fairly limited. This is illustrated in the following table:

	1950	1952	1954	1956
Senior staff	3	6	18	25
Intermediate (grades 6 to 10)	165	460	1,113	1,967
Total Saudi force	10,767	14,819	14,182	13,213

All in all the companies could point to an upward trend in promoting nationals to supervisory and managerial positions. If there is any difference in views between them and the host countries in this respect, it pertains to the pace of the trend and its qualitative features rather than to its existence. Among the producing companies the Consortium stands in the forefront in respect to the advancement of local nationals, both quantitatively and qualitatively. But this should not lead one to condemn the other companies, inasmuch as each company operates in its own sociocultural environment, different from others in terms of educational standards and the availability of skilled personnel. It is obvious that Saudi Arabia, a desert country with a few secondary schools emphasizing religious education, cannot compare with Iran with its long tradition of arts and sciences, its secular educational system, and its universities. Moreover, the company cannot compete with the government when the latter, because of the dearth of talent and the needs of the rapidly expanding bureaucracy, offers its young nationals high-ranking positions, sometimes on a subcabinet level, regardless of their experience.

OIL AND STATE IN THE MIDDLE EAST

In a number of cases it has been the intangible factors rather than the figures, which were at stake. These largely revolve around the basic attitude of the Westerner toward the nationals of the host country. Is he treating them condescendingly or on a footing of equality? Is he genuinely polite and courteous? Does he show racial prejudice? Is he willing to accept the inevitable differences in cultures and customs and respect them? Is he narrowly clannish or is he willing to associate with the nationals? Does he make any effort to learn the local language? Is the management willing to put local nationals in positions of authority over Westerners? Is there genuine desire to understand the people of the host country and their aspirations?

These are the important, and perhaps the decisive, questions upon the right solution of which much depends. Although they call, first of all, for conscientious self-appraisal on the part of every Westerner employed by the companies, management's role in encouraging the right attitudes and mental habits cannot be overstressed. Such encouragement may take institutional forms, such as providing for language instruction for Western employees or organizing special indoctrination courses for newly engaged personnel from overseas. It may also take the form of personal example set by the top executives and emulated down the line. If training is needed for the host countries' nationals to make them useful to the industry, education in international living is no less necessary to Westerners if the industry wants to survive and prosper in the midst of an alien environment.

PART FIVE

A CASE STUDY

CHAPTER XVI

Repercussions of the Suez Crisis

THE Suez crisis of October and November 1956 is an excellent subject for a case study of the repercussions which a political event essentially unconnected with oil may have on the position of the oil industry in the Middle East and in the world at large.

The salient facts of the crisis can be summed up as follows: On October 29, in an action officially described as aiming at the destruction of fedayeen bases,[1] Israel invaded Egyptian territory and in a swift movement occupied the Gaza Strip and large portions of the Sinai Peninsula. On October 30 Britain and France jointly demanded cessation of hostilities within twelve hours, announcing that failure of either side to comply would result in military action on their part to keep the two belligerents separated, with a view to protecting the Suez Canal. Inasmuch as Egypt did not heed the ultimatum, France and Britain on October 31 launched their attack, which consisted in a preliminary bombing of Port Said and a few other Egyptian concentrations, followed by an invasion of the Canal Zone by land troops on November 5.

The attack on Egypt was taken up by a special session of the United Nations General Assembly, which demanded an immediate ceasefire and prompt evacuation of foreign troops from Egypt and provided for the creation of a United Nations Emergency Force (UNEF) to assure the peaceful withdrawal of foreign forces and the policing of the border area between Israel and Egypt. In addition, strong rep-

---
[1] "Voluntary" Egyptian commando units.

resentations were made separately by the United States and Russia, calling upon Israel, Britain, and France to desist from further hostilities and to evacuate Egyptian territory. On December 22 Britain and France completed the withdrawal of their troops from Egypt.

Israel basically agreed to comply with the UN resolutions and began a gradual evacuation. It drew a distinction, however, between Egyptian territory proper and the Gaza Strip, trying to make evacuation of the latter conditional upon recognition of the freedom of navigation in the Gulf of Aqaba and the Suez Canal. Eventually under strong pressure from the United States, which suspended its aid programs to Israel at the time of the invasion, Israel on March 7, 1957, withdrew its forces from the Gaza Strip. This was understood to have been accompanied by verbal reassurances from Washington that America would uphold the principle of free navigation in both the Suez and the Aqaba waterways, but no formal commitment to this effect was announced. With the evacuation of Israel's troops the military phase of the Suez crisis came to an end, and American economic aid to Israel was resumed.

Politically, however, three points still awaited solution. One was the presence of the UNEF on the Egyptian side of the Israeli-Egyptian border and, more particularly, how long it was to remain there. The second was navigation in the Gulf of Aqaba, which, since the destruction of the Egyptian shore batteries by the Israelis in the early days of the invasion, was under the virtual naval control of Israel, a status likely to be perpetuated so long as the UNEF was in control of the Sinai coast opposite the only navigable entrance to the Gulf, the Strait of Tiran. The third was the Suez Canal, which, blocked by Egyptian sabotage action, continued to be closed to navigation. The latter problem had at least three ramifications: freedom of navigation for Israel, international control as opposed to purely Egyptian control, and compensation of the Suez Canal Company nationalized by Egypt in July of 1956.

A well-rounded analysis of the Suez crisis is not the intent of this study. We are concerned only with the effects of this crisis on oil operations. From the point of view of the oil industry the drama began to unfold when the Suez Canal was blocked on October 31 by the scuttling of a number of ships and by demolitions carried out on order of the Egyptian government. The following day, November 1, 1956, at noon Radio Cairo broadcast the official announcement that Egypt had decided to break off all relations with Britain and France. The same broadcast carried two appeals. One, from the Federation of Arab Labor Unions, called on all workers in the Arab world to implement the decisions adopted at the conference of the Federation held the previous August. These provided for sabotage of oil installations and pipelines and for attacks on foreign military bases. The other appeal originated from the rector of Al-Azhar University, calling all Moslems to a Holy War.

REACTIONS TO THE INVASION IN IRAQ AND LEBANON

The events in Egypt made a profound impression on the Arab world. Although Egypt's relations with some Arab states were strained at the time of the invasion, a high degree of unanimity was displayed by Arab governments and public in condemning the aggression. Numerous expressions of solidarity were heard from both official and unofficial quarters, coupled with declarations of readiness to come to the rescue of the attacked nation.[2] But could these protestations be translated into action and, if so, in what way? Paradoxically, of the three invading states, Israel was the most immune to retaliation inasmuch as the only way in which the Arab states, already boycotting it economically, could hurt it was by launching a military counteroffensive. The results of hostilities with this little but relatively strong state were unpredictable.

The situation was different in the case of France and Britain, both possessing considerable assets in the Arab countries. Of these, oil was obviously one of the most conspicuous and therefore most vulnerable targets. Actually neither France nor Britain, jointly or separately, were the sole owners or operators of any oil-producing

[2] Typical in this respect was the statement made in the United Nations General Assembly by Dr. Fadhil al-Jamali, Iraq's representative. On the issue of Palestine and Israel's aggression in Egypt he declared, "All the Arab world is Egypt and all Arab statesmen are Nassers" (*New York Times*, Dec. 7, 1956).

company or pipeline in the Middle East. The Iraq Petroleum Company was an international corporation with nearly 25 per cent American participation. Nevertheless, IPC had always been identified in the minds of the public as a British organization, not only because of British control of almost one-half of its stock, but also because of the predominantly British management. IPC was a "logical" target for possible nationalist retaliation against Franco-British action in Egypt. In view of the fact that IPC's operations extended to three countries, Iraq, Syria, and Lebanon, three different policies toward the company could have been adopted. The danger to IPC consisted in the fact that a hostile stand by any one of these countries was apt to affect the bulk or at least a substantial part of the company's production. Ultimately the policies of the governments in question were determined by three principal factors: (*a*) the economic stake of a given country in uninterrupted production or transit operation; (*b*) its general orientation in international politics; and (*c*) the pressure of public opinion.

In the first place, there was Iraq; it was the only Arab state in the Baghdad Pact; it was politically at odds with Egypt; and it relied to a pronounced degree (69 per cent of budget receipts in 1955) on oil revenues. Its government was conducting a determined policy of alignment with the West against possible Soviet penetration, despite the unpopularity of such a policy with the broad masses of the Arab East. And because it had the will to move against the prevailing current, it was interested in avoiding undue public excitement over internal or external issues so as to maintain much-needed internal stability in the crucial period of reconstruction and development. There was no doubt that as Arabs and nationals of a freshly emancipated country the Iraqi leaders felt keen resentment at the thought of another Arab land being invaded by the superior forces of Israel, Britain, and France. To this could be added the embarrassment of being formally aligned with the West at a time when two major European powers, in seeming co-operation with the Jewish state, were violating the principle of territorial integrity in the Arab East. Moreover the Iraqi government had to face an outburst at home which might have considerably weakened its position. Theoretically the government was presented with two alternatives: either to follow the popular trend, apply retaliatory measures to Franco-British interests, and thus regain its prestige with the Arab masses at home and abroad; or to persist in its policy of friendship with the West regardless of the effects upon Arab public opinion.

Although outwardly commendable as a means of bringing the government and the people together, the first policy, if pushed to the extreme, might have resulted in undercutting the economic basis of the government's authority. If the government decided to repudiate the IPC concession, it would probably experience difficulties and dangers similar to those encountered by Dr. Mossadegh's regime in Iran a few years earlier. Although such a course would have clearly been disastrous economically, it is conceivable that it could have brought the same kind of short-range political advantage as that enjoyed by Dr. Mossadegh in the emotionally surcharged climate of Iran between 1951 and 1953. But even this advantage would not have been absolutely certain; by so drastically reversing their policy Iraqi leaders would have acknowledged its failure and opened the door for popular demand to cede their place to others who had consistently advocated a neutralist and Pan-Arab orientation.

The second alternative—that of continuous co-operation with the West—had this advantage that it would not only assure the continuity of oil revenues, but also serve as proof that a policy of soberness and moderation could spare the country the dangers stemming from adventurous extremism. In this connection it might be observed that, although publicly committed to expressions of solidarity with Egypt as a victim of aggression, Iraqi leaders were not unaware of the provocation of Egypt's policies prior to the crisis, especially as expressed by the seizure of the Suez Canal in July 1956. Actually there never was much hesitation among Iraq's leaders as to which course to follow. Their stated objective was to assure the maximum benefit to their country through a truly independent policy. Consequently they limited their actions to what was morally desirable but politically feasible, by (*a*) declaring their readiness to assist Egypt according to the provisions of the Arab Joint Defense Treaty of June 17, 1950; (*b*) breaking off diplomatic relations with France (but not with Britain); and (*c*) refusing to attend any formal meetings of the Baghdad Pact alliance so long as Britain continued her aggression in Egypt. The differentiation between France and Britain was defensible on the ground of France's alleged collusion with Israel and of her continuous hostility toward Arab nationalism as expressed by the crisis in Algeria. The government experienced a few tense moments when, in response to inflammatory appeals of opposition leaders, city mobs demonstrated in favor of Egypt. These demonstrations took a particularly violent form in Najaf and Mosul. But, determined as it was not to be deflected from its consistent policy, Iraq's government reacted to these outbursts swiftly and decisively, nipping in the bud the possible spread of disorders.[3]

Similarly cool-headed was Lebanon's reaction. This little country also had two alternatives: to identitfy itself with Egyptian-sponsored militant Pan-Arabism or to preserve its special character as a trading entrepôt of the eastern Mediterranean and a half-Christian political bridge between the West and the East. Although the first alternative might have brought it closer to its natural Arab hinterland (surely a desirable objective,) it would also have carried the risk of dissolution of its own national identity, in addition to all the economic pitfalls of such a policy. Thus the issues in Lebanon were even more weighty than in Iraq, because at stake were not only the survival of a government and of a sound economic foundation, but also the preservation of the state as an independent entity. Consequently despite heavy pressure from the Pan-Arab-oriented elements of opposition, the government of President Camille Chamoun and Premier Sami as-Solh did not hesitate to refuse to burn its

[3] Pro-Egyptian and antigovernment demonstrations occurred in the major urban centers of Iraq in early November and late December 1956. Simultaneously the principal opposition leaders presented a petition to the king in which criticism of the existing regime was coupled with a demand for free elections and freedom of political association. In response the government used police to quell the disturbances, suspended the schools for a two-week period, and placed under arrest the chief signers of the petition. Among those arrested were Kamel Chaderchi, leader of the National-Democrat Party; Faiq Sammarrai and Mohammed Sadiq Shanshal, both leaders of the Independence Party; Sami Bashalem, a deputy in the Parliament; Senator Mohammed Shabibi of the United Popular Front; Hussein Jamil, president of the Lawyer's Association; and Abdur Rahman al-Bazzaz, dean of the Baghdad Law School. Subsequently Kamel Chaderchi was sentenced to three years' imprisonment, and four others were placed on probation or on bond, each for one year.

bridges with the West. It neither broke off diplomatic relations with France or Britain nor did it apply any retaliatory measures to IPC.

SABOTAGE OF PIPELINES IN SYRIA

Syria took an altogether different stand. Her government, linked with Egypt by close ideological ties and special military agreements,[4] promptly broke off diplomatic relations with Britain and France. On November 2, 1956, all three pumping stations of IPC's pipeline in Syria (T-2, T-3, and T-4) were blown up, stopping the flow of oil from Iraq to the Mediterranean terminals at Banias and Tripoli. Ostensibly this was the spontaneous action of the workmen, acting on patriotic impulse and in response to the previously mentioned Egyptian appeal to sabotage Western oil installations. However, the Syrian action had all the characteristics of advance planning, thorough preparation, and expert execution. It was inconceivable that the major quantities of explosives needed for such an operation could have been stored and kept by anybody without prompt detection by the rather ubiquitous police and army intelligence. The army's active role in this affair is made clear by the fact that shortly before the sabotage operations the British supervisory personnel at the desert pumping stations were arrested and evacuated by the troops. Circumstantial evidence of official connivance is the fact that the government rejected IPC's request to allow it to repair the damage, persisting in this attitude until the evacuation of Egypt by all the invading forces.[5]

Eventually, on December 20, Syria allowed IPC's engineering team to inspect the pipelines. On March 6, 1957 (four months after the stations were blown up) official permission to start repairs was given. Inasmuch as the demolition of the three stations was complete, the repair work which began on March 9 was bound to be slow. At first, arrangements were made to by-pass the wrecked stations and to pump

[4] Agreement of Oct. 20, 1955, whereby Syrian forces were placed under a joint Syro-Egyptian command headed by General Abdul Hakim Amer, commander in chief of the Egyptian Army.

[5] Although the British and French troops withdrew from Egypt by Dec. 22, 1956, Syria continued to oppose repair work until the evacuation of Israel's forces from Sinai and the Gaza Strip.

oil by the force of the undamaged stations in Iraq. Soon afterward the company began to install new equipment in order to increase the pumping capacity of the Iraqi stations. It hoped eventually to dispense with two of the Syrian stations and to rely on one together with the pressure generated in Iraqi territory.[6] Such a solution possessed the advantage of decreasing the number of possible hazards stemming from any future sabotage action, while also reducing the role of the Syrian section of the pipeline in the over-all technological operation.

Global and Regional Effects of Syrian Sabotage

The effects of the Syrian sabotage action were both global and regional. This action affected the economy and security of Western Europe and, indirectly, of the Free World as a whole. It should be borne in mind that the demolition of the Syrian pumping stations was accompanied by another major act of sabotage in the Arab world, the blocking of the Suez Canal by Egypt in the first days of the Anglo-French invasion. Although Egypt's action differed from the Syrian in that it did not aim specifically at oil enterprises, it was bound to affect the oil traffic vitally inasmuch as in 1956 the latter accounted for 65 per cent of the total tonnage transited through the Canal. Translated into absolute figures the situation was as follows: Both the Suez oil traffic, representing 77 million tons a year and the Syrian pipeline transit amounting to 25 million tons a year came to a complete stop. Together this constituted 102 million tons a year or 2.04 million barrels a day, of which 87.5 million tons a year or 1.75 million barrels a day went to Europe. In 1956 Europe was importing oil at the rate of 125 million tons a year or 2.5 million barrels a day. In the postinvasion period therefore Europe was deprived of 70 per cent of its usual oil supplies. True enough, the oil which normally went through the Suez Canal could be shipped around the Cape. But this meant a delay of at least two weeks. Be-

[6] The old throughput capacity of the pipeline was about 25 million tons a year. Within a few days after the first repairs the pipeline operation could be resumed at the rate of 11 million tons a year. A new type of pump installed in Iraqi territory could increase the throughput capacity to 36 million tons a year, which represented a pressure of 1,000 pounds per square inch, an increase of 30 per cent over the previous pressure.

cause of the longer time needed for tankers to make the journey,[7] supplies via the Cape could amount to barely 60 per cent of the oil directed through Suez. Thus Europe was bound to be short about 45 per cent of its normal supplies. Even this figure is based on the assumption that all the tankers hitherto using the Canal would go to Europe instead of being divided between European (80 per cent) and the Western Hemisphere (20 per cent) destinations, as was the case before the blocking of the Canal.

The damaging effects on the economy of the highly industrialized European countries of such an interruption are obvious (see Chapter II above). European governments were obliged to turn toward other, primarily Western Hemisphere, sources of supply. But this presented three difficulties: (*a*) a drain on their hard currency reserves; (*b*) the difficulty of obtaining oil supplies at short notice; and (*c*) the securing of adequate tanker space for transportation.

The United States government responded to this crisis by appealing to the American oil industry to lessen the European crisis by voluntary action. Soon after the nationalization of the Suez Canal by Egypt in July of 1956 the government had encouraged fifteen leading American oil corporations to create the Middle East Emergency Committee and had exempted it from the provisions of antitrust legislation. This Emergency Committee, presided over by Stewart Coleman of Standard Oil of New Jersey and reactivated after the invasion of Egypt, promptly set to work to fill the gaps in European oil requirements. According to the committee's estimates, 1.25 million barrels a day were needed by the Free World, including 800,000 a day for Europe. Inasmuch as the United States has a surplus productive capacity estimated at about 2 million barrels a day, Europe's requirements could have been met from American domestic wells had it not been for limited pipeline facilities. The committee prepared a plan whereby the domestic output was increased by about 725,000 barrels a day (about 10 per cent of the total United States production), and 50,000 additional barrels a day

[7] A T-2 tanker, the type commonly used after World War II, requires 19 days to sail from the Persian Gulf to Liverpool via the Suez Canal and 34 days via the Cape.

were secured from Venezuela and 25,000 from Canada. Of the total increased output (about 800,000 barrels a day) thus obtained, some 500,000 barrels a day were shipped to Europe and the rest were made available to the East Coast of North America, which before the Suez-Syrian crisis received about 250,000 barrels a day through the Canal and 100,000 a day through the pipelines.

The quantity shipped to Europe did not fully meet its requirements. The deficit of 300,000 barrels a day was equivalent to about 12 per cent of the total European consumption, but this was taken care of by rationing (in Britain and some countries on the Continent) and the elimination of nonessential uses. Thanks to its prompt and efficient action the committee succeeded in sparing Europe a major disaster.[8]

The regional effects of Syrian and Egyptian sabotage were manifold. Before the crisis Middle Eastern oil production was about 190 million tons a year (3.8 million barrels a day). The sabotage actions on the transit routes reduced it by one third, i.e., by about 63 million tons a year, of which 25 million tons were from Kirkuk and 38 million tons from the Persian Gulf area. In purely financial terms the greatest sufferer was Iraq, whose normal output of about 35 million tons a year was reduced by about 37 per cent in 1957. The two Levantine transit countries, Syria and Lebanon, also suffered income reductions. Syria's loss until the reactivation of the IPC pipeline was estimated at $50,000 a day, and Lebanon's, whose terminal at Tripoli was handling 7 million tons a year, was $5,600 a day.

Consequences of Sabotage in Syria: Labor Dispute

Apart from the financial losses to the governments there were other effects, no less important. Most notable were the repercussions in Syria. In the first place, the demolition of the pumping stations resulted in increased unemployment in the country. Having asked for and been denied permission to repair the stations, IPC had no need for the bulk of its labor force. Consequently it offered its workers and employees two alternatives: either to take an unpaid leave of indefinite duration (with the right of re-employment should the

[8] Statistical data in this section have been compiled from a variety of sources, including an article by J. H. Carmical, "Solving Oil Crisis Mostly Up to U.S.," *New York Times*, Dec. 9, 1956.

work be resumed) or to accept dismissal with the usual separation pay. A great majority of workers chose the first alternative. An obvious consequence of this situation was that the Syrian government could be held directly responsible by the discontented workers for causing and prolonging their unemployment. In fact, considerable pressure was put on the authorities by the unionized oil workers to annul IPC's decision concerning dismissals and unpaid leaves and to compel the company to pay wages for the period of enforced idleness. In a number of petitions and manifestoes the oil workers demanded equal treatment with the workers in Iraq and Lebanon, who had been retained in their jobs by IPC despite the work interruption.

In accordance with the provisions of the Syrian Labor Code [9] these complaints were taken up by a Joint Guild Board, a body composed of three workers' representatives, three employer's representatives, and a delegate from the Ministry of Labor and Social Affairs. On February 12, 1957, the board issued a decision enjoining IPC to continue to pay full wages to workers on unpaid leave since November 11, 1956. Although the decision favored the workers, it was received grudgingly by their union. It soon demanded that no difference should be made between those discharged with compensation and those who had opted for unpaid leave. Moreover, the workmen employed by IPC's contractors, particularly by the major contractor, Ahmad Sharabati, asked for identical treatment with those employed directly by the company. IPC appealed the decision of the board to the Superior Arbitration Board.

Although a hearing was set for February 18, 1957, the dispute was soon transferred to the field of politics and legislation. Against the background of an intense press campaign and strong lobbying by the workers' delegation in Damascus to influence the legislative and executive authorities in their favor, the Parliamentary Commission of Social Affairs took up the workers' complaints and early in March (shortly before IPC was granted permission to begin repairs on the pipeline) presented to Parliament a draft bill fully satisfying the

[9] Act no. 279, Labour Code, dated June 11, 1946, in *Official Gazette of the Syrian Republic*, June 13, 1946, no. 25. A translation of the text is in ILO, *Legislative Series 1946—Syria 1, Act: Labour Code*.

workers' demands. On the motion of the Communist deputy, Khaled Bakdash, the bill was placed on the session's agenda. After a brief debate, during which the workers' delegation sat in the galleries, the Parliament on March 7, 1957, passed a 7-article bill, according to which, "in all cases of suspension of work due to *force majeure* beyond the control of employers and workers, the employer shall, upon resumption of his suspended work, immediately reintegrate all the workers who upon such suspension have been discharged or put on leave without pay" (Art. 2). Article 5 made the law retroactive to November 1, 1956, and Article 6 declared it applicable "to concessionary companies whose capital exceeds ten million Syrian Pounds," thus leaving it beyond doubt that IPC was the principal target of this legislation.[10]

Although of dubious legality (if we consider its retroactive effect and its potential conflict with the terms of IPC's concession), the law was not contested by the company. Anxious to resume work as quickly as possible, it was obviously loath to raise legal issues at a time when the Syrian government was becoming amenable to restoration of normal relations. Resumption of operations did not automatically mean return to the *status quo ante*. The three destroyed pumping stations could not be reconstructed at short notice, the period necessary for their rebuilding being estimated at ten to twelve months. Moreover, the company was planning to increase the diminished throughput capacity by installing more powerful pumps in Iraqi territory rather than by rebuilding the Syrian ones. Consequently its operations in Syria were bound to be reduced, and only a fraction of the personnel formerly employed would be needed. The company paid the workers their wages in compliance with the new law and then on March 16 declared that its operations would henceforth not require their services and that it was serving notice to the majority of its re-employed labor force.

What happened next illustrates rather eloquently the pronounced flexibility of legal concepts and procedures in the Middle East, where social, economic, and political considerations tend to overshadow purely legalistic ones. On the basis of the law just passed and the company's prompt compliance with it, it would appear that the

---

[10] The bill was promulgated on March 12–13, 1957, as Law no. 360.

matter was closed. But nothing would be further from the truth. As soon as the company announced its decision to fire a majority of its working force, a hue and cry began to the effect that IPC was harming the workers' interests and violating the law. The latter, said workers' delegates and their political friends, had enjoined the company to reintegrate the workers; hence the company was expected to continue their employment. The fact that it had no need for their services did not diminish the workers' insistence on their presumed rights.

The government was caught between two fires: the law just enacted (with which the company had complied) and the political pressure of the workers. Dominated by anti-Western elements, the cabinet had very little hesitation as to the course to follow. The workers' demands were a paramount consideration, so far as it was concerned. As to the newly passed law, the government said that the company had adopted a mistaken interpretation of it. The law enjoined the company not only to pay the wages for the period of enforced inactivity, but also to re-employ the idle workers. The fact that the company had re-employed them only nominally before dismissing them again constituted a violation of the clearly expressed intention of the law. This was the general argument of the specially appointed ministerial commission (composed of Ministers of Public Works, Finance, and National Economy), which in the next two months held many meetings with IPC representatives. The government's position was untenable, inasmuch as it was against the principles of logic and common sense that a profit-making private corporation should employ a labor force in obvious excess of its needs. As a matter of immediate political advantage, however, the government could hope for some *ad hoc* arrangement with the company whereby the latter would employ (or at least pay wages to) the dismissed workers for an additional period, the length of which would be determined by what was economically acceptable to the company and politically advantageous for the government.

The company had to weigh the disadvantage of further concessions to the government (by agreeing to repeated violations of legal principles) with the advantage of continuing its transit operations in Syria. Within reasonable limits these concessions could be economically justified. The most disturbing element was the unreliability of the other party, to whom written engagements and even its own laws meant little if the political circumstances dictated otherwise. Ever since the Egyptian crisis Syria had been in a state of considerable ferment, with extremist elements asserting their leadership in national affairs. The dispute with IPC clearly transcended the narrow limits of the government-company relationship.

In addition to the ministerial commission, three other distinct elements were influential. These were the unionized workers themselves, the press, and Parliament. The subject was being debated in at least two parliamentary commissions: the Commission for Social Affairs and the Finance Commission. Deputies from Homs, Banias, and Deir ez-Zor, representing principal concentrations of IPC's labor force, began to play an increasingly active role. Consequently the ministerial commission was obliged to negotiate on two fronts simultaneously: with IPC on the one side and with the parliamentary commissions, regional deputies, and workers' delegates on the other.

Toward the end of May a compromise formula was reached. IPC agreed to pay its workers wages for the period ending June 30, 1957 (i.e., almost four more months), after which it was free to dismiss surplus labor. It was characteristic, however, of the political climate in Syria that the matter was not allowed to rest as the subject of a bilateral agreement. Instead the government moved to incorporate the terms of the agreement into a new bill, which was duly passed by Parliament as Law no. 412 in June 1957.[11] The new law meant that Syria was determined to emphasize her sovereignty in her dealings with the company through the instrumentality of unilateral legislative action. It not only impinged upon the terms of the concession but also constituted a novelty in that it was explicitly aimed at the Iraq Petroleum Company instead of at "concessionary companies" as had been the preceding law.

The provisions of the Law no. 412 were as follows:

1. Law no. 360 of March 12, 1957, was abolished.

2. All workers who were in the service of IPC on November 3, 1956, were to be considered as having been on duty until June 30, 1957.

[11] Published in *Al-Jaridat ar-Rasmiyah* no. 27, June 13, 1957.

3. Only after June 30, 1957, is the company permitted to discharge its surplus laborers, with due respect to "Ministerial decisions relating to seniority."

4. Contractors for IPC were also enjoined to pay all workers employed by them on November 3, 1956, for the period of idleness, but this obligation was limited to two months' wages.

5. In addition, the government undertook to pay IPC workers one month's wages and the contractors' workers three months' wages. The payment would be made by IPC, but was to be deducted from the revenue due to the government from the company for the year 1958.

6. Workers dismissed by the company according to the provisions of this law were to enjoy priority in employment opportunities which might arise in future company operations. In case of such re-employment they were to be engaged according to seniority and at the same grade and salary.

7. A decree was to be issued, according to proposals of the Minister of Labor and Social Affairs, "setting out rules of priority of engagement, in the various administrations and institutions under state supervision, of the workers whose services have been dispensed with by the company."

The enactment of the new law marked the end of a dramatic phase in the history of the pipelines in Syria, a phase of direct reaction to the foreign invasion of Egypt in the fall of 1956. Relations between IPC and the Syrian government ostensibly returned to normalcy in June 1957, but certain psychological wounds, mutually inflicted, were not likely to heal quickly. For the first time the Syrians had tasted the fruit of sovereignty in unilaterally dealing with a private corporation whose only fault was that it was partly owned by the nationals of a country guilty of military intervention in a different area. Although their action was unnecessarily destructive (the same result could have been accomplished by a simple government order to shut off the flow of oil) and abounded in measures clearly at variance with the terms of the concession agreement, it gave them a sense of power and satisfaction at having humiliated a once-powerful and seemingly invincible organism. By resorting to such measures they have so gravely shaken the foundations of mutual trust between the company and the country as to discourage and possibly preclude any further expansion of transit operations. Henceforth the company will not consider any governmental decision as final. In fact, while the finishing touches were being put on the settlement of the labor controversy, Syria's Minister of Finance came forward with a new, though not totally unexpected idea considering the prevailing political climate. In his annual report to Parliament on the economic and financial situation of Syria, in which he analyzed a draft of the state budget for the fiscal year 1957, the minister stated that Syria expected to receive from IPC an amount equal to that of the previous year:

> We have considered the oil revenues included in the budget for 1957 as being equal to those estimated in the budget for 1956, because it is within our right to receive them in their totality. The stoppage of the flow of oil is due, in fact, to the cowardly aggression on sister-Egypt and on Syria at the same time and we believe that it is natural, if not evident, that IPC should pay the Syrian government all of this revenue, while adding to it the supplementary sums stemming from the increase in the price of transportation.[12]

By thus introducing a new element of uncertainty into the company-government relationship, Syria's official circles were damaging their country's economy in the long run. This fact was, however, entirely subordinated in their minds to political priorities, a condition common to the Middle East as a whole in this age of ebullient nationalism.

REACTIONS TO THE SUEZ CRISIS IN SAUDI ARABIA AND THE PERSIAN GULF

As a sister Arab country Saudi Arabia was not remiss in expressions of solidarity with invaded Egypt. Following a royal proclamation, general mobilization of the army was ordered, and training camps for volunteers were established. On November 6, 1956, Radio Mecca broadcast the text of three official Saudi announcements. The first declared that because of the armed aggression against Egypt

[12] *Recueil des lois syriennes: Exposé du Ministre des Finances, 1956, Annexe 1—1957* (Damascus, 1957).

by Britain and France the Saudi government had decided to break off diplomatic relations with both aggressor countries. The second proclaimed an embargo on the shipment of Saudi oil and oil products to Britain and France and a ban on the fueling of British and French ships and of all ships carrying oil to these two countries. The third contained a statement that the Saudi army had moved toward Jordan to co-operate with the armies of the sister Arab states.[13]

It was the second of these proclamations which proved of utmost importance to Aramco. In compliance with it, the company was obliged to stop the flow of oil through the underwater pipeline to the Bahrein refinery inasmuch as the latter was located in British-controlled territory. Shipments of oil to Britain and France suffered interruption primarily as the result of the blocking of the Suez Canal. Consequently the Saudi embargo merely added the seal of official action to the general dislocation of oil shipments already manifest. Yet as a result of the Saudi ban buyers of Aramco's oil (i.e., its parent corporations), whose main markets were located in Europe, had to reroute many of their shipments so as to avoid supplying France and Britain. The Saudi ban lasted until the evacuation of Israeli forces from the Gaza Strip. On March 9 it was lifted, and on the same day oil again flowed through a pipeline to the Bahrein refinery.

It was not appropriate for Aramco, as a private company, to take any public stand on the international political crisis. Consequently it withheld official comment. But working as it did in the Saudi Arabian environment, the company could and did express understanding of, and sympathy with, the public reaction to Egyptian events. When a Red Crescent drive was initiated among its Arab and Moslem employees to collect donations for the victims of warfare in Egypt, the company's facilities were made available to the voluntary collectors, and its public relations office served as a channel of information about the results attained. According to a broadcast from Radio Mecca on November 14, Aramco's employees donated SR178,000 to the Egyptian Red Crescent Society. The drive was carried out in the company's three districts, Dhahran, Abqaiq,

[13] The texts are in *Umm al-Qura*, no. 1640, 6 Rabi' II, 1376 (Nov. 9, 1956).

and Ras Tanura. Furthermore, F. A. Davies, chairman of the Board of Directors, declared that the company itself would contribute an amount equal to that collected by the employees. The co-operative spirit displayed by the company on this occasion did a good deal to relieve the tenseness evident among Arab and Moslem employees when the news of the invasion of Egypt reached the Eastern Province. Aramco's willingness to subject its shipping operations to inspection by Saudi officials—the latter were unusually sensitive to the attributes of national sovereignty—paid a dividend in the maintenance of mutual confidence.

By far the most important factor, however, in sparing the company the vexation experienced by the less fortunate corporations was its exclusively American character. This permitted the Saudi government to draw a clear distinction between American and British-French interests and naturally affected the attitudes of the transit countries, in particular Syria. As a result, neither the principal producing operations nor the pipeline transit of Aramco's oil to the Mediterranean were interrupted.

The Persian Gulf principalities presented a much less serene picture. Subjected to varying degrees of British political control, these principalities contained two separate elements: the native rulers and the population. While the rulers were anxious to preserve their normal relations with the British, their subjects did not feel that compulsion and in a few cases actively expressed hostility to Britain. As could be expected under the circumstances, the most adverse reaction to the Suez crisis occurred in the two most advanced principalities, Bahrein and Kuwait, both of which experienced considerable turbulence. In Bahrein the workers of the Bahrein Petroleum Company (Bapco) went on strike on November 4. Simultaneously riots broke out in the capital city of Manama and the adjoining township of Muharraq. This was followed by a general strike that paralyzed government services and the economic life of the islands. The stoppage of work was so widespread that even the nurses in the government hospitals walked out leaving their patients unattended. Appeals to reason and legality broadcast by Radio Bahrein proved of little avail to stop the popular upsurge. Demonstrators set fires in the cities and attacked shops and buildings belonging to

Bahreinis and foreigners alike. Among those attacked by the mobs were the editorial offices of the newspaper *Al-Khalij,* the Catholic Church, and the repair docks. The organization around which the discontented elements rallied was the Committee for National Unity, whose banished leader had chosen Egypt as his place of residence. Although the influence of alien agents, Egyptian or Communist, was more than probable, the riots had enough spontaneity in them to warrant the conclusion that they were an expression of genuine nationalism, provoked by the events in Egypt. After nine days of rioting the police were able to restore order, and laborers resumed their work in the oilfields and the refinery.

In Kuwait the popular reaction was also strong, though differently expressed. There were some demonstrations in the city of Kuwait (in which for the first time women participated alongside men), but they lacked the violence of the Bahrein riots. Notable was the 2-month-long boycott which the Kuwaiti merchants instituted against British firms and customers. They went so far as to refuse to sell anything to British residents in the principality. Coupled with this were voluntary contributions for the benefit of Egypt. In contrast to the orderly demonstrations in the city was the sabotage in the oilfields. Rather paradoxically it did not occur during the first days of invasion but a month later. On the night of December 10 to 11 the oil installations were rocked by about a dozen explosions; some took place at the oil wells, some in the underwater pipelines at Mina al-Ahmadi, and others in the gas line feeding an electric generator and a water distillation plant, both owned by the Kuwait government. Four wells were blown up and one of them caught fire and was not extinguished until four days later. The pipelines were also damaged, causing temporary suspension of the tanker-loading operations. It appears, however, that the saboteurs were not very experienced and that they lacked the knowledge seriously to impair the oil operations. The damaged wells were but a tiny fraction of the 185 wells operated by the Kuwait Oil Company and consequently did not affect the over-all production picture.

The Kuwait sabotage incident and the Bahrein riots were instructive in several respects. They gave Britain a reminder that her position in the Persian Gulf was no longer unchallenged and that in any serious manifestations of hostility Arab nationalism, stimulated and fed by the revolutionary regime in Cairo, would inevitably aim at the most obvious target, the oil installations in the British-protected area. Although acts of sabotage and violence occurred only as a result of major provocation, i.e., the British invasion of Egypt, the delay in Kuwait proved that such actions might be the result of long-range political planning. Britain's treaty relations with the native rulers could no longer be considered as a sufficient safeguard of the imperial *status quo* in the Gulf. From now on the British must consider not only the rulers—most of whom have tangible reasons for maintaining their friendship with Britain—but also the people. And their attitudes are likely to reflect both domestic and international factors. A repressive regime at home and strong anti-Western propaganda emanating from Cairo or from some other center of Arabism are bound to inspire nationalist extremism. A more progressive outlook on the part of the local rulers, both in governmental and economic spheres, together with a lack of serious provocation from the outside, will do a great deal toward converting their frustrated subjects into citizens having a stake in the preservation of the separate status of the sheikhdoms and the maintenance of the working relationship with Britain.

THE TURKISH PIPELINE PLAN

The sudden cutoff of the two principal transit routes for Middle Eastern oil stimulated much thinking in international oil circles with a view to devising alternative routes and better safeguards for the future. As early as mid-March 1957 representatives of eight major oil companies met in London to discuss expansion and improvement of the pipeline network in the Middle East. This meeting was followed by a larger conference of seventeen leading oil corporations held in London on May 13–16. The conference studied a proposal to build a pipeline linking the Iraqi oilfields with the Mediterranean through Turkey. The purpose was to avoid dependence on Syria and Egypt, both of dubious political reliability.

Several variants of the project were taken into consideration. One provided for construction of a pipeline from the Kirkuk oilfields to a terminal in Iskenderun (Alexandretta). Such a line would pass through the Ramandag area in Turkey and then follow the railroad line north of the Syrian border. Another provided for southward extension of the pipeline to Basra and for feeder lines linking Iran, Kuwait, and Saudi Arabia with the principal line. Between these two variants came several other alternatives which depended on the willingness of countries and companies to tie their production with the projected system. Implementation of the first plan would involve the laying of a 30-inch pipe, 675 miles long—the distance between Kirkuk and Alexandretta—and would cost at least $350,000,000. The aim would be to have a pipeline with a minimum 500,000-barrel, and preferably with 800,000-barrel, daily throughput capacity. Should the system be extended to the head of the Persian Gulf, the line would be over 1,000 miles long and would necessitate the laying of either two parallel 30-inch pipes or of a single 42-inch pipe. Its cost might then rise to between $700,000,000 and $840,000,000 and its capacity might be doubled. The conference discussed also the possibility of diverting the terminal of two Iraqi pipelines from Haifa to the Lebanese coast as well as the problem of permanent repairs to the existing IPC pipeline in Syria. These two points were, however, subordinate to the first—that of the Iraqi-Turkish line; upon the decision on that much else depended. The need for increased tanker tonnage was also reviewed, but only incidentally inasmuch as the handling of tankers is a matter for individual companies to decide.

Although the conference adjourned with a basic agreement that construction of the Turkish pipeline was advisable, many vital points were left to further decisions. This was due to the necessity of ascertaining the views of the countries directly concerned, in particular Turkey and Iraq, before definite decisions could be reached. Increased attention was being given both by the companies and by certain Western governments to a system of legal safeguards whereby the new pipelines would gain international status through the conclusion of treaties between the states having a stake in the safe transit of oil. In fact, the problem had already been discussed at a meeting of President Eisenhower and Britain's Prime Minister Harold Macmillan in Bermuda in March 1957. It was mentioned by Secretary of State John Foster Dulles at a news conference held in the latter part of March in Washington. Queried by newsmen, the Secretary confirmed the fact that the matter had been on the agenda of the Bermuda meeting and declared that such a pipeline was as much in need of international status as the Suez Canal.[14]

The reaction in Turkey to the proposed pipeline was positive. The Turks saw no fault in a project which would enhance their importance to, and provide an additional link with, the West. Moreover, the projected pipeline would assure Turkey a steady revenue either from transit fees or from part ownership, the possibility of which was contemplated by the interested oil companies.

Iraq's position was not identical with that of Turkey. As a producing state, heavily dependent on the unhindered transit of its oil, it shared the desire of the companies to assure safer outlets than those in Syria. Moreover, its economic development plans called for an increased production of crude so that, regardless of political developments in Syria, it needed to extend its outlets and transit facilities. As an Arab country, however, Iraq could not disregard the trends of public opinion at home and in the Arab world as a whole. These were definitely hostile to the Turkish project. Foremost among the opponents of the scheme was, of course, Syria. Its press fulminated against it, calling it another example of "imperialist conspiracy" designed to rob the Arabs of their inalienable rights.[15] The controlled Egyptian press spoke in a similar vein, sensing a new blow to Colonel Nasser's revolutionary regime should the pipeline plan be carried out. The semiauthoritarian character of the regime in Iraq made it hard for the Iraqi press to voice strong opinions about the subject, but there was no doubt that the politically conscious strata of the population were opposed to the idea of favoring Turkey over an Arab country as a transit route. Iraq's ruling group, which did not always see eye to eye with the populace, was guided by three considerations: (a) fear lest agreement with the project provoke a public outburst comparable to that which occurred in

[14] *New York Times,* March 26, 1957.

[15] "Conspiracies of Oil Companies against Syria and Egypt and Their Collaboration with the Governments of Turkey and Iran," *Al-Ayyam* (Damascus), Nov. 17, 1957. See also *Barada,* June 18, 1957; *Al-Ayyam,* Aug. 30, 1957; and *Ash-Sham,* Aug. 30, 1957 (all of Damascus).

several Arab countries following the signing of the Baghdad Pact; (b) fear lest adherence provoke Syrian retaliation in the form of renewed stoppage (or sabotage) of the existing pipelines (this fear was partly based on the increasingly pro-Soviet orientation of Syria's foreign policy and on the possible ability of Syria to replace its revenues from the pipeline by some form of aid from the Soviet bloc); and (c) the desire to reduce to a minimum its dependence on foreign-controlled transit routes, whether Syrian or Turkish.

Leading members of the government were determined to make the Baghdad Pact a working concern and consequently were serious in their desire to cultivate the alliance with Turkey. But they knew that political alignments are not eternal. Moreover, they did not quite share the pessimism of the West about the situation in Syria. Although strongly disapproving the pro-Soviet trend in Damascus, they believed that Syria would not have damaged IPC's pipeline had it not been for the major provocation of the tripartite aggression in Egypt. They refused moreover to believe in the permanence of the leftist orientation in Syria, thinking that there might be a change of government at some future date. In their desire to make their outlets as independent as possible of foreign control, they were most interested in assuring the flow of oil from Kirkuk and Mosul as well as from Basra through a terminal on the Persian Gulf. The ideal solution technically would have been to build a pipeline linking the northern oilfields with the natural port in Kuwait, but a political obstacle stood in the path of such a scheme, namely, the refusal of the ruler of Kuwait to negotiate any agreement so long as Iraq did not recognize his northern boundary. Iraq was not prepared to give such recognition, and as a result the Kuwait terminal had to be shelved. The Iraqis came to believe that, economic factors notwithstanding, a scheme to build their own deep-water port on the Persian Gulf was most desirable from a political point of view.

While privately advocating this solution, Iraqi leaders tried to keep the door open to negotiations with Syria with a view to diverting the original Haifa line to a terminal located on the Syrian or Lebanese coasts and, possibly, to increasing the throughput capacity of the existing lines. In connection with the possibility of a new deal with Syria, the Iraqis were anxious to stress two points:

(*a*) that the task of negotiating transit agreements should be carried out not by the concessionary company but by the government of the country whose oil was to be transited through foreign territory and (*b*) that before committing itself to this course of action Iraq would want to have a treaty guarantee from the transit countries that the flow of its oil would be adequately protected. In raising the first point, the Iraqis were introducing a new element or, indeed, a new theory in the international oil business. This theory rested on the assumption that oil, as a principal national resource of a producing country, was too important an asset to be subjected to transit negotiations between a private company and a foreign sovereign power. What was needed was an agreement between two or more sovereign states acting as equals. The second point was a logical sequence of the first. If an agreement between sovereign states was to be negotiated, it could provide for adequate safeguards against hostile acts, whether official or unofficial. In consonance with this second point the government of Iraq declared in a note to the Arab League of April 1957 that if the Arab transit countries wanted it to resist effectively the pressure of oil companies to have a pipeline built through Turkey they should provide for adequate guarantees against sabotage or stoppage of the flow of oil.[16] In line with this policy Iraq's Minister of National Economy, Dr. Nadim Pachachi, paid a visit to Damascus early in October 1957. It was reported that chief among the items discussed was a treaty guarantee for the pipelines, which would pave the way for extension of the network through Syrian territory.

Iraq's desire to secure formal safeguards for transit rights through agreements between sovereign states was in line with a similar desire expressed earlier in the year by the Bermuda conferees and by Secretary Dulles. The difference lay in the emphasis and the scope of the agreements rather than in the basic legal concept. While Westerners spoke of "internationalization" and implied their own participation in such agreements, the Iraqis thought of bilateral or

[16] *Al-'Alam* (Damascus), April 25, 1957. The matter of guarantees was subsequently raised by Iraq at the 4th session of the Arab League Economic Council in Cairo on May 25–June 3, 1957. See *Ash-Sham* (Damascus), May 30, 1957, and *Ar-Rai* (Damascus), June 17, 1957.

multilateral pacts of a strictly regional character, to be concluded by the producing and the transit countries and possibly to be given additional sanction through some general agreement under the auspices of the Arab League.

The Turkish pipeline scheme not only evoked a negative response, but it also stimulated the theory that Arab oil must pass exclusively through Arab lands. This theory has already been mentioned in the discussion of the general problem of unification of Arab oil policies. Inasmuch as this idea found strong supporters, if not actually initiators, among the Saudi petroleum experts, it was obvious that Saudi Arabia could not be counted upon as a backer of the Turkish scheme. The Kuwait government did not take any public stand on the proposed scheme, but its political differences with Iraq made its adherence to the scheme dubious.

IRANIAN-TURKISH AGREEMENT

Iran, whose oil-based economy has been heavily dependent on the Suez Canal, passed through many anxious moments when the Canal was blocked. Consequently it welcomed any initiative that might render it independent of the political happenings in Cairo. As the suggested Iraqi-Turkish line was slow in materializing, the government of Iran decided to initiate bilateral talks with Turkey aiming at construction of a pipeline between the newly discovered oilfields at Qum in central Iran and the Turkish Mediterranean coast. Noteworthy is the fact that what has been merely tentative theory in Iraq, namely, the need for negotiations between two or more sovereign states rather than between a private corporation and a foreign government, has become reality in Iran. To be sure, the two situations were not identical. In Iraq oil is exploited exclusively by a foreign corporation; in Iran the government since 1951 has taken upon itself the task of exploitation and, notwithstanding its agreement with the Consortium covering the southwestern part of the country, it has proceeded independently to develop the oilfield discovered at Qum. Although the construction of the pipeline by the joint efforts of the Iranian and Turkish governments will have no direct bearing upon the operations of the Consortium, indirectly it is bound to affect them and the Consortium may sometime want to become a partner to such an enterprise.

Preliminary agreement to build the pipeline was reached between the two countries early in September 1957. According to later information, financing and construction was to be assumed by the Allen Company of Texas, which was expected to operate the pipeline jointly with the Iranian government. Many technical difficulties would have to be surmounted because the line would pass through mountainous terrain in western Iran and eastern Turkey. Some sectors of the line would have to be heated to prevent freezing of the Qum crude with its high paraffin content. This would increase construction costs and prolong the period of amortization. Early in November 1957 an Iranian delegation arrived in Ankara to work out the details and to agree on the financial aspects of the enterprise. The latter proved to be difficult. The Turks asked for a higher share in the profits than was palatable to the Iranians. As a result, negotiations were suspended, not to be resumed until nearly a year later.

Eventually, on October 18, 1958, Fatin Rustu Zorlu, Turkish Foreign Minister, and Abdullah Entezam, chairman of the National Iranian Oil Company, on behalf of their governments, signed in Ankara an 80-year agreement to construct a 965-mile pipeline linking the Qum area of Iran with a Mediterranean port, presumably Iskenderun or Mersin. Iran was to finance the project, the cost of which was estimated at some $600,000,000. The 32-inch pipeline was to have a capacity of 25 million tons a year (500,000 barrels per day) and was to serve as an outlet for oil extracted by NIOC, with the proviso that if an agreement was made with the Consortium the latter's crude could also be handled. Turkey obtained the right to purchase, at 6 per cent discount, 10 per cent of the oil transported, and in addition it was to receive transit royalties on a percentage basis. The latter were estimated at $10,000,0000 to $20,000,000 annually. Construction was not to start until the development of sufficient reserves at Qum. Work on it, once begun, was expected to take three years.[17] The agreement was subject to ratification by the Parliaments of both countries.

[17] *International Oilman*, Dec., 1958. For the text of this agreement, see *Iran-Presse*, Dec. 1–3, 1958.

Implementation of the agreement will affect not only the Consortium but the Iraq government. Iraq will not be able to remain unmoved by the existence of a pipeline skirting its territory and giving Iranian oil a transit advantage over its own.

PIPELINE SCHEMES IN ISRAEL AND EGYPT

The Egyptian crisis acted as a stimulant to two other projects, both of which deserve brief mention.

Although Israel was thwarted by American and United Nations opposition to its invasion of Egypt, it managed to salvage at least one fruit of its victory, virtual opening of the Gulf of Aqaba. The opening of this passage could have two practical results: first, the creation of a maritime link between Israel and the Indian Ocean; second, the availability of Israel as a transit area between the Red Sea and the Mediterranean, a position hitherto monopolized by Egypt. Anxious to make the most of this opportunity, the Israelis promptly laid a pipeline between Elath and the Mediterranean in the hope of attracting international oil business to this new transit route.

Their original plans envisaged three alternatives. The first was construction of a 16-inch pipeline from Elath to Ashdod Yam on the Mediterranean near Askelon. This would cost $32,000,000. According to Levi Eshkol, Israel's Minister of Finance, France was ready to contribute half of this sum.[18] The second provided for a 32-inch line with a 16-inch extension to Haifa. France and "other parties" were reported willing to assume part of the $80,000,000 cost in this case. The third alternative called for the laying of a 32-inch line from Elath to Beersheba with a 16-inch extension from there to Ashdod Yam. Not one of the three plans was fully implemented. By the end of April 1957 a 135-mile-long, 8-inch pipeline had been completed between Elath and Beersheba; work on the line was then stopped. Further construction depended on assured freedom of navigation in the Gulf and willingness of the oil-producing states east of Aqaba to use the Israeli pipeline. The first condition was fulfilled when on April 7, 1957, an American tanker, *Kern Hills*,

[18] *Oriente Moderno*, April, 1957, p. 215, and *New York Times*, April 10, 1957.

chartered by the Israel National Oil Company, passed the Strait of Tiran and unloaded its 16,700 tons of oil in three storage tanks in Elath, thus establishing a precedent of unmolested passage through the Gulf. This, incidentally, was a situation which had reportedly formed the subject of a "deal" between the United States Department of State and Israel when the latter was urged to evacuate Egyptian territory. It was understood that in return for the evacuation the United States, without any formal commitment on its part, would uphold the principle of the freedom of navigation by letting an American vessel enter the Gulf unchallenged by Egypt (in view of the presence of UNEF in the Strait of Tiran) so as to create a precedent. It is not clear whether Washington was informed in advance of the passage of the *Kern Hills*. Whether or not any such agreement existed, the Israelis were anxious to establish the precedent without delay, and this explains their haste in chartering the American tanker.

Where did the oil come from? It came from Iran, but by subterfuge. According to Iranian sources, the master of the ship declared upon leaving Abadan that her destination was South Africa. During the voyage (unless the captain had previous secret instructions) the destination was changed to Elath.

This brings us to the second condition for success of the Israeli pipeline project, the willingness of producing countries or concessionary companies to have their oil conveyed through Israel's territory. The right of host countries to determine by which routes oil should be shipped is uncertain. Concession agreements do not as a rule contain any provisions on this subject, and under normal circumstances the presumption is that choice of the best routes to assure effective marketing would be the exclusive concern of the operating company. Moreover, the company may not have any say in this matter if it sells its oil to trading companies in the ports of the host country. These trading companies or "offtakers" may have their own marketing and transit arrangements, with which the host country would have no right to interfere. But these considerations are based on the assumption of normal circumstances, and circumstances were not normal in the Middle East in 1957. Apart from the immediate issues created by Israel's invasion of Egypt, there was the long-standing Arab boycott of Israel from which no country or private organization having interests in the area could stand entirely aloof. Regardless of the legal aspects of the problem, it was more than doubtful that the Consortium, largely composed of corporations having heavy investments in Arab countries, would be willing to disregard Arab wishes in this respect.

There remained one further possibility, namely, Iranian disposition on its own of the 12.5 per cent of the output of crude to which it was entitled by its agreement with the Consortium. As a Moslem country Iran was generally inclined to take the Arab side in the chronic Arab-Israeli conflict. As a member of the Baghdad Pact, Iran was reluctant to embarrass its close ally, Iraq. It also was anxious to cultivate the friendship of Saudi Arabia, with whose ruler the shah of Iran had shortly before exchanged state visits. It was, in fact, King Saud's reported protest against the shipment of Iranian oil to Elath which hastened Iran's official moves. On April 16, 1957, the Iranian Ministry of Foreign Affairs announced that the National Iranian Oil Company (NIOC) had received orders forbidding it to sell its oil to Israel. As for exports made by the Consortium, said the statement, the Iranian government had recommended that the latter follow NIOC's example, and it was "very probable" that the recommendation would be observed out of respect for the government's interests.[19]

Iran's attitude might change, of course, as the country does not share the emotional bias of its Arab neighbors toward Palestine. It is conceivable that in case of a real emergency or protracted difficulties with Egypt with regard to the Suez Canal Iran might reconcile itself to use of the Israeli transit route. Consequently Israel's hasty construction of a pipeline of a small diameter served two purposes: first, to carry oil for its own needs, presumably imports from Burma or Indonesia, and, second, to act as a symbol or a reminder that, if worst came to worst, Israel was an alternate route for oil shipments.

Another plan which received publicity in connection with the Egyptian crisis provided for the construction of a 120-mile pipeline paralleling the Suez Canal on its western side and linking Port Taufiq with Port Said. The plan antedated the invasion and was

[19] *Oriente Moderno*, May, 1957, p. 309.

under serious study by Egypt's government. Bids for its construction had been invited. One of the most serious bidders was the Greek shipping magnate, Aristotle Onassis, who late in October 1956 (on the eve of the invasion) declared that his plan, originally submitted in 1955, was under active consideration by the Egyptian government. It called for the construction of either three parallel lines of 32-inch pipe or of two lines of 48-inch pipe, which would have a daily throughput capacity of 150,000 tons or 950,000 barrels. This would amount to about three-quarters of the Canal's normal oil movement, which is around 1.2 million barrels a day. According to Onassis, the cost of moving oil through the pipeline would approximate one-half the tolls received from large tankers transiting the Canal.[20] One of the reasons for initiating the plan was the Canal's limited clearance, which makes it impossible for giant supertankers to pass through it when loaded. The Suez invasion gave this matter a new political dimension.

In reviewing the Onassis plan, the Egyptian government had to consider several factors bearing upon its national interests. On the one hand, was the factor of competition between the pipeline and the Canal. Should the pipeline reduce the revenues Egypt was receiving from the tankers, it would obviously be against Egypt's interest to permit its construction. On the other hand, the pipeline might merely absorb the expansion of the oil traffic, an expansion which was generally expected and which could be proved by the steadily rising figures of supply and demand. Even then there would still remain the question of whether it might not be more advantageous for Egypt to concentrate on deepening and improving the Canal to enable it to handle the new 100,000-ton supertankers. This would make the outside world more dependent on the Canal, and the greater this dependence the greater the strategic role Egypt could hope to assume.

One further factor had to be considered, the competition between Egypt and other countries as a transit route for oil. Should the necessary technical improvements in the Canal not materialize or be unduly delayed, or should Canal tolls differ too much from the costs of pipeline operation, the oil industry would inevitably turn to other

[20] *New York Times*, Oct. 28, 1956.

transit routes. In this respect Turkey, Syria, and even Israel were all potential competitors of Egypt. Moreover, the Onassis plan had the advantage of providing immediate financial means for its implementation as against the more difficult task of securing adequate funds for the deepening and widening of the Canal, which would be a costlier project. It was apparently the financial and competition factors which caused the Egyptian government to react favorably at first to the Onassis proposal. On June 29, 1957, Hassan Ibrahim, chairman of the Egyptian Economic Organization, announced that plans were underway to establish an Egyptian stock company to build and exploit the Suez Canal pipeline. Fifty-one per cent of the shares were to be controlled by the Egyptian government, with the remaining 49 to be owned by Onassis and other oil-shipping companies. Hassan Ibrahim also stressed that the projected pipeline would be "the cheapest possible, cheaper than that which Israel is planning to build." [21] Soon afterward, for reasons not yet fully explained, the Egyptian government broke off the negotiations, and by the fall of 1957 it was generally believed that it had definitely abandoned the plan.

This review of the Suez crisis makes plain its world-wide reverberations. Decisions made by a handful of men in Tel Aviv, Cairo, and Damascus, all capitals of small and weak states, affected the security and well-being of populous and advanced states of Western Europe and of the Free World as a whole. The vulnerability of the West has passed from the realm of theoretical speculation to the realm of reality. The crisis starkly revealed Europe's heavy dependence on the few transit channels, while testing the West's capacity to deal effectively with a dangerous situation. The lesson should not perhaps be regretted. It stimulated much thinking as to means to avert similar contingencies in the future by devising alternate transit routes and taking other precautions of a political and an economic nature. The oil industry, the principal sufferer, received a strong warning of its vulnerability to politically motivated hostility, causing it to be more alert to the sociopolitical aspects of its role and function in an area undergoing revolutionary transition. With regard to the Middle

[21] *New York Times,* June 30, 1957. See also "Construction d'un pipeline entre Suez et Port Said," *L'Orient* (Beirut), June 29, 1957.

Eastern states, Syria's behavior is particularly worthy of attention. It seems to prove that the concept of Arab solidarity has ceased to be a mere slogan and can be translated into action. It also seems to prove that countries passing through the ferment of nationalist awakening no longer abide solely by consideration of economic self-interest.

Conclusion

THE preceding sixteen chapters have reviewed the manifold aspects of the relationship between oil and state in the Middle East. Is there a moral to the story? We would hesitate to narrow down a complex set of problems to a simple, crisp formula; that could be achieved only through gross oversimplification of the issues at stake. Yet some generalizations are not only permissible but expected in a study of this kind. From the totality of relationships discussed three aspects stand out as of paramount importance: (*a*) the vital role that Middle Eastern oil is playing in the economy of Western Europe and, consequently, of the Free World; (*b*) the actual and potential contribution that the oil industry is making to the welfare and economic progress of the Middle East; and (*c*) the insecurity in which the predominantly Western-owned oil companies have to carry out their functions.

If the first two aspects represent the true facts of the situation, there is a mutuality of interest between the West and the Middle East with regard to the continuous and smooth operation of the existing oil enterprises. But if this is so, why should there be insecurity? The evidence of the previous chapters indicates that the insecurity results from a number of challenges to which the companies are subjected. These challenges are political, legal, economic, and social in character.

The basic challenge is undoubtedly political. It stems from the over-all political climate prevailing between the West and the

Middle East. If this climate is one of friendship and mutual trust, there is good reason to expect Western-owned enterprises to be spared hostility and molestation. If the reverse is the case, it is likely that the larger the enterprise the more conspicuously will it stand out as a possible target of anti-Western policies. The oil companies may therefore be exposed to punishment and adversity resulting from crises not of their making. The Palestine War of 1948 and the Suez crisis of 1956 are two cases in point. In both the oil industry suffered dislocations, penalties, and threats born out of developments over which it had little or no control. This statement may be challenged by persons at home and abroad who ascribe to the big oil corporations a major influence in formulating the governmental policies of the United States and Great Britain. Analysis of the domestic power and influence of these companies is outside the scope of this study. The point the author wishes to make is that, even if a considerable degree of such influence is assumed, it would constitute only one of the multiple forces trying to exert pressure on the government in a politically competitive Western society. Consequently, unless one adopts a purely Marxist approach to the phenomena of capitalism and imperialism, one is bound to refute a simple equation between government policies and company interests.

Is there any way in which the companies can protect their position, short of acquiring a decisive voice in the policy-making processes of their own countries? Paradoxical as it may seem, it would appear that the best protection for the companies lies in a rather negative virtue, namely, in their ability to dissociate themselves as much as possible from the official policies of their home governments. In this respect British practice has differed considerably from American. In all three predominantly British-managed companies, namely, those in Iran, Iraq, and Kuwait, the identification of the company with its home government has been considerable, although the intensity of this phenomenon has fluctuated with circumstances. The closest identification was in Iran. During the 1951 dispute the British government went so far as to argue at the World Court that the concession agreement was in reality an international convention between Britain and Iran. American practice, most vividly exempli-

CONCLUSION

fied in Saudi Arabia, has been to stress the independent and purely commercial character of the oil enterprise. This has certainly been more in consonance with prevailing attitudes in the Middle East itself, as attested by recent resolutions of the Arab League enjoining its members not to grant concessions to companies wholly or partly owned by foreign governments.

In other areas the companies can enhance and protect their position by their own policies despite the manifold challenges to which they are subjected. One area where both challenge and opportunity to improve their position occur concerns the concession agreement. Here the challenge is linked to the whole conceptual framework of the companies' status. Is it fair and proper that, in contrast to the practice in the West, the companies should operate on the basis of long-term concessions? Is it just that a single company or group should have a concession covering virtually the whole area of a middle-sized country? And how valid is the charge, so frequently voiced by nationalists, of a state-in-state relationship? In the preceding chapters we have been trying to evaluate the validity of such criticisms. No simple and easy solution can be offered in answer.

As for the basic legal problem—that of the validity of concessions in international law—the companies have no choice but to insist on their unquestioned binding force and to promote at home and abroad the notion of the sanctity of a contract between a foreign person and a sovereign. There is no middle road on this issue, and for the companies it is a point on which they cannot afford to be flexible without undermining the basis of their existence. But to state this is not to equate it with rigidity in the companies' policies in their totality. Matters such as the term of a concession agreement and an exclusive claim to a huge territory, to give some striking examples, can and should be treated with flexibility if the companies do not want to be out of tune with the march of the times. In other words, though they are perfectly justified in insisting on the letter of their agreements, they should be willing to revise the terms of the latter from time to time, so as to keep in touch with the sociopolitical reality underlying all legal arrangements. Too much discrepancy between reality and law is bound to produce tension, which may end in an explosion. The wise lawmaker will never disregard the conditioning elements which alone are capable of giving the law respect and durability. Contracts cannot safely be divorced from their social background.

The same flexibility should apply to the whole question of the state-in-state relationship. Although there is no doubt that a high degree of self-sufficiency is required for companies to conduct their operations efficiently in a technologically retarded country, alertness to the changes the country is undergoing is mandatory. The following episode is illustrative. One of the companies operating in the Middle East possessed its own airfield located near a major city of the host country. The commercial airline used a public airport inconveniently located and poorly equipped to handle the ever-increasing local passenger traffic. As a result local people envied the oil company officials, who could easily reach common destinations, while they had either to suffer privations or to renounce their trips altogether. Then one of the company executives suggested that the local commercial airline be permitted to use the company airport. His suggestion encountered considerable opposition on the part of conservative colleagues, who predicted dire consequences. The executive held firmly to his view and succeeded in having his proposal adopted. His initiative bore fruit: the services of the local airline were considerably improved and its passengers were truly appreciative of the company's willingness to abandon its erstwhile isolation.

What has been said about the need for flexibility applies, albeit with reservations, to the financial terms of the concessions. No person can deny that there is a difference between the early pioneer days when the presence of oil had not yet been proved and the present era when oil is being abundantly and profitably extracted. True enough, the companies have recognized this difference by renegotiating the financial terms of their agreements, once in the 1930's and the second time in the 1950's. The question here seems to revolve not so much around willingness to revise the terms as around the timing and the manner of the revision. Imaginativeness and generosity have always paid dividends in the increased good will of the host country. The search for the right formula for the pipelines is a case in

CONCLUSION

point. Why should the transit countries be oblivious to the profits made by the producing countries and the companies out of the pipeline operation? Is it not legitimate for them to ask: "How much is the right-of-way through our territory worth to you?" Fortunately for the companies their somewhat rigid early insistence on the "transportation company" concept has given way to recognition that the transit states' plea was not wholly without merit. This has undoubtedly strengthened their position in the transit states. Having more to lose from interruption of the flow of oil than formerly, these states will now be less inclined toward recklessness in their oil policies.

The newly promoted concept of an integrated operation has presented another serious economic challenge to the established companies. As was pointed out in Chapter IV, strong resistance to this concept may be expected on the part of the older companies, notwithstanding its acceptance by the Japanese newcomers. It is not easy to say where right lies in this matter. It is understandable that the host countries should want to have a share in at least the outside operations that influence the extraction and refining carried out within their territory. The companies too have good reasons for refusing to surrender any part of profits based on their transportation and marketing arrangements, which have been organized independently of, and often antedate, the main extractive operation. There is, moreover, another weighty consideration which seems to militate in favor of the companies' stand on this issue. In the uneven relation existing between a sovereign state and a foreign company, the latter is compelled to seek extra safeguards in case the former should be tempted to abuse its sovereign power. Control of the marketing outlets is such a safeguard, and it seems unfair that host governments should deny the companies this minimum ingredient of security.

The possibility of the abuse of sovereign power brings us to political challenges to the companies' position. We have already discussed the political difficulties stemming from the general course of intergovernmental relations, which, as we have pointed out, are largely beyond the companies' control. At this juncture we shall focus our attention on those political challenges with which the companies can cope, at least in part. The main challenge, as we see it, results from the fact that the companies have to transact their business with the legally constituted governments of the host countries. The more satisfactory the relationship, the greater the stake the companies have in the preservation of the government which treats them fairly. The dilemma, to be sure, is not of exclusive concern to the companies. Western governments have often faced the painful problem of close collaboration with, and a stake in, foreign governments whose character and *modus operandi* have left much to be desired. In the case of the companies, the matter has been complicated by the circumstance that the host governments have often been authoritarian, old-fashioned, and, sometimes, repressive. A few years ago such an association implied future danger; today, with the revolution in Iraq and a rather abrupt change from the old to the new in Syria, the dangers are actualities.

How are the pitfalls of such a relationship to be avoided? There seems to be no pat formula for such situations, and on the intergovernmental level this dilemma has indeed taxed the ingenuity of the policy-making bureaus of Western capitals. Only a general guiding principle can be offered. This is that the companies should do their utmost to avoid political identification with any government actually in power in the host countries. Achievement of this principle will not be easy, and actual details must be left to the intelligence, tact, and imaginativeness of company executives. No matter how difficult the task, however, it would seem worth trying, especially in this era of revolutionary change in the Middle East.

When we speak of revolutionary change, we should realize that its implications and ramifications go far beyond the contingency of a coup which results in a displacement of one government by another in a given country. The change is both broader and deeper. It is reflected in the rise of new social strata who want more employment, speedier advancement, greater participation in the socio-economic life of their countries, and a higher degree of recognition. The white-collar employees of the companies represent a not unimportant segment of these new strata. To ignore the changes in their mentality and aspirations is to be oblivious to the deep currents in the society in the midst of which the companies exist. For-

CONCLUSION

tunately, however, this is a sector where the companies can do much to meet the demands of the changing scene if they conduct a well-conceived human relations policy. The advancement of national employees to positions of responsibility commensurate with their skills is one of the most important items for the companies to attend to.

Would that mean the eventual replacement of the Western personnel by nationals of the host countries? And if so, would this not remove one more company safeguard, the safeguard of having the exclusive know-how of industrial operations? To ask the first question is to pose the problem of the fundamental purpose of an oil company. Is it a business enterprise primarily representing an investment in a profitable venture, or should it be treated as a career organization? We would suggest that while the business aspect is decisive, the two are not mutually exclusive. The career aspect must not be neglected if healthy continuity of the operation is to be maintained. A company employing Western personnel should assure opportunities for advancement and reasonable security, failing which the personnel's morale will suffer and so will the operations. But to state this is not to deny the need for careful study of each case when a retirement, a resignation, or a termination of contract produces a vacancy that could possibly be filled by a national employee. As for the safeguard of a Western-staffed management, this should not be treated as an isolated factor but in relation to the over-all position of the company in the host country. If the promotion of nationals to high managerial positions will create a better feeling toward the company, there will be less need to seek additional and somewhat artificial safeguards.

When dealing with the problem of human relations, one must not overlook the dilemmas posed by native labor and its actual or attempted organization. Thus far the oil-producing countries have tended to discourage unionization as incompatible with their forms of government, and the companies, without much regret, have conformed to these policies. The governments' attitude has undoubtedly been based on fear lest organized labor become a political force capable of undermining the existing governmental systems. This fear has not been wholly unfounded. Indeed, it has found corroboration in the fact that labor unions in Iran and the Arab countries have shown a rather alarming proclivity to degenerate into militant political organizations. They not only do not hesitate to subordinate industrial to political questions, but they also become the instruments of outside groups—whether Communist or Pan-Arab—whose primary objectives have little or nothing to do with labor's welfare. Yet it is perhaps legitimate to ask whether the disturbing proclivities of the unions have not been, at least partly, stimulated by the negative and overly conservative attitudes of the host governments. Experience with labor unionism in a more progressive country, such as Lebanon, has not been altogether unsatisfactory. Even in Syria, during periods of moderate governments company experience with unions has not been wholly negative. This means that when trouble occurred in other periods (such as the sabotage of the pipelines and the resulting labor complications in 1956–1957), it was due not so much to the existence of unionism as to specific government policies.

What has been said thus far would seem to postulate a more progressive attitude on the part of the host governments toward labor questions. But where do the companies fit into this picture? Assuming that their managements accept the thesis that a slow and orderly evolution toward unionism is preferable to underground conspiracies leading to outbursts of violence, can they implement this thesis without incurring the accusation of interference in the domestic affairs of the host countries? They probably cannot if their actions are taken in obvious disregard of the host government's wishes and policies. If, however, their relations with the host government admit of frank consultation, there is no reason why their own and the government's views should not be harmonized in a search for the best possible solution.

In thus reviewing the challenges to the companies' security and in suggesting certain ways in which they can be met, we have concentrated on the companies and their policies. It may be proper at this juncture to remind ourselves that an oil enterprise represents a mutuality of interests between the companies and the host governments. Consequently the burden of making it successful should not be placed exclusively on the companies. Although the desire of

CONCLUSION

the host governments to maximize the benefits from their countries' natural resources is legitimate and understandable, it should be pointed out that these resources, while "natural," were hidden (and sometimes unsuspected) until a Western oilman invested his skill and capital in their discovery. This he did in the expectation of a fair or, perhaps, a spectacular return. So long as he did not acquire his concession by fraud or with the aid of bayonets, it is hard to find anything reprehensible in a gamble he was willing to take. To protect himself, he secured a contract believing that it represented an unquestioned binding engagement. If such an engagement is lightheartedly breached or unilaterally repudiated by the host government, the basis of international economic intercourse is shaken.

There is no doubt that the host governments, especially in this era of revolutionary nationalism, are subjected to two great temptations. One is to find a solution to their economic problems by nationalizing foreign-owned enterprises. The other is to make such enterprises a target of political hostility, with an eye to the popularity this may gain for the government in power. A dispassionate scrutiny reveals the fallacy of both approaches. Expropriation of foreign enterprises, no matter whether with or without some pledged compensation, is inevitably going to be resisted by the victims and their home governments, as the experience in Mossadegh's Iran has shown. And although the prevailing trend in Western capitals is to shun the use of force under such circumstances, measures of economic boycott likely to be instituted against the violators of contracts may prove so onerous as to defeat the original objective of expropriation. It is harder to prove the fallacy of the second motivation—that of internal political advantage gained by bullying a foreigner. Many nationalist leaders in Africa and Asia have founded their careers on such methods and exploits. But the law of diminishing returns seems to apply here as in many other spheres of human activity. People can be whipped up into a frenzy of enthusiasm for the spectacular deeds of their leaders, but in the long run it is ability to fulfill people's needs that counts. Furthermore, there is a connection, however intangible, between a country's reputation in the family of nations and the ultimate happiness of its citizens. Adventures in lawbreaking may sometimes bring immediate gains, especially if there is some confusion about the enforcement machinery. But they may boomerang, provoking retaliatory measures, which sometimes may exceed in their violence the original acts of transgression.

The deriving of benefits from the exploitation of natural resources is a two-way affair. Host governments are more likely to increase their gains by negotiation than by unilateral action. As for the companies, their primary objective is to maintain their concessions and thus to assure the continuity of their operations. There are many elements of insecurity in their position, but the companies have many ways and means of increasing their margin of safety. This they can do, first and foremost, by avoiding undue rigidity, which, unless relaxed, will build up sociopolitical pressures in the host countries to a bursting point. Flexibility should not, however, be equated with softness and indecision. On the contrary, it should denote a firm and intelligent determination to do one's best to bridge such gaps between law and reality as the march of time inevitably creates.

Appendix Tables

I. Crude oil production in the Middle East
(Thousands of barrels per day and thousands of tons per year)

Country	1955 Bpd	1955 Tons	1956 Bpd	1956 Tons	1957 Bpd	1957 Tons	1958 Bpd	1958 Tons
Bahrein	30.1	1,505	30.0	1,500	32.5	1,625	40.6	2,030
Egypt	35.0	1,750	32.0	1,600	52.3	2,615	65.2	3,260
Iran	320.0	16,000	530.0	26,500	689.0	34,450	808.2	40,410
Iraq	690.0	34,500	633.0	31,650	438.6	21,930	756.2	37,810
Israel	.0	0	0.6	30	1.2	60	2.0	100
Kuwait	1,100.0	55,000	987.0	49,350	1,250.0	62,500	1,376.0	68,800
Neutral Zone	24.6	1,230	41.0	2,050	78.0	3,900	84.1	4,205
Qatar	114.0	5,700	127.0	6,850	147.0	7,350	162.0	8,100
Saudi Arabia	951.0	47,550	1,000.0	50,000	893.9	44,695	991.3	49,565
Turkey	4.5	225	6.4	320	6.3	315	6.5	325
Total Middle East	3,269.2	163,460	3,387.0	169,850	3,588.8	179,440	4,292.1	214,605

Source: *Oil and Gas Journal*, last issues of December for the years 1955, 1956, 1957, and 1958.

APPENDIX TABLES

II. Crude oil production of major world areas, 1958

	Barrels per day	Tons per year
United States	6,459,300	322,965,000
Total Western Hemisphere	10,138,300	506,915,000
The Middle East	4,292,100	214,605,000
Other outside Soviet orbit	822,140	41,107,000
Total Free World	15,187,340	759,367,000
The Soviet bloc	2,522,000	126,100,000
Total world	17,709,340	885,467,000

Source: Oil and Gas Journal, Dec. 29, 1958.

III. Estimated oil revenues of certain countries in the Middle East (U.S. dollars)

Country	1957	1958
Bahrein	$ 11,000,000	$ 11,000,000
Iran	214,000,000	246,000,000
Iraq	136,900,000	235,000,000
Kuwait	365,000,000	415,000,000
Qatar	44,500,000	57,000,000
Saudi Arabia	285,900,000	300,000,000
Total	$1,057,300,000	$1,274,000,000

Compiled from a variety of sources, including William S. Evans, *Petroleum in the Eastern Hemisphere* (The First National City Bank of New York, April, 1959).

APPENDIX TABLES

IV. Proved reserves of crude oil and refining capacity in the Middle East and North Africa in 1958

Country	Reserves, thousands of bbls.	Refining capacity, 1,000 bpd
Middle East		
Aden	—	120.0
Bahrein	230,000	186.5
Egypt	400,000	74.7
Iran	33,000,000	493.0
Iraq	25,000,000	55.8
Israel	50,000	87.0
Kuwait	60,000,000	220.0
Lebanon	—	24.0
Neutral Zone	6,000,000	50.0
Qatar	2,500,000	0.6
Saudi Arabia	47,000,000	189.0
Turkey	70,000	6.9
North Africa		
Algeria	3,500,000	—
Libya	50,000	—
Morocco	8,500	2.1

Source: *Oil and Gas Journal*, Dec. 29, 1958.

V. Tanker tonnage of major tanker-owning countries (in dead-weight tons)

National registry *	End of 1957	End of 1958
Liberia	8,671,000	11,200,000
Panama	3,449,000	3,900,000
United States	5,784,000	6,100,000
United Kingdom	7,624,000	8,000,000
Norway	7,624,000	8,500,000
Soviet bloc	783,300	821,000

Sources: For 1957 (except the Soviet bloc), *Sun Oil Report* (Philadelphia, 1958); for 1958, unofficial estimate; for the Soviet bloc, *Western Form Shipping Report*, no. 123, December, 1958.

* Exclusive of government fleets, with the exception of the Soviet bloc.

APPENDIX TABLES

VI. Tanker tonnage controlled by five major American oil companies at the end of 1957 (in dead-weight tons)

Standard Oil Company (New Jersey)	2,569,000
Gulf Oil Corporation	1,175,000
The Texas Company	795,000
Socony Mobil Oil Company	745,000
Standard Oil Company of California	443,000
Caltex *	938,000
Stanvac *	455,000
Total	7,120,000

Source: Sun Oil Report (Philadelphia, 1958).

* Caltex is owned by Standard of California and Texas; Stanvac is owned by Standard (New Jersey) and Socony Mobil.

Bibliographical Note

THE history of the oil industry in the Middle East has been treated in two nearly encyclopedic works: Brigadier S. L. Longrigg, *Oil in the Middle East: Its Discovery and Development* (London and New York, 1954) and Benjamin Shwadran, *The Middle East, Oil, and the Great Powers* (New York, 1955). Longrigg's book, with its attention to technical and economic detail, reflects the author's intimate knowledge of the industry, born of his long association with Iraq Petroleum Company, and may be considered as a major *apologia* of the oil enterprise in the Middle East. Shwadran's emphasis is on the political aspects, foreign and domestic, and the general tone of his work implies considerable criticism of both the oil companies and the host governments. Earlier than these two books was a shorter volume by Raymond F. Mikesell and Hollis B. Chenery, *Arabian Oil* (Chapel Hill, N.C., 1949), which dealt mostly with American oil interests. A good general description, with historical background, is contained in a pamphlet, *Middle East Oil Development,* published by the Arabian American Oil Company (4th ed.; n.p., 1956).

Apart from these general works, individual oil-producing countries or special problems have been treated in monographic studies. Such is Alan W. Ford, *The Anglo-Iranian Oil Dispute of 1951–1952: A Study of the Role of Law in the Relations of States* (Berkeley and Los Angeles, 1954). The Anglo-Iranian oil dispute has also stimulated highly partisan expressions of views, notable among the latter being Nasrollah Saifpour Fatemi, *Oil Diplomacy: Powderkeg in*

Iran (New York, 1954) and L. P. Elwell-Sutton, *Persian Oil: A Study in Power Politics* (London, 1955). While the former reflects the nationalist bias of its author, the latter contains a scathing, though technically undocumented, condemnation of the oil company involved in the dispute. Saudi Arabia's oil development has been competently described by Roy Lebkicher in *Aramco and World Oil* (New York, 1952).

Examination of Western governmental policies concerning Middle Eastern oil is outside the scope of this study. Those interested in the formulation of American policies might well consult two monographs: *Petroleum and American Foreign Policy* by Herbert Feis (Stanford, Calif., 1944) and *American Security and Foreign Oil* by Bernard Brodie (Foreign Policy Reports; New York, 1948). Herbert Feis's *Seen from E.A.: Three International Episodes* (New York, 1947) also has a chapter dealing with American oil policies. A longer and authoritative treatment of the subject is contained in Halford L. Hoskins, *Middle East Oil in United States Foreign Policy* (Legislative Reference Service, Library of Congress, Public Affairs Bulletin No. 89; Washington, 1950). British policies and attitudes toward Middle Eastern oil have been analyzed, rather critically, by Michael Brooks in *Oil and Foreign Policy* (London, 1949) and, in a more sympathetic vein, by Sir Olaf Caroe in *Wells of Power, the Oilfields of South-West Asia: A Regional and Global Study* (New York, 1951).

In addition to these specialized works, certain general studies on the Middle East contain sections on Western oil policies. Halford L. Hoskins' *The Middle East: Problem Area in World Politics* (New York, 1954) stands out as one of the most mature analyses of the strategic aspects of Middle Eastern oil and its role in United States foreign policy. A lucid and thought-provoking treatment of the subject is also presented by John C. Campbell in *Defense of the Middle East: Problems of American Policy* (New York, 1958).

The most recent addition to the literature on oil is David H. Finnie, *Desert Enterprise: The Middle East Oil Industry in Its Local Environment* (Cambridge, Mass., 1958). Compact and readable, it provides a perceptive analysis of the position and policies of the oil companies in their present sociopolitical setting in the Middle East.

BIBLIOGRAPHICAL NOTE

The nature of the present study has made it necessary, for the most part, to have recourse to primary materials. These have included surveys made by the United Nations and intergovernmental organizations, government and company publications, texts of laws, treaties, and agreements, statistics, newspaper articles and pamphlets, records of various congresses, reports by experts and consultants, and books written in native languages of the Middle East. These sources have been supplemented by frequent interviews with company executives, government officials, and political leaders in the Middle East.

For data on the role of oil in the European economy, the author has relied largely on four studies published by the Organization for European Economic Co-operation: *Europe's Growing Needs of Energy* (Paris, 1956), *Le Pétrole: Perspectives européennes* (Paris, 1956), *Maritime Transport: A Study by the Maritime Transport Committee* (Paris, 1957), and *Europe's Need for Oil: Implications and Lessons of the Suez Crisis* (Paris, 1958).

Many useful data on production, refining, transportation, investment, and national budgets of the Middle Eastern states may be found in the periodic volumes of the United Nations, *Economic Developments in the Middle East*. The economic development problems of Iraq and Iran have been exhaustively treated in a number of reports presented by the World Bank, special engineering organizations, and individual consultants. On Iraq, authoritative data are contained in International Bank for Reconstruction and Development, *The Economic Development of Iraq* (Baltimore, 1952) and in a report submitted to the Iraq Development Board by Lord Salter, *The Development of Iraq, A Plan of Action* (n.p., 1955). Detailed information on Iran may be found in Overseas Consultants, Inc., *Report on Seven Year Development Plan for the Plan Organization of the Imperial Government of Iran,* 5 vols. (New York, 1949). No comparable studies have as yet been made of Saudi Arabia and Kuwait, and information on their economic development has to be gathered from occasional company reports and magazine articles.

For the description and analysis of the concession agreements, recourse to the original texts has been necessary. These can be found in a variety of sources. Official government gazettes constitute the most important direct source. These may be supplemented by a number of other sources containing the translated texts of agreements, notable among which are J. C. Hurewitz, *Diplomacy in the Near and Middle East, A Documentary Record: 1535–1914*, 2 vols. (Princeton, 1956), *Platt Oilgram News Service* (New York and Chicago), and *Petroleum Times* (London).

Legal provisions adopted by the host countries concerning their oil resources and the granting of concessions are found both in fundamental laws and special legislative acts. For the former, it is useful to consult Helen Miller Davis, *Constitutions, Electoral Laws, Treaties of States in the Near and Middle East* (Durham, N.C., 1953). For the latter, many texts may be found, not only in official government gazettes, but also in *World Petroleum Legislation* (New York), a periodical publication. The legal status of the concession agreements has been treated both in general works on international law and in special monographs. Standard works, such as L. Oppenheim, *International Law*, 4th ed. (London, New York, Toronto, 1928), H. W. Briggs, *The Law of Nations: Cases, Documents and Notes* (London, 1938), and J. L. Brierly, *The Law of Nations* (4th ed.; Oxford, 1949), contain discussions of the contractual relationship between a state and a private alien. An illuminating recent study on this subject, "The Sanctity of Contract between a Sovereign and a Foreign National," was presented by Lowell Wadmond at the meeting of the American Bar Association in London in 1957. The Suez crisis of 1956 stimulated much debate on the subject of concessions and the international obligations of states. Excellent presentations in this connection have been made by Quincy Wright, Clyde Eagleton, and A. L. Goodhart in *Tensions in the Middle East*, a symposium edited by Philip W. Thayer (Baltimore, 1958).

The descriptions of the companies' government relations organizations have been based almost entirely on interviews with company officials and on such company documents as have been made available to the author.

Questions pertaining to maritime and desert boundaries in the general area of the Persian Gulf and the adjacent territories are nowhere treated comprehensively in a single study. Treaties and official government pronouncements are the main source of information. Many indispensable texts may be found in C. U. Aitchison, *A Collection of Treaties, Engagements and Sanads Relating to India and the Neighboring Countries* (5th ed.; Calcutta, 1933), vol. XI; *The American Journal of International Law;* Saudi Arabia's official gazette, *Umm al-Qura* (Mecca); and *The Persian Gulf Gazette,* an official journal of the British Residency in the Persian Gulf. The positions of Great Britain and Saudi Arabia in the Buraimi dispute have been stated in voluminous memoranda submitted to the arbitration tribunal: *Arbitration Concerning Buraimi and the Common Frontier between Abu Dhabi and Saudi Arabia: Memorial Submitted by the Government of the United Kingdom of Great Britain and Northern Ireland* (n.p., 1955), and *Arbitration for the Settlement of the Territorial Dispute between Muscat and Abu Dhabi on One Side and Saudi Arabia on the Other: Memorial of the Government of Saudi Arabia,* 3 vols. (n.p., 1955). As for the Oman dispute, both the Oman delegation in Cairo as well as the Arab Information Office in New York have been releasing pamphlets and bulletins presenting the case of the imam of Inner Oman as against Great Britain and the sultan of Muscat.

Data on pipelines have been secured from three principal sources: official texts of pipeline agreements, the press in the transit countries, and direct information from the companies.

Multifarious projects for international control of the Middle Eastern oil industry may be traced in a variety of documents. Especially worthy of attention are *Reports of the Proceedings* of the postwar congresses of the International Co-operative Alliance, an organization which, more consistently than any other, has advocated some form of international supervision of the oil industry. The Arab League has released a number of documents pertaining to oil and economic matters, of which the decisions and resolutions of the High Economic Council are particularly relevant to this study.

In his research on public opinion, the author has relied heavily on the press in the Middle East. Books written on the subject of oil in Persian and Arabic were also used to some extent. For the oil industry's reaction, various media of publicity have been reviewed, including the company employee magazines, such as *Ahl an-Naft* in Iraq and *Qafilat az-Zait* in Saudi Arabia.

BIBLIOGRAPHICAL NOTE

As to labor problems, it has been necessary to consult the labor legislation of the producing and transit countries. Although official government gazettes and compilations were the main source for this, the *Legislative Series* published by the International Labour Office has proved useful in providing English texts of certain laws. The actual course of labor relations in the countries under review has never been subjected to a thorough and comprehensive study, and field research was essential. A worthy attempt to inquire into the labor situation under the AIOC regime in Iran was made by a special ILO mission. Its report was entitled *Labour Conditions in the Oil Industry in Iran* (Geneva, 1950). No further studies of this kind have appeared since. For data on the Pan-Arab labor movement, the author has had to rely on personal interviews and the local press. Extensive documentation has, of course, been available on the companies' personnel, wage, and benefit policies in the annual reports published by Aramco, IPC, and the Iranian Consortium.

Index

Abadan, 11, 41
Abu Dhabi, 128
  boundaries, 143
  forces enter Buraimi, 147
  submarine areas arbitration, 131, 132
Aden, 41
Aden Protectorate, 22, 141
Advertising, institutional, 242 ff.
Aflaq, Michel, 158
Agip Mineraria, 12, 82 ff.
Ahl an-Naft magazine, 245 ff.
AIOC:
  claim to Kuwait, 20
  concession, features of, 64
  housing, 301
  Iraq holdings, 16
  nationalization, 10 ff., 109
  payments to Iran, 1911–1951, 76
  payments to Iran, 1933, 67
  payments to Iran, 1950, 39 n.
  personnel, 312
  publicity of concessions, 205, 206
  public relations, 236
  training, 307
  wages, 296
Aitchison, C. U., 369
Ajman, 128
Algeria, 23
Amerada Petroleum Corp., 24
American Petroleum Institute, 171
Aminoil, 22

Anglo-American oil agreements, 169 ff., 179
Anglo-Iranian Oil Co., *see* AIOC
Anglo-Persian Oil Co., *see* AIOC, British Petroleum Co.
Appurtenance theory, 129
Aqaba, 140
Aqaba, Gulf of, 133, 320, 345 ff.
Arab Development Bank, 183, 184
Arabian American Oil Co., *see* Aramco
Arab League:
  Bludan resolutions, 188
  boycott of Israel, 188
  Buraimi dispute, 147
  no concessions to foreign governments, 84, 191
  oil experts committee, 189 ff.
  oil policy, 187 ff.
  Palestine crisis, 187, 188
  petroleum office, 189 ff.
  pipeline guarantees, 342
  regional development, 182 ff.
Arab Oil Congress, 192, 195, 196 ff.
Arab Renaissance Party, 158; *see also* Ba'ath
Arab Socialist Renaissance Party, 225, 279
Aramco:
  borderland exploration, 144
  concession, features of, 64 ff.
  concession granted, 17 ff.

371

INDEX

Aramco (*cont.*)
  contractors, 231, 232
  economic development, 234, 235
  exploration program criticized, 210
  government relations, 113 ff.
  health services, 299
  housing, 232, 303
  personnel policies, 313
  pipeline agreements, 154 ff.
  pipeline management, 159
  profit-sharing formula, 68, 71, 76, 109, 220 n.
  publicity of concessions, 206
  public relations, 237 ff.
  purchasing policy, 230, 231
  saving schemes, 305
  Syria, dependence on, 108
  training, 309
  turnover, 297
  wages, 295
Aramco Overseas Co., 35
Ardalan, Dr. Ali Gholi, 135
Asquith of Bishopstone, Lord, 132

Ba'ath Party, 279
Bahra Agreement, 140 n.
Bahrein:
  agreement with Saudi Arabia, 133
  concession grant, 21
  continental shelf, 127
  dispute with Iran, 134 ff.
  employment figures, 41
  reactions to Suez crisis, 336
  refining, 42
  revenue from oil, 39
Bakhtiari tribe, 122, 123
Banias, 26, 31
Bapco, 21
Barawi, Dr. Rashid al-, 222
Basrah Petroleum Co., 13, 17
Berenger-Long Agreement, 15, 167
Black, Eugene R., 184
Bludan resolutions, 188
Blue Line, 138, 142, 144
Boggs, Whittemore, 133 n.
Boundaries, maritime, 132 ff.
Brewer, Sam Pope, 81 n.

Brierly, J. L., 368
Briggs, H. W., 368
British Oil Development Co., 16
British Petroleum Co., 10, 13; *see also* AIOC
Brodie, Bernard, 366
Brooks, Michael, 366
Bullard, Sir Reader, 146, 147
Buraimi dispute, 145 ff.
Burgan field, 20, 22
Bustani, Emile, 179, 196, 197

California Arabian Standard Oil Co., 17
Campbell, John C., 366
Canada, 13, 22
  continental shelf, 130
Caroe, Sir Olaf, 366
Chenery, Hollis B., 365
China, 290, 291
Churchill, Winston, 11
Clemenceau, Georges, 14
Coleman, Stewart, 327
Communist Party, 205, 285
Compagnie Française des Pétroles, 10, 13, 18
Concession agreements:
  growth, 9 ff.
  legal status, 94 ff.
  pattern, 63 ff.
  validity, 353
Consortium:
  agreement with Iran, 64
  ban of sales to Israel, 347
  composition, 10, 11
  conciliation committees, 264
  housing, 301
  personnel, 313, 314
  production guarantees, 75
  public relations, 239 ff.
  training, 307
  wages, 296
Consumption of oil:
  in Europe, 28 ff.
  in Middle East, 41 ff.
Continental Oil Co., 24
Continental shelf, 65

INDEX

Continental shelf (cont.)
  convention on, 130
Cowden, Howard A., 176, 177
Cox, Major Percy, 122

Dammam conference, 145
D'Arcy, William Knox:
  concession, features of, 64 ff.
  concession of, 10
D'Arcy Kuwait Co., 19
Davies, F. A., 336
Davis, Helen Miller, 368
Denmark, 34
Development plans, regional, 177 ff.
Development programs, in Middle East, 45 ff.
Dihigo, Dr. Ernesto, 146
Doha (Qatar) arbitration, 131
Dubai, 128
Dulles, John Foster, 340, 342

Eagleton, Clyde, 105, 368
Ebtehaj, Abol Hassan, 59, 60
Economic grievances, 209 ff.
Egypt:
  constitution, 89
  consumption of oil, 42
  legislation on oil, 92, 93
  pattern of oil industry, 20
  pipelines, 156, 157
  refining, 42
  revenue from Suez Canal, 45
  tankers, 42, 43
  see also Suez Canal, Suez crisis, United Arab Republic
Eisenhower, Dwight D., 185 ff., 339
Elwell-Sutton, L. P., 366
Employment figures, 40, 41
Entezam, Abdullah, 344
Eshkol, Levi, 345
Europe:
  consumption trends, 28 ff.
  cost of imports, 34, 35
  refining capacity, 30
  tanker fleets, 31 ff.

Faisal, Emir, 145

FALU, 281 ff.
  call for sabotage, 321
Fanfani, Amintore, 184
Fatemi, Nasrollah Saifpour, 220, 365
Federation of Arab Labor Unions, *see* FALU
Feis, Herbert, 366
Finnie, David H., 366
Ford, Alan W., 365
France:
  consumption of oil, 35
  invades Egypt, 319
  Iraq concession, 14 ff.
  refining, 30
  tanker fleet, 34
  see also Compagnie Française des Pétroles
Fuad Line, 142, 145

García-Amador, Señor, 97
Germany:
  consumption of oil, 35, 36
  industrial recovery, 109
  Iraq concession, 14, 15
Getty Oil Co., 22
Ghaleb ibn Ali, Imam of Oman, 149
Goodhart, A. L., 106, 368
Great Britain:
  agreement with U.S., 169 ff.
  Bahrein dispute, 134 ff.
  boundary agreements, 137 ff.
  consumption of oil, 35
  dominant in Bahrein, 21
  dominant in Persian Gulf, 24
  invades Egypt, 319
  Iraq concession, 14 ff.
  navy shifts to oil, 11
  Oman rebellion, 148 ff.
  tanker fleet, 34
Greece, 35
Gulbenkian, C. S., 13
Gulf Oil Corp., 6, 19

Hadda Agreement, 140
Haifa, 26, 41
Hamza, Fuad Bey, 142, 143
Haruni, Yusuf Mustafa al-, 221 n.

373

INDEX

Hasan, Mahmoud, 146
Hay, Sir Rupert, 145 n.
Hendryx, Frank, 196, 197
Hilu, Yusuf Khattar, 221
Holland:
    consumption of oil, 35
    tanker fleet, 34
Holmes, Major Frank, 19, 21
Hoskins, Halford L., 366
Hourani, Akram, 158
Housing, 300 ff.
Hurewitz, J. C., 368
Hussain, Adil, 221 n.
Hussein, King, 285

Ibn Saud, King, 17
Ibrahim, Hassan, 349
ICFTU, 287, 288
Imperialism, charges of, 216 ff.
Industrial grievances, 212 ff.
Integrated operations, 78 ff.
    evaluation, 355
Integration, social, policy of, 230 ff.
International Bank of Reconstruction and Development, see World Bank
International Co-operative Alliance, 173 ff.
International Co-operative Petroleum Association, 175, 176
International Court of Justice, 101, 102
International law, 94 ff.
Investment in oil facilities:
    in Europe, 30
    in Middle East, 38
Investments, protection of, 103
IPAC, 13
IPC:
    concession, features of, 64 ff.
    concessions in Arabian Peninsula, 142
    consultation committees, 268
    government relations, 117 ff.
    health services, 299
    housing, 233, 302
    payments to transit states, 43, 44
    personnel, 313, 315
    pipeline agreements, 1953–1958, 162 ff.
    pipelines owned, 26
    pipelines repaired, 325 ff.
    pipelines' throughput, 31
    publicity of concessions, 206, 207
    public relations, 237 ff.
    Qatar concession owned, 22
    saving schemes, 305
    taxed in Lebanon, 71
    training, 308
    wages, 295
Iran:
    and Arab Oil Congress, 196
    arbitration rejected, 102
    concessions, growth of, 10 ff.
    constitution, 89
    consumption of oil, 42
    continental shelf, 128
    development plan, 53 ff.
    dispute over Bahrein, 134 ff.
    employment figures, 40, 254
    industrial relations, 261 ff.
    labor laws, 256
    Mossadegh law, 1944, 90
    nationalization law, 90
    offshore agreements, 82 ff.
    Petroleum Act 1957, 91, 128
    refining, 41
    regional development plans, 181, 182
    revenue from oil, 38, 39, 40, 67, 76
    tankers, 43
    transit through Israel, 346 ff.
    tribal protection, 121
    Turkish pipeline plan, 338 ff., 343 ff.
    see also AIOC, Consortium
Iranian Oil Participants, Ltd., see Consortium
Iranian Workers Party, 225
Iran Party, 220, 225
Iraq:
    boundaries, 140
    concession granted, 13 ff.
    constitution, 89
    consumption of oil, 42
    continental shelf, 128, 129
    development plan, 49 ff.
    employment figures, 40, 254
    industrial relations, 266 ff.
    labor laws, 254
    losses due to Suez war, 328

INDEX

Iraq (cont.)
  reactions to Suez crisis, 321 ff.
  refining, 41, 53
  revenue from oil, 38, 39
  tankers, 43
  Turkish pipeline plan, 340
  see also IPC
Iraq Petroleum Company, see IPC
IRCAN, 13
Iricon Agency, 10
Israel:
  boycott of, 188, 190, 195
  charges of influence, 217
  Eisenhower plan, 186
  invades Egypt, 319 ff.
  pipelines, 192
  pipeline schemes, 345 ff.
  refining, 41
  regional development, 178, 181
  strikes oil, 23
  tankers, 43
  see also Haifa, Suez crisis
Istiqlal Party of Iraq, 225
Italy:
  concession in Iran, 12
  consumption of oil, 35, 36
  government in oil business, 84
  supplies to Israel, 190
  tankers, 34

Jablonski, Wanda, 81 n.
Jabri, Majdeddin, 158
Japan, enters oil business, 110
Japan Petroleum Trading Co., 22, 84 ff.
Jemayyel, Pierre, 211
Jordan:
  constitution, 89
  see also Pipelines, Transit countries, Transjordan

Kamel, Fathi, 283, 286
Kern Hills, 345
Khadduri, M., 95 n.
Khadra, Abdullah, 225
Khanaqin, 16
Khanaqin Oil Co., 14
Kharg and Kargo islands, 13
Khatib, Subhi, 279, 283

Khuzistan, 11
Kikhya, Rushdi, 158
Koran injunctions, 95
Kuwait:
  boundaries, 137, 139, 140
  concession granted, 19 ff.
  continental shelf, 128
  economic development, 46 ff.
  employment figures, 40
  excluded from Red Line area, 16
  Japanese agreement, 84 ff.
  reactions to Suez crisis, 337
  regional development, 197
  revenue from oil, 38, 39, 47, 182
  tankers, 43
Kuwait Oil Co., 19 ff.

Labor legislation:
  in Iran, 256
  in Iraq, 254
  in Lebanon, 259
  in Saudi Arabia, 258
  in Syria, 260
Lahham, Aref al-, 158
League of Nations, 15
Lebanese Congress, 225
Lebanon:
  constitution, 89
  employment figures, 41
  industrial relations, 276 ff.
  losses due to Suez war, 328
  oil administration, 111
  pipeline agreements' revisions, 160 ff.
  public relations, 204
  reactions to Suez crisis, 321 ff.
  refining, 42
  revenue from pipelines, 44, 160
  tax on pipelines, 71, 163
  see also Pipelines, Transit countries
Lebkicher, Roy, 366
Legal status of concessions, 94 ff.
Legislation, oil, 12, 90 ff.
Lessani, Senator A., 220
Liberia, 34
Libya:
  oil discovered, 23
  respects contracts, 196
Liebesny, H. J., 95 n.

375

INDEX

Literature on oil, Arab and Iranian, 219 ff.
Little, Arthur D., Inc., 50
Lloyd George, David, 14
Long-Berenger Agreement, 15, 167
Longrigg, Brig. S. L., 365

Macmillan, Harold, 339
Makki, Hussein, 220
Masjid-i-Suleiman, 11
Middle East Emergency Committee, 327 ff.
Mikesell, Raymond F., 365
Miller, Sir Edington, 50
Mohammerah, Sheikh of, 121, 122
Morrison-Knudson Co., 54
Mossadegh, Dr. Mohammed, 177, 204, 220, 221
Mosul, 14, 15
Mosul Petroleum Co., 13, 16
Muscat-Oman:
    boundaries, 141
    forces enter Buraimi, 147
    IPC concession, 22
    rebellion in Oman, 148 ff.

Nabulsi, Suleiman, 285
Nasser, Gamal Abdul, 223, 224, 282, 284
National-Democratic Party of Iraq, 225, 226
National Iranian Oil Company, see NIOC
Nelson, Wesley, 50
Neutral Zone, Iraqi-Saudi, 140
Neutral Zone, Kuwait-Saudi Arabian, 22
NIOC:
    as state oil agency, 10 ff., 111
    forbidden to sell to Israel, 347
    nonbasic operations, 72, 73, 239, 240
    offshore agreements, 82 ff.
    Qum operation, 73
    statute, 90
    see also Consortium, Iran
Norway:
    consumption of oil, 35
    tanker fleet, 34

OEEC, 29 ff., 184, 367
Offshore, see Submarine areas
Ohio Oil Co., 24
Oman (Inner), 147, 148 ff.; see also Muscat-Oman
Onassis, Aristotle, 348
Open Door principle, 15, 21
Oppenheim, L., 368
Oqair conference, 138, 139, 140
Overseas Consultants, Inc., 54, 367

Pachachi, Dr. Nadim, 342
Palestine war, 352
Panama, 34
Pan American Petroleum Corp., 12, 82 ff.
Pan-Arab activities, see Arab League, FALU
Paris arbitration, 132
Payments, pattern of, 67, 68
Pella, Giuseppe, 184
Persian Gulf, 22
Persian Gulf principalities, 5, 127 ff.
Pipelines:
    division of profits evaluated, 355
    Egyptian plans, 347 ff.
    expansion, 31
    international guarantees, 340, 342
    Iranian-Turkish, 12
    Israeli projects, 345 ff.
    mandated areas' agreements, 153, 154
    percentages of oil transited through, 26
    sabotage of, 289, 325 ff.
    throughput capacity, 31
    Turkish pipeline plan, 338 ff.
    see also IPC, Jordan, Lebanon, Suez crisis, Syria, Tapline, Transit countries
Pirnia, Hussein, 220
Political grievances, 215 ff.
Pricing, 77, 80
Production statistics, 37, 361, 362
Public Affairs Institute, 180
Public opinion:
    and Tapline agreement, 208
    importance of, 4, 203 ff.
Public relations, 235 ff.

INDEX

Qafilat az-Zait magazine, 246 ff.
Qasem, Anis, 196
Qatar:
  boundaries, 137, 142, 143
  continental shelf, 128
  employment figures, 41
  IPC concession, 22
  revenue from oil, 39
  submarine areas arbitration, 131
Qatar Petroleum Co., 22
Qazzaz, Said, 267
Qudsi, Nazem, 158
Qum, 11, 12, 343, 344

Radcliffe, Lord, 131
Ras al-Khaimah, 128
Red Line agreement, 16, 17, 18, 167, 172
Refining:
  desired by host countries, 80
  in Arabian-Japanese agreements, 85, 86
  in Europe, 30
  in Italy, 84
  in Middle East, 41 ff.
REPAL, 23
Reserves of crude, 363
Reuter, Baron de, concession, 88 n.
Revenues from oil, 38, 39, 40, 362
Riyadh Line, 143
Royal Dutch–Shell group, 10, 13
Russia:
  dominates WFTU, 262
  oil agreement with Iran, 222
  penetrates Middle East, 262
  supplies Israel, 191
  tanker fleet, 34
  see also Soviet bloc
Rusta, Reza, 261 ff.
Ryan, Sir Andrew, 142

Said, Nasser, 285
Said, Nuri as-, 266
Said bin Taimur, Sultan, 149
Salman, Mohammed, 189 ff.
Salter, Lord, 50, 367
San Remo Conference, 15, 167
Sapphire Petroleums, 13, 82 ff.

Saud, King (Crown Prince), 133, 271, 347
Saudi Arabia:
  Arabian pipeline proposal, 192, 193, 194, 195
  Aramco 1950 agreement, 68
  Bahrein, agreement with, 133
  concession granted, 17 ff.
  continental shelf, 126, 127
  employment figures, 40
  industrial relations, 268 ff.
  Japanese agreement, 84 ff.
  labor laws, 258
  pipeline problem, 165 ff.
  reactions to Suez crisis, 334 ff.
  regional development, 182
  revenue from oil, 38, 39
  tax laws, 70, 71
  territorial waters, 126, 127, 133
  see also Aramco, Tapline
Shakir, Major Amin, 223
Shammar tribe, 124
Sharabati, Ahmad, 157, 329
Sharjah, 128
Shell, 188; *see also* Royal Dutch–Shell
Shwadran, Benjamin, 365
Sib, treaty of, 148
Sidon, 214, 225
Sidon price dispute, 80, 81
SIRIP, 12
Socialist Progressive Party of Lebanon, 225
Socony Mobil Oil Co., 10, 13, 17, 188
Solh, Sami as-, 324
Soviet bloc, 34, 287, 290, 291
Soviet Union, *see* Russia
Standard Oil Co. (New Jersey), 10, 13, 17, 23
  and Anglo-American agreement, 171
Standard Oil Co. of California, 10, 17, 21
Standard Oil Co. of Indiana, 12
State-in-state question:
  criticism, 218, 222
  evaluation, 354
  refutation, 229 ff.
Submarine areas:
  general problems, 126 ff.

INDEX

Submarine areas (cont.)
  Iranian, 12, 13
  new Iranian agreements, 82 ff.
  off Neutral Zone, 22
Suez Canal, 31
  percentages of oil transited through, 26
  tonnage transited, 44, 45
  traffic figures, 326
Suez crisis, 27, 37, 43, 249, 319 ff., 352
Sweden, 34, 35
Syria:
  coups d'état, 101
  desire to keep pipelines, 226
  employment figures, 41
  financial claims, 334
  industrial relations, 279
  labor dispute with IPC, 328 ff.
  labor laws, 280, 332 ff.
  legislation, 93
  losses due to Suez war, 328
  oil administration, 111
  pipeline agreements revised, 160 ff.
  public relations, 204
  refining, 42
  revenue from oil, 43, 44
  sabotage of pipelines, 325 ff.
  Tapline ratification, 155 ff.
  see also Pipelines, Suez crisis

Taif, Treaty of, 141
Tankers, 31 ff.
  Arab Tanker Co. proposal, 192, 195
  tonnage statistics, 363, 364
Tapline:
  first agreements, 154
  payments to Lebanon, 44
  percentage of Saudi oil handled, 26
  profit-sharing formula, 164 ff.
  pumping stations, 115
  Syrian concession, 155 ff.
  taxed by Lebanon, 71
  throughput capacity, 31
  see also Aramco, Pipelines, Sidon price dispute, Transit countries
Tariki, Abdullah, 81, 194 n., 196, 197
Taxation:
  by home governments, 75, 76
  by host governments, 66
  exemption criticized, 212
  exemption principle, 70
  Lebanon and pipelines, 163
  United States tax, 19
Technical assistance:
  general, 120
  to Saudi Arabia, 117, 120
Territorial claims:
  desert borders, 137 ff.
  submarine areas, 126 ff.
Texas Co., 10, 17, 21
Thayer, Philip W., 368
Transit countries, 25, 41, 43 ff.
Transit routes, 31 ff.
Tribal protection:
  in Iran, 121
  in Iraq, 124
Tripoli (Lebanon), 26
Transjordan:
  boundaries, 140, 141
  see also Jordan
Trucial Coast, 22, 141
Truman proclamation, 126, 129
Tudeh Party, 261 ff.
Turkey:
  constitution, 89
  consumption of oil, 42
  legislation on oil, 23, 92
  pipeline agreement with Iran, 343 ff.
  pipeline plans, 12, 338 ff.
  refining, 42
  tankers, 43
Turki ibn Abdullah ibn Ataishan, Emir, 145, 146
Turkish Petroleum Co., 14 ff.

Umm al-Qaiwain, 128
United Arab Republic:
  constitution, 89
  pipeline negotiations, 166
United Nations:
  crusade for control of oil, 173 ff.
  enforcement of law, 105
  law of the sea, 130
  Suez crisis, 319 ff.
United States:
  agreement with Britain, 169 ff.

INDEX

United States (*cont.*)
  aid to Iraq, 50
  Arab development proposals, 185
  continental shelf doctrine, 126
  defends Gulf Oil in Kuwait, 20
  defends oil interests in Iraq, 15
  Federal Trade Commission's report, 78 n., 176
  Suez crisis, 327 ff.
  tanker tonnage, 34

Venezuela, 193, 196, 210
Vidal, F. S., 233 n.
Violet Line, 138, 142

Visscher, Dr. Charles De, 146

Wadmond, Lowell, 96 n., 368
WFTU, 174, 288
World Bank, 54, 184, 186
Wright, Quincy, 368

Yasin, Sheikh Yusuf, 146
Yazbek, Yusuf Ibrahim, 221 n.
Yemen, 141
Yunes, Zaidan, 286

Zaim, Col. Husni az-, 158, 159, 218
Zorlu, Fatin Rustu, 344